Natural Fiber-Reinforced Composites

Natural Fiber-Reinforced Composites

Thermal Properties and Applications

Edited by Senthilkumar Krishnasamy, Senthil Muthu Kumar Thiagamani, Chandrasekar Muthukumar, Rajini Nagarajan, and Suchart Siengchin

WILEY-VCH

Editors

Dr. Senthilkumar Krishnasamy
Department of Materials and Production Engineering
The Sirindhorn International Thai-German Graduate School of Engineering (TGGS)
King Mongkut's University of Technology North Bangkok
1518 Wongsawang Road, Bangsue
Bangkok, 10800
Thailand

Dr. Senthil Muthu Kumar Thiagamani
Department of Mechanical Engineering
Kalasalingam Academy of Research and Education
Krishnankoil, 626 126
Anand Nagar, Tamil Nadu
India

Dr. Chandrasekar Muthukumar
School of Aeronautical Sciences
Hindustan Institute of Technology & Science
Padur, Kelambakkam
Chennai, 603103, Tamil Nadu
India

Prof. Rajini Nagarajan
Department of Mechanical Engineering
Kalasalingam Academy of Research and Education
Krishnankoil, 626 126
Anand Nagar, Tamil Nadu
India

Prof. Suchart Siengchin
Department of Materials and Production Engineering
The Sirindhorn International Thai-German Graduate School of Engineering (TGGS)
King Mongkut's University of Technology North Bangkok
1518 Wongsawang Road, Bangsue
Bangkok, 10800
Thailand

Library of Congress Card No.: applied for

British Library Cataloguing-in-Publication Data
A catalogue record for this book is available from the British Library.

Bibliographic information published by the Deutsche Nationalbibliothek
The Deutsche Nationalbibliothek lists this publication in the Deutsche Nationalbibliografie; detailed bibliographic data are available on the Internet at <http://dnb.d-nb.de>.

Print ISBN: 978-3-527-34883-1
ePDF ISBN: 978-3-527-83155-5
ePub ISBN: 978-3-527-83157-9
oBook ISBN: 978-3-527-83156-2

Typesetting Straive, Chennai, India

Printed on acid-free paper

10 9 8 7 6 5 4 3 2 1

Contents

Preface

We are glad to present the book entitled "Natural Fiber-Reinforced Composites: Thermal Properties and Applications" to thermal analysts, materials scientists, materials engineers, material chemists, and researchers working in the field of bio-composites.

This book focuses on exploring the thermal properties of hybrid composites reinforced with natural fibers. As per literature, the thermal properties of these composites could be analyzed by the techniques such as differential scanning calorimetry (DSC), thermomechanical analysis (TMA), thermogravimetric analysis (TGA), and dynamic mechanical analysis (DMA). In lieu of this, all the chapters are designed to cover a broad audience with the aforementioned techniques on hybrid composites reinforced with natural fibers.

This book consists of 18 chapters and is structured as follows: Chapters 1, 2, and 3 present the overview of thermal properties of natural fiber reinforced thermoset and thermoplastic hybrid composites. Chapters 4 to 11 discuss the thermal properties of hybrid composites made up of different natural fibers such as sugar palm, sisal fiber, flax fiber, pineapple leaf fiber, grass/cane fiber, banana fiber, kenaf fiber and hemp fiber. Chapters 12 to 16 provide insights on the effect of nanofillers (e.g. graphene, nanoclay, CNT, metal oxide) on thermal properties of natural fiber reinforced hybrid composites. Chapter 17 discusses the effects of chemical treatments of hybrid composites on thermal properties. Chapter 18 provides an overview of physical, mechanical, and thermal properties of hybrid polymer composites.

We thank our parents and sincerely appreciate the publisher, the typesetting professional associated with this book, and thank all the contributing authors for their valuable time and efforts in submitting their work to this book.

1

Thermal Characterization of the Natural Fiber-Based Hybrid Composites: An Overview

Chandrasekar Muthukumar[1], Senthilkumar Krishnasamy[2,3], Senthil Muthu Kumar Thiagamani[3], Rajini Nagarajan[3], and Suchart Siengchin[4]

[1]*Hindustan Institute of Technology & Science, Department of Aeronautical Engineering, Kelambakkam, Chennai 603103, Tamilnadu, India*
[2]*King Mongkut's University of Technology North Bangkok, Center of Innovation in Design and Engineering for Manufacturing (CoI-DEM), 1518 Wongsawang Road, Bangsue, Bangkok 10800, Thailand*
[3]*Kalasalingam Academy of Research and Education, Department of Mechanical Engineering, Krishnankoil 626126, Tamil Nadu, India*
[4]*King Mongkut's University of Technology North Bangkok, The Sirindhorn International Thai-German Graduate School of Engineering (TGGS), Department of Materials and Production Engineering, 1518 Wongsawang Road, Bangsue, Bangkok 10800, Thailand*

1.1 Introduction

Hybrid composite is fabricated by adding two or more fibers into a single polymer system [1]. The resulting material has a unique feature that combines the advantages of each fiber. Since different fibers are added together, the benefits of one particular type of fiber property could be compensated with the other fiber lacking a specific property. The performance of hybrid composites could be influenced by many factors [2–7]:

i. Fiber length
ii. Fiber loading
iii. Fiber orientation
iv. Fiber layer sequence
v. Fiber/matrix interfacial bonding
vi. Failure strain of fiber

The hybrid effect is termed as an apparent synergistic improvement of properties due to different fibers in a single matrix system. The selection of fibers and their properties is of main importance to achieve the enhanced properties for the hybrid composites. Besides the physical, chemical, and mechanical stabilities of fiber, the

Natural Fiber-Reinforced Composites: Thermal Properties and Applications, First Edition.
Edited by Senthilkumar Krishnasamy, Senthil Muthu Kumar Thiagamani, Chandrasekar Muthukumar, Rajini Nagarajan and Suchart Siengchin.

matrix system also defines the strength of the hybrid composites. The different types of hybrid composites are characterized as follows [8–12]:

i. Tow by tow: the fibers are mixed up randomly or regularly.
ii. Sandwich hybrid composites: one material is sandwiched between two different layers.
iii. Inter-ply or laminated: two or more fiber layers are alternatively stacked regularly.
iv. Intimately mixed fibers: various types of fibers are mixed up randomly.

Though the hybrid composites have many advantages, the prime challenges are replacing the synthetic fiber-reinforced composites using biocomposites. Biocomposites exhibit functional and structural stability during storage and degrade upon disposal into the environment. "Engineered natural fiber" is one of the exciting concepts to obtain the enhanced strength in the biocomposites, which involves the blending of the leaf and stem fibers. The correct blending of these two fibers exhibits optimum balance in mechanical properties, resulting in balanced stiffness–toughness properties [13–15].

The mechanical and physical characteristics of the natural fiber are influenced by many factors: (i) maturity of the plant fiber, (ii) harvesting time and region, (iii) soil condition, (iv) rain, (v) sun, etc. Since the natural fibers are nonabrasive and hypoallergenic, they could be processed efficiently. Amongst the various properties of natural fibers, the low density and the cellular structure allow them to exhibit better thermal properties. However, the amorphous hemicellulose on the fiber surface can be a potential threat to the better interfacial bonding between the matrix and the fiber, thereby reducing the properties. Hence, the mechanical and thermal properties of the biocomposites could be further enhanced through chemical treatments [16]. Natural fiber has cellulose, hemicellulose, and lignin susceptible to degradation on exposure to elevated temperature [17–19]. Thus, many studies exploring the thermal properties of the biocomposites have been published over the years [20–22]. By botanical type, the natural fibers are classified into six major types (Table 1.1).

Table 1.1 Classification of the natural fibers.

Seed	Bast	Leaf	Core	Grass and seed	Others
Kapok	Jute	Banana	Flax	Canary	Roots
Coir	Ramie	Pineapple	Kenaf	Barley	wood
Cotton	Flax	Curaua	Hemp	Wheat	
Oil palm	Hemp	Sisal	Jute	Grass	
Rice	Kenaf	Abaca		Corn	

1.2 Thermal Characterization

The thermal analyses encompass a family of techniques that would share a common feature, whereby any material's response could be measured through heating or cooling. Thus, a significant connection is held between the temperature and the physical property of the materials. The most common thermal techniques that have been used by researchers and by industrial organizations for thermal characterization are thermomechanical analysis (TMA), thermogravimetric analysis (TGA), differential scanning calorimetry (DSC), and dynamic mechanical analysis (DMA). These techniques are not only used for measuring the physical properties with respect to the temperature changes but also used in the following areas: (i) to substantiate mechanical properties and thermal history of the biocomposites, (ii) to estimate the service life of composites in different environments, and (iii) as one of the quality control approaches in polymers and their manufacturing industries. Figure 1.1 shows some of the essential thermal analysis techniques and the characteristics measured [23–25]. In terms of research, thermal behavior of the biocomposites has been investigated by varying fiber volume fractions [26–28], varying fiber layering patterns [29, 30], using different types of chemical treatments [31, 32], adding different kinds of fillers [19, 33, 34], and using polymer blends [35, 36].

For instance, Table 1.2 presents some of the experimental works carried out on thermal properties using different natural fibers.

1.2.1 DMA

Figure 1.2 presents the step-by-step process involved in the DMA of the polymers and polymer-based composites. Output parameters such as storage modulus (E' or G'), loss modulus (E'' or G''), and damping factor ($\tan \delta$) obtained as the function of temperature are shown in Figure 1.3a. As the polymer or composite is heated in the temperature range with the simultaneous application of oscillatory load, it undergoes displacement or strain where some energy gets stored in the material,

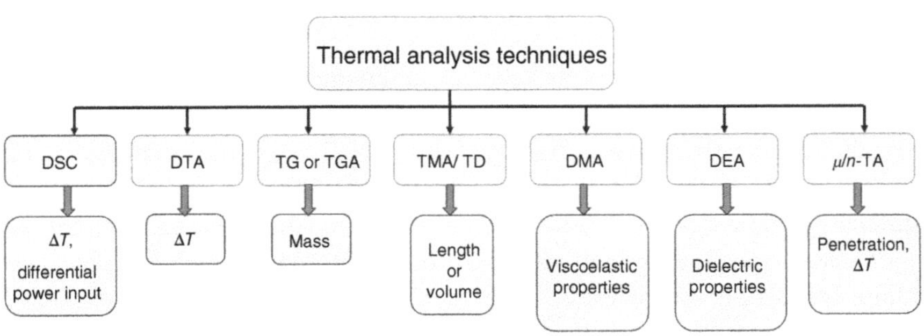

Figure 1.1 Various thermal analysis techniques and their applications. DFA, dielectric analysis.

Table 1.2 Reported thermal based works of natural fiber-reinforced hybrid composites.

Hybrid composites	Details of study	References
	Thermoset polymers	
Flax/sugar palm/epoxy	DMA	[6]
Flax/woven aloe vera/epoxy	TGA, DMA	[20]
Sisal/cattail/polyester	Thermal conductivity	[37]
Date palm/coir fiber/epoxy	TGA	[38]
Sisal/jute/sorghum/polyester	TGA	[39]
Coir/*Luffa cylindrica*/epoxy	DMA	[40]
Bamboo/kenaf/epoxy	TGA, DMA, DSC	[41]
Ramie/sisal/epoxy Sisal/curaua/epoxy	TGA, DSC	[42]
Flax/aloe vera/hemp/epoxy	TGA, DMA	[43]
Kenaf/pineapple leaf fiber/phenolic	TGA	[44]
	Thermoplastic polymers	
Sugar palm/roselle/polyurethane	TGA	[45]
Jute/bamboo/polyethylene	DSC, TGA	[46]
Sugar palm/roselle/polyurethane	TGA	[47]
Seaweed/sugar palm fiber/thermoplastic sugar palm starch agar	TGA	[48]
Coir/pineapple leaf fiber/polylactic acid (PLA)	TGA	[49]
Coir/pineapple leaf fiber/ PLA	TGA, TMA	[50]
	Biodegradable polymers	
Sisal/hemp/bioepoxy	DMA, TGA	[29]
Kenaf/sisal/bioepoxy	TGA, DSC, DMA	[51]
Sisal/hemp/bioepoxy	TGA	[52]

while some energy is dissipated as heat due to the internal friction. The resultant strain measured by applying the oscillatory load is represented as loss modulus, storage modulus, and phase angle or damping factor. The ability of the tested material to store the energy is termed as the storage modulus while the tendency of the material to dissipate heat energy is termed as the loss modulus. Storage modulus represents the stiffness of a polymer or composite and is often related to Young's modulus. Loss modulus is related to the molecular chain motions such as transition and relaxation within the polymer during the heating process and applied load. Tan δ is a dimensionless number obtained through the ratio of loss modulus to the storage modulus. Lower tan δ indicates higher stiffness and better interfacial bonding between fiber and polymer matrix, which restricts the molecular mobility within the polymeric chains.

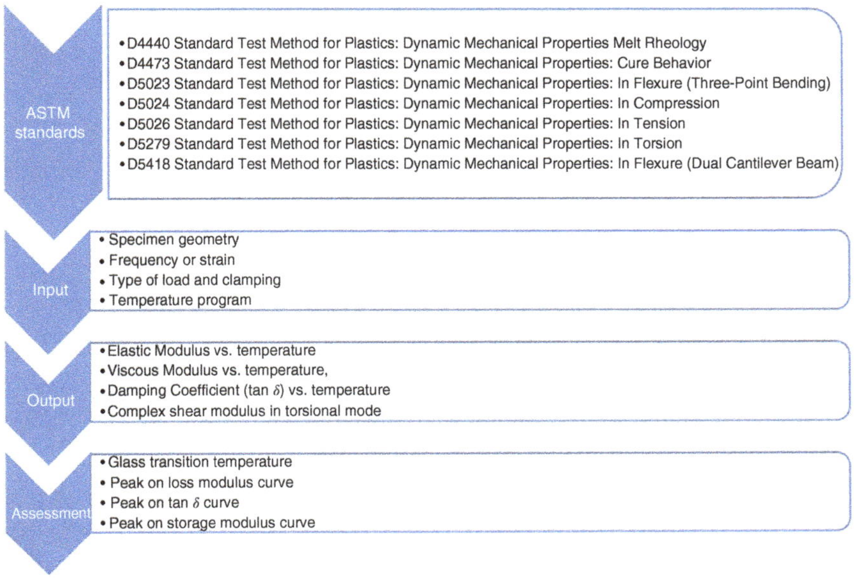

Figure 1.2 Thermal characterizations of the polymer and polymer-based composite through DMA, step-by-step process.

Polymers are viscoelastic and can be classified into crystalline, amorphous, and semicrystalline (has both crystalline and amorphous characteristics) depending upon the composition. It is because of this characteristic that polymers or polymer-based composite undergoes phase change during the simultaneous application of the load and heating process (Figure 1.3b). Figure 1.3a shows the typical data obtained from DMA. Glass transition temperature (T_g) is the tangent obtained in the phase change region between glassy state and rubbery state. T_g can be below the melting temperature for a polymer, which has both crystalline and amorphous characteristics. The material tends to get softer rather than melting at T_g. DMA is particularly useful in identifying the cross-linking density of the polymer, as shown in Figure 1.3b. It can be noticed that polymers with a high cross-linking density have higher T_g and greater loss modulus and storage modulus, while it is vice versa for polymers with low cross-linking density [53].

1.2.2 TMA

TMA is a common technique used for investigating the dimensional change of material under the combination of temperature and a fixed load. Figure 1.4 presents the step-by-step process involved in the TMA of the polymers and polymer-based composites. Dimensional change of material (at nanoscale) under the influence of temperature and load can be measured in various testing modes shown in Figure 1.5. Changes in the free volume of material depending upon the heat absorption or heat release with respect to the temperature can also be determined with this technique.

Figure 1.6a–c shows that the T_g measurement for a polymer or a polymer composite can be derived from the TMA, DSC, and DMA.

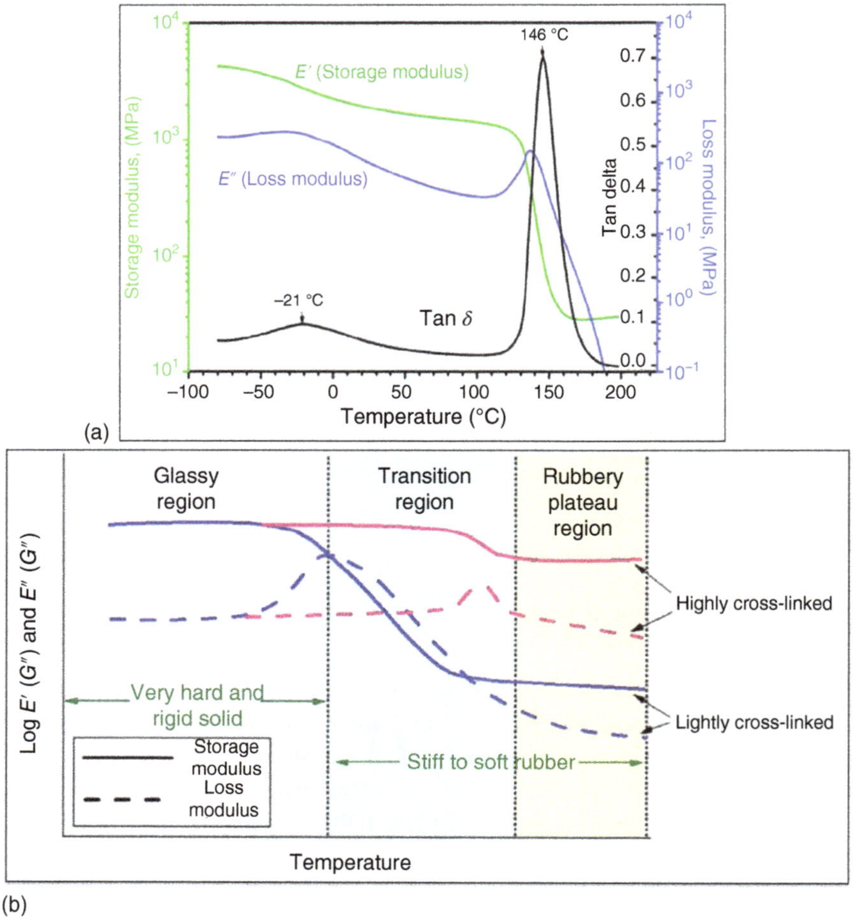

Figure 1.3 Thermal characterization of polymer and polymer-based composite with DMA. (a) Typical curve. (b) Viscoelastic characteristics of the polymer. Source: Saba et al. [53].

1.2.3 DSC

Figure 1.7 presents the step-by-step process involved in the DSC of the polymers and polymer-based composites. In DSC, the sample is heated around 30 to the elevated temperature beyond 300 °C with the constant supply of liquid nitrogen in a controlled chamber. Heat flow from the sample is measured as a function of the temperature shown in Figure 1.8. The changes in crystalline properties (T_g), melting temperature (T_m), and cold crystallization temperature (T_c) due to the introduction of two or more natural fibers in the hybrid composite can be evaluated.

1.2.4 TGA

Figure 1.9 presents the step-by-step process involved in the TGA of the polymers and polymer-based composites. It is an effective technique for evaluating thermal

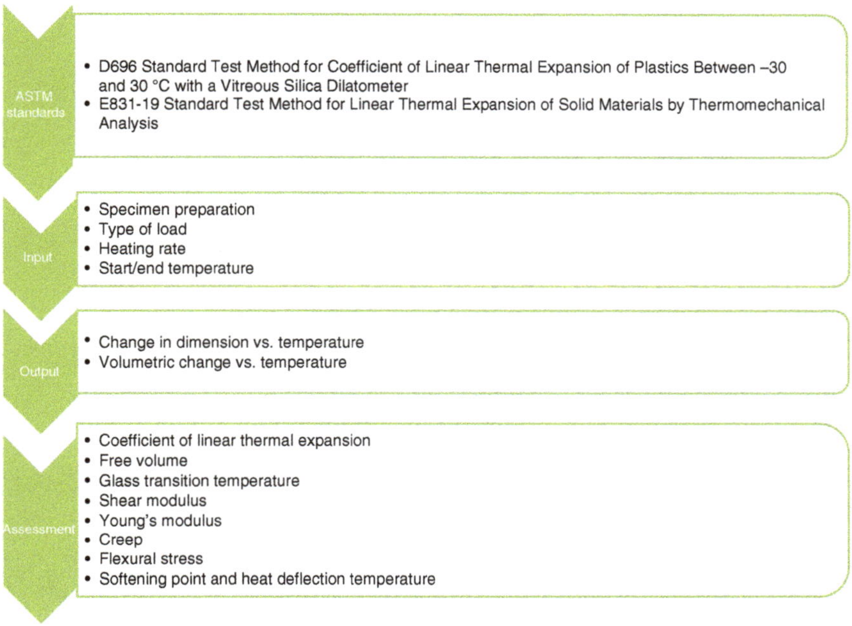

Figure 1.4 Thermal characterizations of the polymer and polymer-based composite through TMA, step-by-step process.

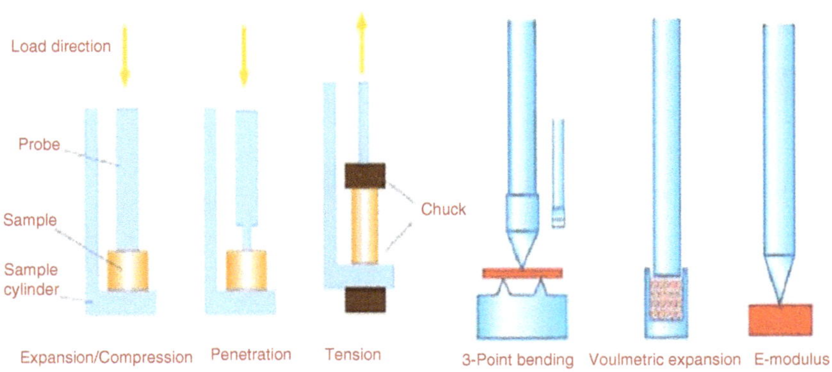

Figure 1.5 Test modes in TMA. Source: Saba and Jawaid [54].

decomposition characteristics of the polymers and polymer composite reinforced with the natural fibers or the synthetic fibers. It provides the quantitative mass change of the sample due to the heating under the controlled atmosphere. A natural fiber obtained from the plants and trees is made up of the constituents such as cellulose, hemicellulose, lignin, pectin, wax, moisture, and ash. The percentage of constituents can vary from one fiber to another, which has a significant influence on the thermal decomposition characteristics of natural fiber and their composites. Also, these fiber constituents are volatile and can decompose at elevated temperatures.

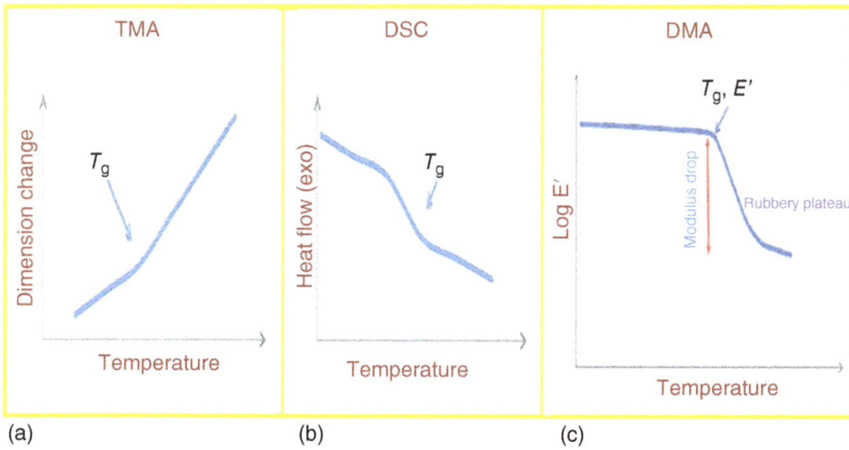

Figure 1.6 T_g measured from the various thermal characterization techniques: (a) TMA, (b) DSC, and (c) DMA. Source: Saba and Jawaid [54].

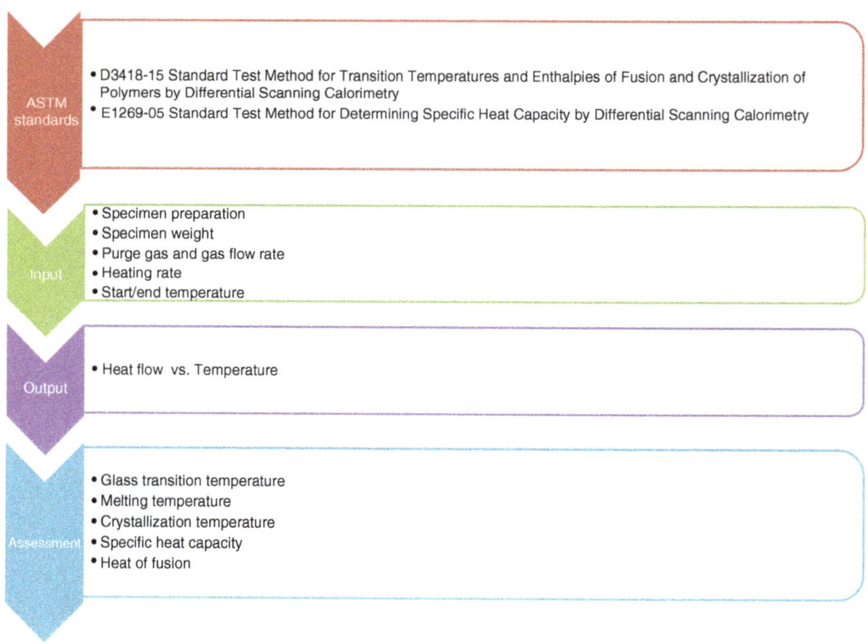

Figure 1.7 Thermal characterizations of the polymer and polymer-based composite through DSC, step-by-step process.

A few milligram of sample is placed in the TGA chamber and heated from room temperature to as high as 700 °C at a defined ramp rate in the presence of nitrogen to prevent oxidation inside the chamber. The thermal stability of a polymer-based composite is usually assessed from the thermogram (TG curve) and the derivative thermogram (DTG curve) obtained from the TGA, as shown in Figure 1.10. Parameters

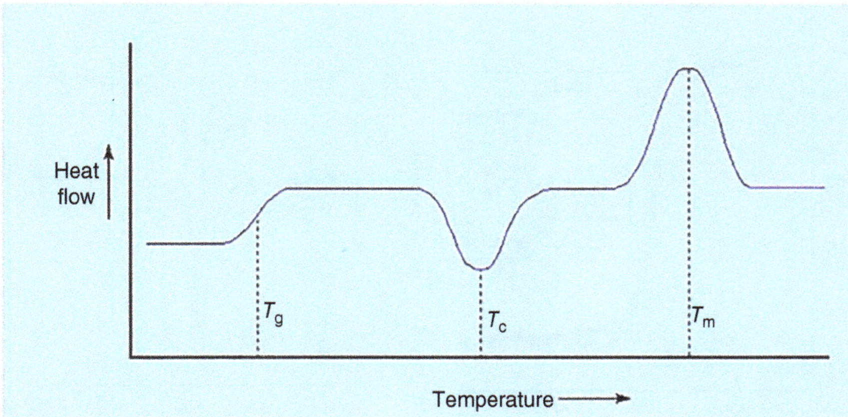

Figure 1.8 Thermogram from the DSC. Source: Chandrasekar et al. [55].

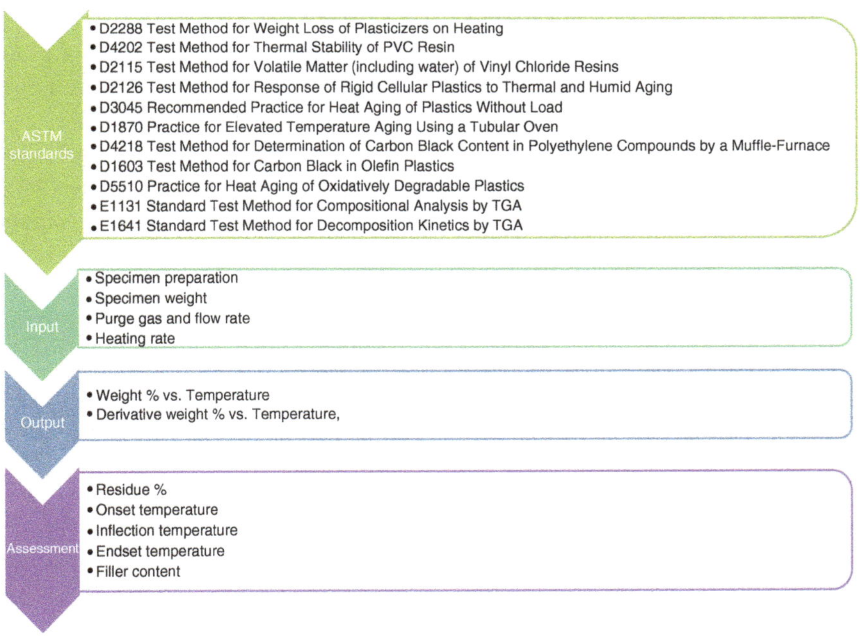

Figure 1.9 Thermal characterizations of polymer and polymer-based composites through TGA, step-by-step process.

such as the onset, endset, inflection temperature, and residue percentage at the end of the heating process in the TGA chamber are usually compared to identify changes due to the reinforcement percentage and type of fiber. Degradation temperature at 5%, 10%, 20%, 40%, and 80% weight loss along with the residue can also be discussed.

In the case of a polymer, thermal decomposition usually occurs in single stage, whereas, for the natural fiber, thermal decomposition occurs in two or three stages

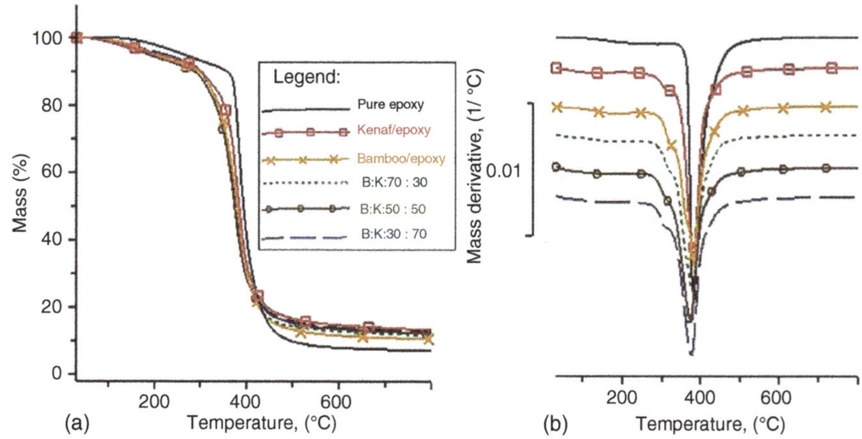

Figure 1.10 A typical TGA curve of the hybrid composite with kenaf and bamboo fibers. (a) Thermogram. (b) Derivative thermogram. Source: Chee et al. [41].

depending on the fiber constituents. Initial mass loss between 50 and 150 °C is due to the evaporation of moisture in the fiber. The weight loss at a temperature range between 150 and 300 °C is associated with the decomposition of hemicellulose and lignin. The final weight loss between 300 and 700 °C is attributed to the decomposition of cellulose. Since the fiber constituents vary from one fiber to another, TGA has proved to be an excellent tool for determining the changes in thermal decomposition characteristics of the hybrid polymer composite reinforced with two or more natural fibers. Thermal stability is also evaluated by residue percentage at the end of the heating process. The higher the residues, the better the thermal stability of the composite.

1.3 Conclusion

Thermal characterization of the hybrid composites using various commercially available techniques such as DMA, TMA, DSC, and TGA has been discussed. The following are the conclusions:

- DMA is useful in determining the creep properties and interfacial interactions of the composites and measuring their stiffness, material behavior with respect to the phase transitions, damping, and relaxation processes in a range of frequencies and temperatures.
- TMA helps in defining the material structure with respect to the dimensional and volumetric change, surface roughness, molecular structure, cure, and cross-linking polymerization under both static and dynamic loads.
- DSC is considered as one of the primary tools for thermodynamic analysis and cure kinetics. It gives useful information on the phase transitions upon heating and quantifies the glass transition temperature, melting temperature, and crystallization temperature related to the polymers and polymer-based biocomposites.

- TGA has been widely used to illustrate the thermal stability of the composites, which provides the quantitative mass change of the sample due to heating. It also provides vital information on the decomposition characteristics of the constituents of the composites at elevated temperatures.

The forthcoming chapters of this book would give extensive information on the above-discussed thermal characterization techniques with respect to different natural fibers and polymers targeted for various applications.

Acknowledgment

This study was financially supported by the King Mongkut's University of Technology North Bangkok (KMUTNB), Thailand, through grant no. KMUTNB-BasicR-64-16 and through grant no. KMUTNB-64-KNOW-07.

References

1 Naveen, J., Jawaid, M., Amuthakkannan, P., and Chandrasekar, M. (2018). Mechanical and physical properties of sisal and hybrid sisal fiber-reinforced polymer composites. *Mechanical and Physical Testing of Biocomposites, Fibre-Reinforced Composites and Hybrid Composites*: 427–440. https://doi.org/10.1016/B978-0-08-102292-4.00021-7.
2 Senthilkumar, K., Kumar, T.S.M., Chandrasekar, M. et al. (2019). Recent advances in thermal properties of hybrid cellulosic fiber reinforced polymer composites. *International Journal of Biological Macromolecules* 141: 1–13.
3 Asim, M., Paridah, M.T., Chandrasekar, M. et al. (2020). Thermal stability of natural fibers and their polymer composites. *Cellulose* 174: 175.
4 Nayak, S.Y., Sultan, M.T.H., Shenoy, S.B. et al. (2020). Potential of natural fibers in composites for ballistic applications – a review. *Journal of Natural Fibers*: 1–11.
5 Thiagamani, S.M.K., Krishnasamy, S., and Siengchin, S. (2019). Challenges of biodegradable polymers: an environmental perspective. *Applied Science and Engineering Progress* 12: 149.
6 Chandrasekar, M., Shahroze, R.M., Ishak, M.R. et al. (2019). Flax and sugar palm reinforced epoxy composites: effect of hybridization on physical, mechanical, morphological and dynamic mechanical properties. *Materials Research Express* 6 (10): 1–11.
7 Chandrasekar, M., Siva, I., Kumar, T.S.M. et al. (2020). Influence of fibre inter-ply orientation on the mechanical and free vibration properties of banana fibre reinforced polyester composite laminates. *Journal of Polymers and the Environment*: 1–12.
8 Matykiewicz, D. (2020). Hybrid epoxy composites with both powder and fiber filler: a review of mechanical and thermomechanical properties. *Materials (Basel)* 13: 1802.

9 Ravishankar, B., Nayak, S.K., and Kader, M.A. (2019). Hybrid composites for automotive applications – a review. *Journal of Reinforced Plastics and Composites* 38: 835–845.

10 Mochane, M.J., Mokhena, T.C., Mokhothu, T.H. et al. (2019). Recent progress on natural fiber hybrid composites for advanced applications: a review. 13 (2): 159–198.

11 Senthilkumar, K., Saba, N., Rajini, N. et al. (2018). Mechanical properties evaluation of sisal fibre reinforced polymer composites: a review. *Construction and Building Materials* https://doi.org/10.1016/j.conbuildmat.2018.04.143.

12 Guna, V., Ilangovan, M., Ananthaprasad, M.G., and Reddy, N. (2018). Hybrid biocomposites. *Polymer Composites* 39: E30–E54.

13 Dong, C. (2018). Review of natural fibre-reinforced hybrid composites. *Journal of Reinforced Plastics and Composites* 37: 331–348. https://doi.org/10.1177/0731684417745368.

14 Li, M., Pu, Y., Thomas, V.M. et al. (2020). Recent advancements of plant-based natural fiber–reinforced composites and their applications. *Composites. Part B, Engineering* 108254.

15 Thakur, V.K., Thakur, M.K., and Pappu, A. (2017). *Hybrid Polymer Composite Materials: Properties and Characterisation*. Woodhead Publishing.

16 Thiagamani, S.M.K., Nagarajan, R., Jawaid, M. et al. (2017). Utilization of chemically treated municipal solid waste (spent coffee bean powder) as reinforcement in cellulose matrix for packaging applications. *Waste Management* 69: 445–454. https://doi.org/10.1016/j.wasman.2017.07.035.

17 Shahroze, R.M., Ishak, M.R., Sapuan, S.M. et al. (2019). Effect of silica aerogel additive on mechanical properties of the sugar palm fiber-reinforced polyester composites. *International Journal of Polymer Science* 2019: 1–4.

18 Shahroze, R.M., Ishak, M.R., Salit, M.S. et al. (2018). Effect of organo-modified nanoclay on the mechanical properties of sugar palm fiber-reinforced polyester composites. *BioResources* 13: 7430–7444.

19 Shahroze, R.M., Ishak, M.R., Salit, M.S. et al. (2019). Sugar palm fiber/polyester nanocomposites: influence of adding nanoclay fillers on thermal, dynamic mechanical, and physical properties. *Journal of Vinyl & Additive Technology* 26 (3): 236–243.

20 Arulmurugan, M., Selvakumar, A.S., Prabu, K., and Rajamurugan, G. (2020). Effect of barium sulphate on mechanical, DMA and thermal behaviour of woven aloevera/flax hybrid composites. *Bulletin of Materials Science* 43: 58.

21 Sarasini, F., Tirillò, J., Sergi, C. et al. (2018). Effect of basalt fibre hybridisation and sizing removal on mechanical and thermal properties of hemp fibre reinforced HDPE. *Composites* https://doi.org/10.1016/j.compstruct.2018.01.046.

22 Monteiro, S.N., Calado, V., Rodriguez, R.J.S., and Margem, F.M. (2012). Thermogravimetric behavior of natural fibers reinforced polymer composites – an overview. *Materials Science and Engineering A* 557: 17–28.

23 Hatakeyama, T. and Hatakeyama, H. (2006). *Thermal Properties of Green Polymers and Biocomposites*. Springer Science & Business Media.

24 Menczel, J.D. and Prime, R.B. (2009). *Thermal Analysis of Polymers: Fundamentals and Applications*. Wiley.

25 Menczel, J.D., Judovits, L., Prime, R.B. et al. (2009). Differential scanning calorimetry (DSC). *Thermal Analysis of Polymers: Fundamentals and Applications*: 7–239.

26 Nurazzi, N.M., Khalina, A., Chandrasekar, M. et al. (2020). Effect of fiber orientation and fiber loading on the mechanical and thermal properties of sugar palm yarn fiber reinforced unsaturated polyester resin composites. *Polimery* 65: 35.

27 Krishnasamy, S., Muthukumar, C., Nagarajan, R. et al. (2019). Effect of fibre loading and $Ca(OH)_2$ treatment on thermal, mechanical, and physical properties of pineapple leaf fibre/polyester reinforced composites. *Materials Research Express* 6: 085545. https://doi.org/10.1088/2053-1591/ab2702.

28 Senthil Muthu Kumar, T., Rajini, N., Obi Reddy, K. et al. (2018). All-cellulose composite films with cellulose matrix and Napier grass cellulose fibril fillers. *International Journal of Biological Macromolecules* 112: 1310–1315. https://doi.org/10.1016/j.ijbiomac.2018.01.167.

29 Krishnasamy, S., Thiagamani, S.M.K., Muthukumar, C. et al. (2019). Effects of stacking sequences on static, dynamic mechanical and thermal properties of completely biodegradable green epoxy hybrid composites. *Materials Research Express.* 6: 105351. https://doi.org/10.1088/2053-1591/ab3ec7.

30 Chandrasekar, M., Ishak, M.R., Sapuan, S.M. et al. (2018). Fabrication of fibre metal laminate with flax and sugar palm fibre based epoxy composite and evaluation of their fatigue properties. *Journal of Polymer Materials* 35: 463–473.

31 Liu, K., Zhang, X., Takagi, H. et al. (2014). Effect of chemical treatments on transverse thermal conductivity of unidirectional abaca fiber/epoxy composite. *Composites. Part A, Applied Science and Manufacturing* 66: 227–236.

32 Arrakhiz, F.Z., Elachaby, M., Bouhfid, R. et al. (2012). Mechanical and thermal properties of polypropylene reinforced with Alfa fiber under different chemical treatment. *Materials and Design* 35: 318–322.

33 Senthil Muthu Kumar, T., Rajini, N., Siengchin, S. et al. (2019). Influence of *Musa acuminate* bio-filler on the thermal, mechanical and visco-elastic behavior of poly(propylene) carbonate biocomposites. *International Journal of Polymer Analysis and Characterization* 24: 439–446. https://doi.org/10.1080/1023666X.2019.1602910.

34 Senthil Muthu Kumar, T., Senthilkumar, K., Chandrasekar, M. et al. (2019). Characterization, thermal and dynamic mechanical properties of poly(propylene carbonate) lignocellulosic *Cocos nucifera* shell particulate biocomposites. *Materials Resreach Express* 6 https://doi.org/10.1088/2053-1591/ab2f08.

35 Ahmetli, G., Deveci, H., Soydal, U. et al. (2012). Epoxy resin/polymer blends: improvement of thermal and mechanical properties. *Journal of Applied Polymer Science* 125: 38–45.

36 Hassan, E., Wei, Y., Jiao, H., and Muhuo, Y. (2013). Dynamic mechanical properties and thermal stability of poly(lactic acid) and poly(butylene succinate) blends composites. *Journal of Fiber Bioengineering and Informatics* 6: 85–94.

37 Mbeche, S.M. and Omara, T. (2020). Effects of alkali treatment on the mechanical and thermal properties of sisal/cattail polyester commingled composites. *PeerJ Materials Science* 2: e5.

38 Dixit, S., Joshi, B., Kumar, P., and Yadav, V.L. (2020). Novel hybrid structural biocomposites from alkali treated-date palm and coir fibers: morphology, thermal and mechanical properties. *Journal of Polymers and the Environment* 28: 2386–2392.

39 Kumar, S.S. (2020). Effect of natural Fiber loading on mechanical properties and thermal characteristics of hybrid polyester composites for industrial and construction fields. *Fibers and Polymers* 21: 1508–1514.

40 Krishnudu, D.M., Sreeramulu, D., and Reddy, P.V. (2020). A study of filler content influence on dynamic mechanical and thermal characteristics of coir and luffa cylindrica reinforced hybrid composites. *Construction and Building Materials* 251: 119040.

41 Chee, S.S., Jawaid, M., Sultan, M.T.H. et al. (2019). Accelerated weathering and soil burial effects on colour, biodegradability and thermal properties of bamboo/kenaf/epoxy hybrid composites. *Polymer Testing* 79: 106054.

42 Pereira, A.L., Banea, M.D., Neto, J.S.S., and Cavalcanti, D.K.K. (2020). Mechanical and thermal characterization of natural intralaminar hybrid composites based on sisal. *Polymers (Basel)* 12: 866.

43 Murugan, M.A., Kumar, A.S.S., and Dineshkumar, C. (2020). Analysis of thermal, dynamic and mechanical properties of hybrid aloevera/hemp FRE bio-composites. *Mater. Today Proc.* 22: 970–975.

44 Asim, M., Jawaid, M., Paridah, M.T. et al. (2019). Dynamic and thermo-mechanical properties of hybridized kenaf/PALF reinforced phenolic composites. *Polymer Composites* 40: 3814–3822.

45 Radzi, A.M., Sapuan, S.M., Jawaid, M., and Mansor, M.R. (2019). Water absorption, thickness swelling and thermal properties of roselle/sugar palm fibre reinforced thermoplastic polyurethane hybrid. *Composites*: 1–7.

46 Liew, F.K., Hamdan, S., Rahman, M.R. et al. (2020). Thermo-mechanical properties of jute/bamboo/polyethylene hybrid composites: the combined effects of silane coupling agent and copolymer. *Polymer Composites* 41 (11): 4830–4841.

47 Radzi, A.M., Sapuan, S.M., Jawaid, M., and Mansor, M.R. (2019). Effect of alkaline treatment on mechanical, physical and thermal properties of roselle/sugar palm fiber reinforced thermoplastic polyurethane hybrid composites. *Fibers and Polymers* 20: 847–855.

48 Jumaidin, R., Sapuan, S.M., Jawaid, M. et al. (2017). Thermal, mechanical, and physical properties of seaweed/sugar palm fibre reinforced thermoplastic sugar palm starch/agar hybrid composites. *International Journal of Biological Macromolecules* 97: 606–615. https://doi.org/10.1016/j.ijbiomac.2017.01.079.

49 Siakeng, R., Jawaid, M., Ariffin, H., and Sapuan, S.M. (2018). Thermal properties of coir and pineapple leaf fibre reinforced polylactic acid hybrid composites. *IOP Conference Series: Materials Science and Engineering* 368 https://doi.org/10.1088/1757-899X/368/1/012019.

50 Siakeng, R., Jawaid, M., Asim, M. et al. (2020). Alkali treated coir/pineapple leaf fibres reinforced PLA hybrid composites: evaluation of mechanical, morphological, thermal and physical properties. *Express Polymer Letters* 14: 717–730.

51 Yorseng, K., Rangappa, S.M., Pulikkalparambil, H. et al. (2020). Accelerated weathering studies of kenaf/sisal fiber fabric reinforced fully biobased hybrid bioepoxy composites for semi-structural applications: morphology, thermo-mechanical, water absorption behavior and surface hydrophobicity. *Construction and Building Materials* 235: 117464. https://doi.org/10.1016/j.conbuildmat.2019.117464.

52 Senthilkumar, K., Ungtrakul, T., Chandrasekar, M. et al. (2020). Performance of sisal/hemp bio-based epoxy composites under accelerated weathering. *Journal of Polymers and the Environment*: 1–13.

53 Saba, N., Jawaid, M., Alothman, O.Y., and Paridah, M.T. (2016). A review on dynamic mechanical properties of natural fibre reinforced polymer composites. *Construction and Building Materials* 106: 149–159. https://doi.org/10.1016/j.conbuildmat.2015.12.075.

54 Saba, N. and Jawaid, M. (2018). A review on thermomechanical properties of polymers and fibers reinforced polymer composites. *Journal of Industrial and Engineering Chemistry* 67: 1–11. https://doi.org/10.1016/j.jiec.2018.06.018.

55 Chandrasekar, M., Ishak, M.R., Sapuan, S.M. et al. (2017). A review on the characterisation of natural fibres and their composites after alkali treatment and water absorption. *Plastics, Rubber and Composites* https://doi.org/10.1080/14658011.2017.1298550.

2

Thermal Properties of Hybrid Natural Fiber-Reinforced Thermoplastic Composites

A. Vinod, Yashas Gowda, Senthilkumar Krishnasamy, M.R Sanjay, and Suchart Siengchin

Natural Composites Research Group Lab, The Sirindhorn International Thai-German Graduate School of Engineering (TGGS), King Mongkut's University of Technology North Bangkok, 1518 Pracharat 1 road, Wongsawang, Bangsue, Bangkok, 10800, Thailand

2.1 Introduction

The prolonged use of synthetic materials is causing harmful effects on the environment through production and disposal. The advent of hazards on the environment has influenced many industrialists and researchers to adopt ecofriendly materials like natural fibers as reinforcements in thermoplastic polymer composites. Over the past few decades, due to the industrial revolution, many natural fibers like jute, sisal, and hemp are used to develop components in the automotive, household, and civil constructions. This extensive use of natural resources has created a demand and an urge to identify new potential resources [1–5]. In this regard, many new natural fibers like *Tridax procumbens, Saccharum bengalense, Parthenium hysterophorus,* and *Catharanthus roseus* are identified as potential reinforcements in polymer matrices [6–9]. The plant fibers are considered as better reinforcements due to their remarkable properties like low density, biodegradability, availability, and processability. Plant fiber-reinforced lightweight structures in automobiles improved its performance by improving the fuel economy and reducing the landfill on disposal. The natural fibers have other good properties like excellent electrical, thermal, and acoustic insulation properties [9–11].

Despite their advantages, there are certain deficiencies like lower stiffness, lower thermal stability, strength degradation with respect to time, water sensitivity, lower impact resistance, and easily prone to bacterial and fungal attacks. In many cases, a single type of reinforcement cannot offer the desirable properties required for the application. To overcome the issue, hybrid composites are developed, which is a combination of two or more materials like natural–natural or natural–synthetic [12]. The reinforcements in hybrid thermoplastic polymer composite may be in the form of fillers, mats, or fibers. The hybridization of natural fiber composites provides an option for achieving a blend of high strength to weight ratio, good thermal stability,

Natural Fiber-Reinforced Composites: Thermal Properties and Applications, First Edition.
Edited by Senthilkumar Krishnasamy, Senthil Muthu Kumar Thiagamani, Chandrasekar Muthukumar, Rajini Nagarajan and Suchart Siengchin.

and durable components when compared to a single type of reinforcements. This approach of hybridization mainly helps in achieving cost-effectiveness and reduces the use of synthetic materials which pollutes the environment [13].

The thermal behavior of the composite is a vital property that defines the functionality of the materials in temperature-sensitive environments. Applications of composites in automobiles, aerospace, and industrial insulation are mostly prone to temperature fluctuation. It is essential to study the thermal aspects of the hybrid composite to develop a more sustainable and durable material that is suitable for commercialization [14]. Various techniques involving the thermal study of natural fiber hybrid composites are discussed in the forthcoming sections 2.2.1 to 2.2.4.

2.2 Thermal Properties

Thermal properties like thermal stability, glass transition temperature (T_g), storage modulus (E'), loss modulus (E''), and melt flow properties of composites, must be evaluated before using in versatile temperature-fluctuating applications [15, 16]. These properties help to tailor the properties of thermoplastic polymer composites to enhance their thermal aspects for suitability and sustainability in the application. Currently, there are various thermoplastic polymers like polypropylene (PP), polyethylene (PE), polystyrene (PS), polylactic acid (PLA), polyamide, and polyhydroxyalkonates (PHA), which are commonly used as a matrix in natural fiber hybrid composites [17, 18]. Thermoplastic polymers are very much sensitive to temperature, which results in phase transition in elevated temperatures. The phase transition temperatures of the thermoplastic polymers can be altered by reinforcing them with various natural fibers or with the combination of natural and synthetic fibers. These kinds of reinforcement block the heat flow and restrict the polymer chain movement by elevating the glass transition temperature (T_g), which subsequently raises the temperature-withstanding ability. These reinforcements play a vital role in controlling certain factors such as heat diffusion, thermal insulation capacity, melt flow, phase transition temperature, etc. Hence, it is very crucial to study the thermal properties of composites. A detailed survey about the thermal properties of hybrid natural fiber-based thermoplastic composites is given in the forthcoming sections 2.2.1 to 2.2.4. The current work will be very beneficial for the researchers, industries, scientists, and academicians to know about the recent advances in testing procedures, characterization principles, and thermal properties of recently identified hybrid thermoplastic composites.

2.2.1 Thermogravimetric Analysis (TGA)

TGA analysis is performed to measure the rate of change in mass with a function of temperature and time in a controlled environment and to predict the thermal stability and the composition of the composite. Instruments like Mettler Toledo (Model: TGA 2 SF) and Perkin Elmer (Model: TGA 8000 N5320011, N5320010, TGA 4000 system) are commonly used to perform the analysis [19]. A sample of 5 mg is placed in

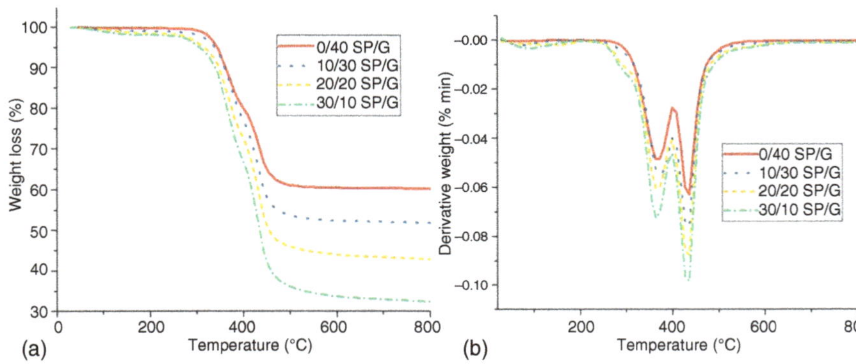

Figure 2.1 (a) TGA sugar palm/glass fiber composite (b) DTG Sugar palm/glass fiber composite. Source: Atiqah et al. [21].

an aluminum crucible within a furnace. The test is carried out in a controlled environment at a heating rate of 10 °C min^{-1} within the range of 30–1000 °C depending upon the composite's composition. The resulting thermogram is the graphical representation of temperature vs. mass, which provides information about thermal stability, compositions of initial samples, and the intermediate compounds formed during heating. These result from physical phenomena like phase transitions, absorption and desorption, solid–gas reactions, and decomposition with respect to temperature. The TGA analysis also provides derivative thermogravimetry charts (DTG), which is the first derivative of thermogravimetry. Furthermore, it helps in the investigation of reaction kinetics, applied kinetics, oxidative degradation, and oxidative stability [20]. The highest peak of the DTG curve at any temperature gives the rate of loss in mass. In case of overlapping reactions, there are difficulties in predicting the initial and final degradation temperatures using TG curves, while the DTG curves provide precise information on initial and final degradation temperatures. A model TGA and DTG results are presented in Figure 2.1.

Figure 2.1a,b shows the thermogravimetry results of sugar palm/glass fiber-reinforced thermoplastic polyurethane hybrid composites. DTG curves from Figure 2.1b show that the increased addition of SPF (30/10 SPF/GF) to the TPU matrix increased the weight loss at temperature T_{max} 435 °C. Furthermore, it reduced the decomposition temperature of the TPU hybrid composites. It is concluded that the reduction in GF and increase in the SPF did not affect the thermal stability of composites [21]. A similar behavior is observed in jute–glass epoxy thermoset composites [22]. When glass fibers are hybridized with bamboo fibers in the polypropylene matrix, the resulting thermograms reveal an increase in thermal stability for 15% bamboo fiber/15% glass fiber/2% maleic anhydride coupling agent, due to increased bonding and the higher thermal stability of glass fibers [23]. Various studies showed that the hybridization of synthetic fibers with natural fibers provides desirable thermal stability, and the use of synthetic materials is reduced by the substitution of natural fibers. Moreover, the resulting thermal properties are comparable with those of synthetic reinforcements. Similarly, when glass fibers and short hemp fibers are hybridized in polypropylene, the TGA results reveal an increase in thermal stability with respect to the increase in glass fibers, due to the higher thermal stability of synthetic glass. The thermal stability of composite with 15% glass and 25% hemp is 474 °C with 18.5% residue [24].

Sugar palm fibers are hybridized with roselle fiber in the TPU matrix. The results indicated an increase in thermal stability for 50/50 RF/SPF. Hybridization also results in an improvement in the initial degradation phase and final degradation temperature [25]. In recent research, when areca fibers are hybridized with moringa fibers in polyethylene terephthalate matrix, the thermal stability is increased with a final degradation temperature at 640 °C, which was much higher than the single fiber-reinforced composites. The alkali treatment also helps in increasing thermal stability by inducing interactions with the matrix [26]. In a similar research, the hybridization of the PLA matrix using alkali-treated kenaf (KF) (15%) and aloe vera fibers (15%) improves thermal stability (343 °C), which is comparatively higher than that of single fiber-reinforced PLA composites [27]. Hence, the thermal stability is improved through the hybridization of natural–natural or natural–synthetic fibers in thermoplastic polymer matrices.

2.2.2 Differential Scanning Calorimetry (DSC)

Differential scanning calorimetry (DSC) is used to measure the heat flow to and from the sample as a function of temperature. The analysis is carried out by heating, cooling, or holding the sample at a constant temperature with respect to a reference sample [28]. The energy absorbed or released by the sample is measured in milliwatts. The procedure is carried out by placing 3–10 mg of samples in an aluminum crucible. The samples are later heated from room temperature to the required temperature at a rate of 10–20 °C min^{-1} in a controlled nitrogen or oxygen environment. Both powdered and flat samples can be used for analysis. For flat samples, care should be taken that the sample surface be at full contact with the crucible to enable proper heat conduction. DSC analysis is most commonly used to analyze the glass transition T_g, cold crystallization, recrystallization, melting, oxidation induction time, and decomposition. Furthermore, it can be used to analyze specific heat capacity, crystallinity, melting behavior, purity of crystalline non-polymeric substances, vaporization, sublimation, desorption, solid–solid transitions, polymorphism, oxidative stability, and compositional analysis [29]. A model DSC curve of polyhydroxy butyrate-*co*-valerate (PHBV) matrix hybridized with talc and wood fiber is presented in Figure 2.2. The results suggest a reduction in the melting temperature (T_m) due to the addition of talc, due to heterogeneous crystal morphology shifting toward the homogenization of the spherulite growth. The crystallization temperature (T_c) increased from 103 to 107 °C in the presence of talc, thereby indicating the increase in the crystallization rate of PHBV during the cooling stage which shows good interaction between the talc and PHBV.

Singh et al. [30] hybridized HDPE using oil-palm fiber and clay. The DSC results concluded that the crystallinity of the composites depends on the dispersion of the fillers, which acts as a nucleating agent in the polymer matrices. Furthermore, it was observed that the enthalpy of melting ΔH_m was decreased from 183 to 143, and

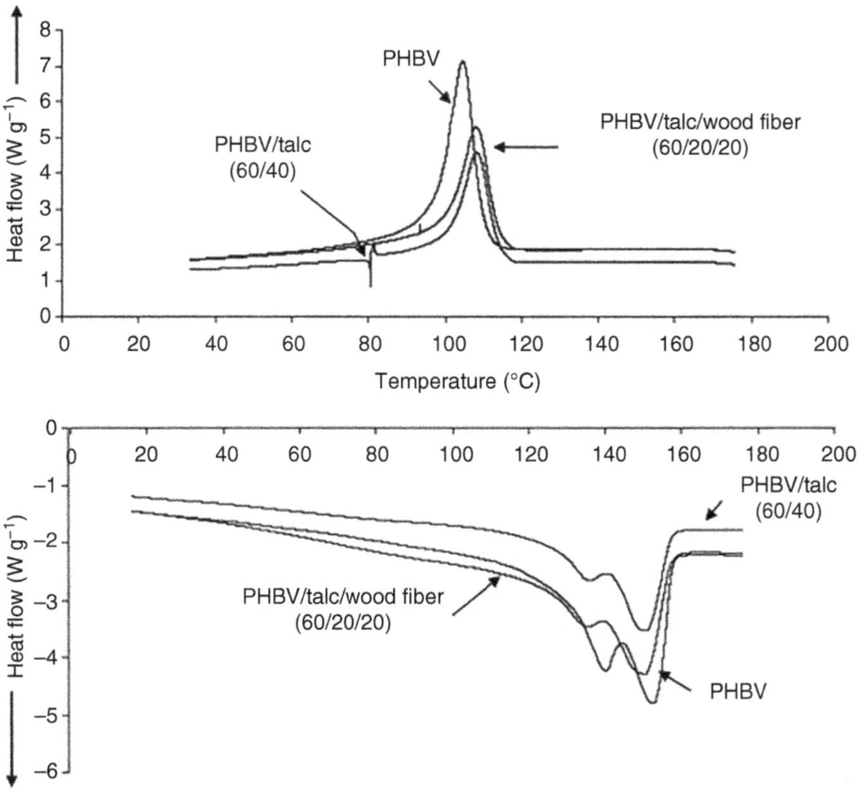

Figure 2.2 DSC curves of PHBV and its composites. Source: Singh et al. [30].

137 J g^{-1} for composites reinforced with 25 wt% of fiber and clay, respectively, when compared to neat HDPE. This indicated the addition of 25% of fillers enhances the thermal stability of the HDPE [31]. Mariam et al. investigated the effect of hybridization on recycled polypropylene using wood floor and glass fibers and concluded that there was no significant difference in the melting temperature (T_m). Inducing wood floor fillers hindered the crystallinity %, which then increased due to the addition of glass fibers. Further, it can be observed that the addition of wood floor at a certain percentage can reduce the manufacturing cost without hindering the performance of the composite [32]. Thermoplastic starch can be hybridized with wood fiber and calcium carbonate filler, and based on the DSC curves the glass transition temperature and melting point were not significantly different. However, there was a slight increase in T_g and melting point due to the effect of hybridization [33].

2.2.3 Thermomechanical Analysis (TMA)

Thermomechanical analysis (TMA) is used to measure the dimensional deformation in the samples during heating or cooling. In other words, it is used to determine the dimensional stability of the composites [34]. Most commonly, TMA analysis is used to measure the coefficient of thermal expansion (CTE), glass transition temperature

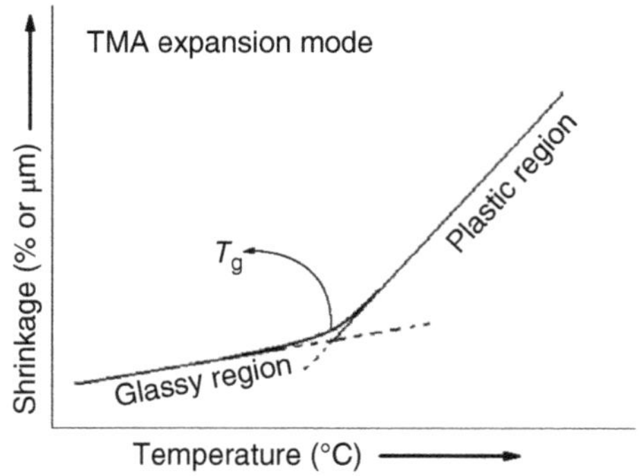

Figure 2.3 Model curve of TMA.

T_g, young's modulus, and the swelling behavior of composites in solvents. The experiment is carried out by holding flat surface samples in two silica plates and expansion quartz probes. Extreme care must be taken that the sample must have flat contact with silica plates to obtain better results. The samples are later placed in a furnace, and the analysis is carried out in nitrogen, oxygen, or ambient air from 30 to 200 °C, in consideration of the composite at a heating rate of 10 °C min^{-1}. Normally 1 N load is applied over the sample to measure the dimensional changes [35].

Figure 2.3 clearly shows solid transition phases of a composite through TMA curves. Idumah and Hassan [36] hybridized polypropylene using kenaf fiber and graphene nanoparticles (GNPs) and concluded that the inclusion of 3% GNP in polypropylene/kenaf/MAPP (maleic anhydride polypropylene)/GNP decreases the CTE; however, CTE was increased above 3 Phr loading of GNP. Asim et al. [37] investigated the thermal, physical, and flammable properties of silane-treated kenaf (KF)/pineapple leaf fiber (PALF) phenolic formaldehyde (PF) hybrid composite. It is observed that the composite that contained (50% PF, 35% PALF, 15% KF) and (50% PF, 15% PALF, 35% KF) showed excellent structural stability due to the good interfacial bonding caused by restrictions of intermolecular movement of polymer chains during elevated temperatures. The chemical treatment on fibers in hybrid composite establishes good structural stability in higher temperatures. Siakeng et al. [38] hybridized PLA using coir and PALFs and found that the composite with 70% PLA, 21% coir, and 9% PALF had less deformation in the plastic region than neat PLA. Lee and Kim [39] investigated the wood/polypropylene/clay nanocomposites and reported that the composite having 1 Phr of montmorillonite coisite20A (nano clay) showed the lowest thermal deformation than the composites without nano clay because the good dispersion of clay particles had good interaction with the polypropylene matrix, thus improving the T_g and also the structural stability of composites at elevated temperatures. Cho and Lee [40] investigated the hybridization effect on silk/wool/polybutylene succinate (PBS) composites. The discarded fibers were processed by electron beam irradiation in various doses 1, 5, 10, 30, and 50 kGy. The PBS was loaded with 30% fibers. TMA results showed less CTE for 50 kGy treated fiber showed less CTE than other PBS composites: 0.87×10^{-4}/°C in the range 40–60 °C and 1.28×10^{-4}/°C in the field 60–80°C. It was due to enhanced adhesion as a result of fiber treatment that restricted molecular mobility in the matrix at high temperatures.

2.2.4 Dynamic Mechanical Analysis (DMA)

The dynamic mechanical analysis (DMA) is used to study the viscoelastic behaviors like storage modulus E', loss modulus E'', and tan delta (tan δ) of the developed composites as a function of time, temperature, and frequency [41]. Storage modulus defines the energy stored by the sample in its elastic limit, loss modulus is the energy dissipated by the sample in the form of heat or the viscous part, and the tan δ is the damping factor of the material. Properties like T_g, amorphous softening, elastic modulus, shear modulus, and mechanical damping can be predicted through DMA. Furthermore, it is used to predict complex modulus, complex/dynamic viscosity, creep compliance, and relaxation modulus [42]. Various modes of deformation analysis in DMA include dual/single cantilever, three-point bending, shear sandwich, compression, and tensile and submersible clamping. A sample DMA curve is presented in Figure 2.4.

Figure 2.4 shows an improvement in loss modulus T_g when compared to other PP composites. The tan δ is also reduced due to the increase in stiffness of the composite resulting from the addition of fillers, while the neat PP has the highest damping factor due to the absence of reinforcements, due to the free movement of

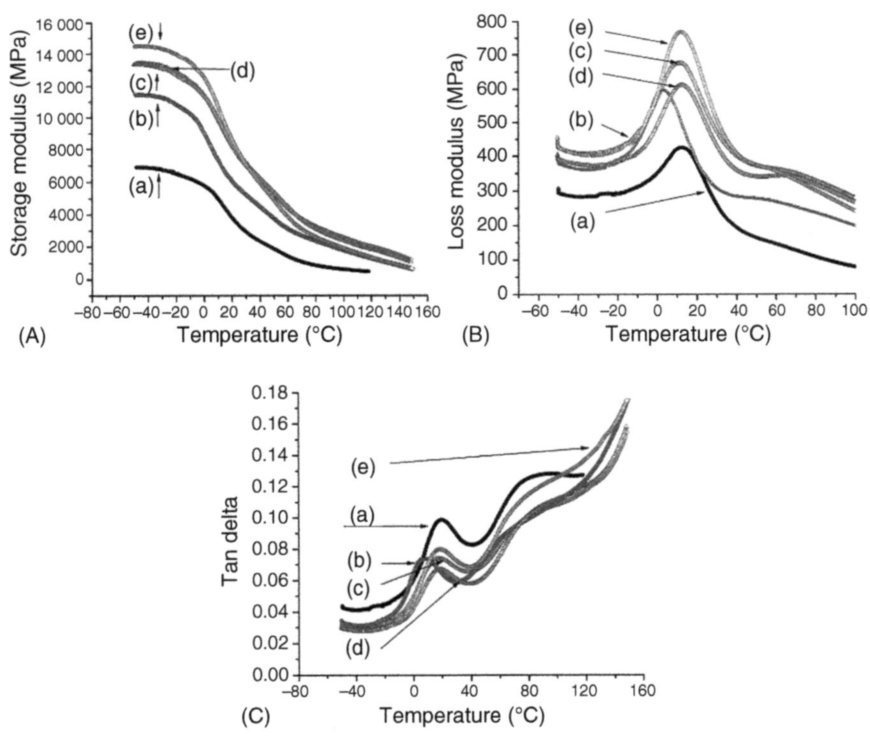

Figure 2.4 Temperature dependence of (A) storage modulus, (B) loss modulus, and (C) tan delta of polypropylene (PP) and PP-based composites: (a) neat PP, (b) PP/newspaper (70%/30%), (c) PP/newspaper/talc 1 (60%/30%/10%), (d) PP/newspaper/talc 2 (60%/30%/10%), and (e) PP/newspaper/ talc 3 (60%/30%/10%). Talc1-Silverline 002, Talc 2-Nicron 403, and Talc 3-Mistron CB. Source: Huda et al. [43].

molecular chains [43]. In one study, Atiqah et al., hybridized sugar palm fiber with glass fiber in the polyurethane matrix, showing that the storage modulus and loss modulus increased with an increase in the percentage of sugar palm fiber compared to glass fibers. However, there was a decrement in tan δ because of an increase in stiffness of the composite [21]. In another research, curaua composites were hybridized using glass fiber in unsaturated polyester resin and it was concluded the increase in the glass fiber content increased its storage and loss modulus, which was due to the increased polymer bond of glass fibers compared to that of curaua fibers [44]. Recently, bamboo fibers have been hybridized with glass fibers and treated with MAPP. The loss modulus and storage modulus are found to increase for MAPP-treated composite, but tan δ is reduced because of increased stiffness due to MAPP treatment [23]. Similarly, when sisal fibers are hybridized with glass fibers in the polypropylene matrix, the thermoplastic hybrid composite with 15% sisal + 15% glass fiber + PP + 2% MAPP shows a higher storage and loss modulus with reduced damping factor [45]. Gerezgiher et al. hybridized thermoplastic starch with wood fiber and calcium carbonate fillers. The DMA result revealed an increase in T_g, storage modulus, and loss modulus and these were higher than those of pure TPS and single reinforced composites [33].

The dynamic mechanical properties of the thermoplastic hybrid composites depend on the type of polymer used, the type of reinforcement, and also the properties of the reinforcements. The enhancement of the T_g in thermoplastic hybrid composites depends on the bonding behavior of the reinforcements and the matrix. In most cases, the improvement in the mechanical properties is reflected in the enhancement of the storage and loss modulus. This is caused by the improvement in the bonding of reinforcement with matrix and an increase in stiffness of the composite. The increase in stiffness of the composite results in the decreased damping factor that is noted from tan δ curves. The increment in the damping factor is mainly due to the presence of voids and improper bonding in hybrid composites.

2.2.5 Melt Flow Index (MFI)

Melt flow index (MFI) is a critical property responsible for the polymer flow under the melting point with the application of standard load (3.2 kg). MFI is the flowability of the thermoplastic materials. The mass flow rate of the polymer is measured for 10 minutes and compared with the standard polymer. A standard temperature of 230 °C is chosen, and MFI is calculated by mass in grams flowing per 10 minutes. The test is mostly carried out according to ASTM D1238-86 or ISO R1133 standard. An orifice or a die with a specific size is used and placed in a heated barrel, and a piston with a load on top is inserted in the heated barrel to push the polymer out of the die or orifice [46] (Figure 2.5).

One study hybridized date palm wood floor with glass fibers in a recycled polypropylene matrix. It is observed that the MFI is decreased with an increase in the addition of wood floor and further increased replacement of wood floor with glass fiber [32]. Similarly, when nylon 6 is reinforced with flax, hemp, and kenaf fibers, the viscosity of the matrix was increased due to the presence of 20% hemp in nylon 6 when compared to other composites [48]. When Oil palm empty fruit

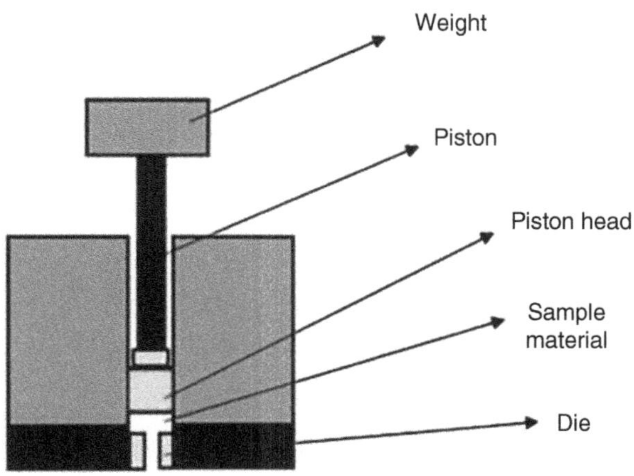

Figure 2.5 Basic structure of melt flow index tester. Source: Singh et al. [47].

bunch fibers were hybridized with glass fibers in recycled polypropylene matrix using coupling agents Polybond 3200 and fusabond P613. It is found the MFI of composite with 60% recycled polypropylene, 28% oil palm fruit bunch fiber, 12% glass fiber with 2.2% of both Polybond 3200 and fusabond P613 exhibited MFI of 1.80 g/10 min, which is much lower than pure polypropylene. The viscosity of the matrix is increased more when reinforced with natural and glass fibers and further with the inclusion of coupling agents, because of increased interaction of fibers with the matrix [49].

MFI is a very vital parameter in case thermoplastic composites are used in temperature-sensitive applications. The viscosity of the thermoplastic polymers is very complicated due to their chemical structure, composition, and processing environment. Furthermore, the viscous behavior of thermoplastics was highly influenced by factors like temperature, pressure, and shear rate. Moreover, the viscous behavior or MFI can be controlled by reinforcements in the thermoplastic matrices.

2.3 Conclusions

The current comprehensive study gives extensive data on the thermal properties of hybrid thermoplastic composites through TGA, DSC, TMA, DMA, and MFI. TGA analysis indicates that the thermal stability of the hybrid composites depends on the type of fiber, chemical composition, and thermal stability of individual fiber used in the composite. The use of surface modification techniques and the addition of coupling agents improve the thermal stability in composites. The DSC analysis provides clear data about the reaction within the composites due to the temperature rise. It is also influenced by the fiber–matrix interactions. The DMA analysis is very important in investigating the amount of energy stored or lost when the chemically modified fibers were used as reinforcements. It helps to study the matrix interaction in filler reinforced hybrid composites. The TMA analysis is widely performed on hybrid composites to study thermal expansion. In conclusion, thermal expansion is reduced with an increase in matrix adhesion to the fibers. MFI indicates that the viscosity of the thermoplastic polymer is increased with the addition of reinforcements. The viscosity is further increased with the further increase in the reinforcement.

References

1 Vinod, A., Sanjay, M.R., Siengchin, S., and Parameswaranpillai, J. (2020). Renewable and sustainable biobased materials: an assessment on biofibers, biofilms, biopolymers and biocomposites. *Journal of Cleaner Production*: 120978. https://doi.org/10.1016/j.jclepro.2020.120978.

2 Bazan, P., Nosal, P., Kozub, B., and Kuciel, S. (2020). Biobased polyethylene hybrid composites with natural fiber: mechanical, thermal properties, and micromechanics. *Materials (Basel)* 13: 2967.

3 Abdellaoui, H., Raji, M., Essabir, H. et al. (2018). Mechanical behavior of carbon/natural fiber-based hybrid composites. *Mechanical and Physical Testing of Biocomposites, Fibre-Reinforced Composites and Hybrid Composites*: 103–122. https://doi.org/10.1016/B978-0-08-102292-4.00006-0.

4 Sanjay, M.R. and Siengchin, S. (2018). Natural fibers as perspective materials. *International Journal of Applied Science and Technology* 11 (4): 233.

5 Sanjay, M.R., Siengchin, S., and Dhakal, H.N. (2020). Green-composites: ecofriendly and sustainability. *Applied Science and Engineering Progress* 13 (3): 183–184.

6 Vijay, R., Singaravelu, D.L., Vinod, A. et al. (2019). Characterization of alkali-treated and untreated natural fibers from the stem of *Parthenium hysterophorus*. *Journal of Natural Fibers*: 1–11. https://doi.org/10.1080/15440478.2019.1612308.

7 Vijay, R., Singaravelu, D.L., Vinod, A. et al. (2019). Characterization of novel natural fiber from *Saccharum bengalense* grass (Sarkanda). *Journal of Natural Fibers*: 1–9. https://doi.org/10.1080/15440478.2019.1598914.

8 Vijay, R., Lenin Singaravelu, D., Vinod, A. et al. (2019). Characterization of raw and alkali treated new natural cellulosic fibers from *Tridax procumbens*. *International Journal of Biological Macromolecules* 125: 99–108. https://doi.org/10.1016/j.ijbiomac.2018.12.056.

9 Vinod, A., Vijay, R., Singaravelu, D.L. et al. (2019). Characterization of untreated and alkali treated natural fibers extracted from the stem of *Catharanthus roseus*. *Materials Research Express* 6: 085406. https://doi.org/10.1088/2053-1591/ab22d9.

10 Joseph, J., Munda, P.R., Kumar, M. et al. (2020). Sustainable conducting polymer composites: study of mechanical and tribological properties of natural fiber reinforced PVA composites with carbon nanofillers. *Polymer-Plastics Technology and Materials* 59: 1088–1099. https://doi.org/10.1080/25740881.2020.1719144.

11 Yorseng, K., Rangappa, S.M., Pulikkalparambil, H. et al. (2020). Accelerated weathering studies of kenaf/sisal fiber fabric reinforced fully biobased hybrid bioepoxy composites for semi-structural applications: morphology,

thermo-mechanical, water absorption behavior and surface hydrophobicity. *Construction and Building Materials* 235: 117464. https://doi.org/10.1016/j.conbuildmat.2019.117464.

12 Kumaran, P., Mohanamurugan, S., Madhu, S. et al. (2019). Investigation on thermo-mechanical characteristics of treated/untreated *Portunus sanguinolentus* shell powder-based jute fabrics reinforced epoxy composites. *Journal of Industrial Textiles*: 1528083719832851. https://doi.org/10.1177/1528083719832851.

13 Jothibasu, S., Mohanamurugan, S., and Vinod, A. (2018). Influence of chemical treatments on the mechanical characteristics of areca sheathflax fibres based epoxy composites. *Rasayan Journal of Chemistry* 11: 1255–1262. https://doi.org/10.31788/RJC.2018.1133096.

14 Athith, D., Sanjay, M.R., Yashas Gowda, T.G. et al. (2018). Effect of tungsten carbide on mechanical and tribological properties of jute/sisal/E-glass fabrics reinforced natural rubber/epoxy composites. *Journal of Industrial Textiles* 48: 713–737. https://doi.org/10.1177/1528083717740765.

15 Atagur, M., Seki, Y., Pasaoglu, Y. et al. (2020). Mechanical and thermal properties of *Carpinas betulus* fiber filled polypropylene composites. *Polymer Composites* 41: 1925–1935. https://doi.org/10.1002/pc.25508.

16 Gan, P.G., Sam, S.T., Abdullah, bin, M.F., and Omar, M.F. (2020). Thermal properties of nanocellulose-reinforced composites: a review. *Journal of Applied Polymer Science* 137: 48544. https://doi.org/10.1002/app.48544.

17 Grünewald, J., Parlevliet, P., and Altstädt, V. (2017). Manufacturing of thermoplastic composite sandwich structures: a review of literature. *Journal of Thermoplastic Composite Materials* 30: 437–464. https://doi.org/10.1177/0892705715604681.

18 Borah, J.S. and Kim, D.S. (2016). Recent development in thermoplastic/wood composites and nanocomposites: a review. *Korean Journal of Chemical Engineering* 33: 3035–3049. https://doi.org/10.1007/s11814-016-0183-6.

19 Asim, M., Paridah, M.T., Chandrasekar, M. et al. (2020). Thermal stability of natural fibers and their polymer composites. *Iranian Polymer Journal* 29: 625–648. https://doi.org/10.1007/s13726-020-00824-6.

20 Duan, Q., Jiang, T., Xue, C. et al. (2020). Preparation and characterization of starch/enteromorpha/nano-clay hybrid composites. *International Journal of Biological Macromolecules* 150: 16–22. https://doi.org/10.1016/j.ijbiomac.2020.01.283.

21 Atiqah, A., Jawaid, M., Sapuan, S.M. et al. (2018). Thermal properties of sugar palm/glass fiber reinforced thermoplastic polyurethane hybrid composites. *Composite Structures* 202: 954–958. https://doi.org/10.1016/j.compstruct.2018.05.009.

22 Braga, R.A. and Magalhaes, P.A.A. (2015). Analysis of the mechanical and thermal properties of jute and glass fiber as reinforcement epoxy hybrid composites. *Materials Science and Engineering: C* 56: 269–273. https://doi.org/10.1016/j.msec.2015.06.031.

23 Samal, S.K., Mohanty, S., and Nayak, S.K. (2009). Polypropylene–bamboo/glass fiber hybrid composites: fabrication and analysis of mechanical, morphological, thermal, and dynamic mechanical behavior. *Journal of Reinforced Plastics and Composites* 28: 2729–2747. https://doi.org/10.1177/0731684408093451.

24 Panthapulakkal, S. and Sain, M. (2007). Injection-molded short hemp fiber/glass fiber-reinforced polypropylene hybrid composites – mechanical, water absorption and thermal properties. *Journal of Applied Polymer Science* 103: 2432–2441. https://doi.org/10.1002/app.

25 Radzi, A.M., Sapuan, S.M., Jawaid, M., and Mansor, M.R. (2019). Water absorption, thickness swelling and thermal properties of roselle/sugar palm fibre reinforced thermoplastic polyurethane hybrid composites. *Journal of Materials Research and Technology* 8: 3988–3994. https://doi.org/10.1016/j.jmrt.2019.07.007.

26 Nayak, S., Khuntia, S.K., Mohanty, S.D. et al. (2020). An experimental study of physical, mechanical and morphological properties of alkali treated moringa/areca based natural fiber hybrid composites. *Journal of Natural Fibers*: 1–12. https://doi.org/10.1080/15440478.2020.1758282.

27 Ramesh, P., Prasad, B.D., and Narayana, K.L. (2020). Characterization of kenaf/aloevera fiber reinforced PLA-hybrid biocomposite. *Advances in Applied Mechanical Engineering*: 1061–1067.

28 Agarwal, J., Mohanty, S., and Nayak, S.K. (2020). Influence of cellulose nanocrystal/sisal fiber on the mechanical, thermal, and morphological performance of polypropylene hybrid composites. *Polymer Bulletin* https://doi.org/10.1007/s00289-020-03178-4.

29 Barczewski, M., Mysiukiewicz, O., Andrzejewski, J. et al. (2020). Thermo-mechanical and mechanical behavior of hybrid isotactic polypropylene glass fiber reinforced composites (GFRC) modified with calcium carbonate ($CaCO_3$). *Polymer Engineering and Science* https://doi.org/10.1002/pen.25404.

30 Singh, S., Mohanty, A.K., and Misra, M. (2010). Hybrid bio-composite from talc, wood fiber and bioplastic: fabrication and characterization. *Composites: Part A* 41: 304–312. https://doi.org/10.1016/j.compositesa.2009.10.022.

31 Essabir, H., Boujmal, R., Ouadi, M. et al. (2016). Mechanics of materials mechanical and thermal properties of hybrid composites: oil-palm fiber/clay reinforced high density polyethylene. *Mechanics of Materials* 98: 36–43. https://doi.org/10.1016/j.mechmat.2016.04.008.

32 Almaadeed, M.A., Kahraman, R., Khanam, P.N., and Madi, N. (2012). Date palm wood flour/glass fibre reinforced hybrid composites of recycled polypropylene: mechanical and thermal properties. *Journal of Materials* 42: 289–294. https://doi.org/10.1016/j.matdes.2012.05.055.

33 Gerezgiher, A.G., Tamási, K., Ibrahim, J.F.M. et al. (2020). Thermoplastic corn starch reinforced with pine wood fibre and calcium carbonate precipitate filler. *Journal of Physics Conference Series* 1527: 012042. https://doi.org/10.1088/1742-6596/1527/1/012042.

34 Ayyagari, S., Al-Haik, M., Ren, Y., and Nepal, D. (2020). Effect of nano-reinforcement topologies on the viscoelastic performance of carbon nanotube/carbon fiber hybrid composites. *Nanomaterials* 10: 1213. https://doi.org/10.3390/nano10061213.

35 Abrisham, M., Panahi-Sarmad, M., Mir Mohamad Sadeghi, G. et al. (2020). Microstructural design for enhanced mechanical property and shape memory behavior of polyurethane nanocomposites: role of carbon nanotube, montmorillonite, and their hybrid fillers. *Polymer Testing* 89: 106642. https://doi.org/10.1016/j.polymertesting.2020.106642.

36 Idumah, C.I. and Hassan, A. (2016). Characterization and preparation of conductive exfoliated graphene nanoplatelets kenaf fibre hybrid polypropylene composites. *Synthetic Metals* 212: 91–104. https://doi.org/10.1016/j.synthmet.2015.12.011.

37 Asim, M., Paridah, M.T., Saba, N. et al. (2018). Thermal, physical properties and flammability of silane treated kenaf/pineapple leaf fibres phenolic hybrid composites. *Composite Structures* 202: 1330–1338. https://doi.org/10.1016/j.compstruct.2018.06.068.

38 Siakeng, R., Jawaid, M., Ariffin, H., and Sapuan, S.M. (2018). Mechanical, dynamic, and thermomechanical properties of coir/pineapple leaf fiber reinforced polylactic acid hybrid biocomposites. *Polymer Composites* 40: 2000–2011. https://doi.org/10.1002/pc.24978.

39 Lee, H. and Kim, D.S. (2010). Preparation and physical properties of wood/polypropylene/clay nanocomposites. *Journal of Applied Polymer Science* 116: 2769–2776. https://doi.org/10.1002/app.

40 Cho D. and Lee H.Y. (2015). Waste silk/wool hybrid biocomposites with PBS matrix: processing, properties and E-beam treatment effect. *International Conference on Composite Materials*.

41 Azammi, A.M.N., Sapuan, S.M., Ishak, M.R., and Sultan, M.T.H. (2020). Physical and damping properties of kenaf fibre filled natural rubber/thermoplastic polyurethane composites. *Defence Technology* 16: 29–34. https://doi.org/10.1016/j.dt.2019.06.004.

42 Nurazzi, N.M., Khalina, A., Sapuan, S.M. et al. (2020). Thermal properties of treated sugar palm yarn/glass fiber reinforced unsaturated polyester hybrid composites. *Journal of Materials Research and Technology* 9: 1606–1618. https://doi.org/10.1016/j.jmrt.2019.11.086.

43 Huda, M.S., Drzal, L.T., Mohanty, A.K., and Misra, M. (2007). The effect of silane treated- and untreated-talc on the mechanical and physico-mechanical properties of poly(lactic acid)/newspaper fibers/talc hybrid composites. *Composites Part B: Engineering* 38: 367–379. https://doi.org/10.1016/j.compositesb.2006.06.010.

44 Ornaghi, H.L. Jr., da Silva, H.S.P., Zattera, A.J., and Amico, S.C. (2011). Hybridization effect on the mechanical and dynamic mechanical properties of curaua composites. *Materials Science and Engineering A* 528: 7285–7289. https://doi.org/10.1016/j.msea.2011.05.078.

45 Nayak, S.L. and Mohanty, S. (2010). sisal glass fiber reinforced PP hybrid composites: effect of MAPP on the dynamic mechanical and thermal properties. *Journal of Reinforced Plastics and Composites* 29: 1551–1568. https://doi.org/10.1177/0731684409337632.

46 Shayan Asenjan, M., Sabet, S.A.R., and Nekoomanesh, M. (2020). Mechanical and high velocity impact performance of a hybrid long carbon/glass fiber/polypropylene thermoplastic composite. *Iranian Polymer Journal* 29: 301–307. https://doi.org/10.1007/s13726-020-00794-9.

47 Singh, R., Kumar, R., and Ahuja, I.S. (2017). Thermal analysis for joining of dissimilar polymeric materials through friction stir welding. In: *Reference Module in Materials Science and Materials Engineering*, 1–15. Elsevier https://doi.org/10.1016/B978-0-12-803581-8.10390-X.

48 Ozen, E., Kiziltas, A., Kiziltas, E.E., and Gardner, D.J. (2013). Natural fiber blend – nylon 6 composites. *Polymer Composites* 34: 544–553. https://doi.org/10.1002/pc.

49 Rivai, M., Gupta, A., Islam, M.R., and Beg, M.D.H. (2014). Characterization of oil palm empty fruit bunch and glass fibre reinforced recycled polypropylene hybrid composites. *Fibers and Polymers* 15: 1523–1530. https://doi.org/10.1007/s12221-014-1523-y.

3

Thermal Properties of the Natural Fiber-Reinforced Hybrid Polymer Composites: An Overview

Jyotishkumar Parameswaranpillai[1], Senthilkumar Krishnasamy[2], Suchart Siengchin[2], Sabarish Radoor[2], Roshny Joy[3], Jinu Jacob George[3], Chandrasekar Muthukumar[4], Senthil Muthu Kumar Thiagamani[5], Nisa V. Salim[6], and Nishar Hameed[7]

[1] *Mar Athanasios College for Advanced Studies Tiruvalla (MACFAST), School of Biosciences, Pathanamthitta, 689101, Kerala, India*
[2] *King Mongkut's University of Technology North Bangkok, The Sirindhorn International Thai-German Graduate School of Engineering (TGGS), Materials and Production Engineering, 1518 Pracharaj 1, Wongsawang Road, Bangsue, Bangkok, 10800, Thailand*
[3] *Cochin University of Science and Technology, Department of Polymer Science and Rubber Technology, Kochi, 682022, India*
[4] *Hindustan Institute of Technology & Science, School of Aeronautical Sciences, Kelambakkam, Chennai, Tamil Nadu, 603103, India*
[5] *Kalasalingam Academy of Research and Education, Department of Mechanical Engineering, Krishnankoil, Tamil Nadu 626126, India*
[6] *Swinburne University of Technology, Faculty of Science, Engineering and Technology, Hawthorn, VIC, 3122, Australia*
[7] *Swinburne University of Technology, School of Engineering, Hawthorn, VIC, 3122, Australia*

3.1 Introduction

Plastic is an important discovery that has improved the quality of human life. In the past few decades, polymers played a significant role in the world's economy because of their light weight, recyclability, and good thermomechanical properties [1]. However, the non-biodegradability of polymer is a primary environmental concern. It is a hazardous material, polluting the environment by generating a vast amount of non-biodegradable waste. Unfortunately, there is no proper way of treating and recycling of polymer. Less than 9% of the plastic is recycled, and the rest is landfilled or thrown into the environment, polluting land, and sea. Hence, biodegradable biopolymers are recently considered as replacements for traditional plastics [2]. Natural fibers are currently used as alternative to synthetic fibers to reinforce polymer composites for advanced applications [3–5]. They are extracted from animals, minerals, and plants [6, 7], and provide strength and stiffness to the composites [8, 9].

Animal fibers are mainly obtained from the hairy mammal, sheep wool, chicken feathers, and spiders. The animal fibers have more tensile strength among the other

Natural Fiber-Reinforced Composites: Thermal Properties and Applications, First Edition.
Edited by Senthilkumar Krishnasamy, Senthil Muthu Kumar Thiagamani, Chandrasekar Muthukumar, Rajini Nagarajan and Suchart Siengchin.

natural fibers [10]. The main constituent of animal fibers is protein. The protein group can be subdivided into α-keratin fibers and fibroin fibers. Hair, wool, and feathers come under α-keratin fiber, whereas silk and spider web are under fibroin fibers. Mineral fibers are naturally produced and are made from mineral resources. Ceramic and asbestos are well-known examples of mineral fiber. However, the applications of asbestos are banned due to the health risk in humans [11]. Among the natural fibers, plant fibers are the most abundant and cheap. Hemp, jute, flax, sisal, abaca, agave, pineapple, kenaf, ramie, and banana are the most used fibers for various industries [12]. The plant fibers are also known as cellulosic fibers since the major constituent is cellulose and have promising mechanical properties. The high strength of natural fibers is due to the hydrogen bonds and other linkages in the cellulosic content present in the natural fibers [1]. However, the performance of the fibers depends on biotic and abiotic factors such as (i) part of the plant from which the fiber is obtained, (ii) the geographical location of plant growth, and (iii) harvest time [7, 13].

The fibers are extracted from the plants by manual extraction, retting, and mechanical extraction, where the latter is the most preferred method for the industrial production of plant fibers [14]. Plant fibers are commonly used as reinforcement for both thermoplastic and thermoset polymers [15, 16]. Recently, natural fibers are used with biodegradable polymers such as polylactic acid (PLA), polycaprolactone (PCL), and poly(3-hydroxybutyrate-*co*-3-hydroxyvalerate) (PHBV) [3, 8]. However, the plant fibers are hydrophilic, while most of the polymers are hydrophobic. Thus, the plant fibers have poor adhesion with polymer matrices. In improving the compatibility of natural fibers with polymer matrices, the fibers were surface-treated with chemical reagents, plasma, laser, ozone, ultraviolet (UV) irradiation, corona treatment, and polymer coating [17, 18]. The factors that affect the properties of the composites are the size of the fiber, fiber diameter, aspect ratio, extraction method, and cellulose content. The disadvantages of natural fiber-reinforced composites are their anisotropic properties, moisture adsorption, and low thermal resistance [19].

Even though chemical and physical treatments improve composites' overall performance, there are drawbacks such as limited mechanical properties, excess water adsorption, and poor thermal properties compared with synthetic fiber-based composites. These deficiencies of natural fiber-reinforced composites can be overcome using new methods such as compatibilization and hybridization. In the compatibilization, a coupling agent compatible with both natural fibers and polymers are preferred to use [20]. Studies have shown that the compatibilizer is required to have adequate enhancements in the properties of the natural fiber-based polymer composites [21]. In the hybridization, two or more natural fibers or natural and synthetic fibers/fillers are used to improve the performance [22–25]. There is a new method called self-hybridization, in which the same fibers with different aspect ratios are used for property improvement [26]. The property of the hybridized composites depends on the fiber loading, fiber length, fiber orientation, fiber/matrix bonding, and stacking sequence in the composites [27]. Recently, most of the studies concentrate on hybridizing natural and synthetic fibers rather than hybrid composites made up of two plant fibers [28, 29]. This is because the properties of the hybrid composites complement with what is lacking in the other fiber and hence justifiable [30, 31].

3.2 Thermal Properties of Natural Fiber Composites

3.2.1 Thermogravimetric Analysis (TGA)

Thermogravimetric analysis (TGA) is used to measure the change in weight percentage of the sample as a function of temperature. Purge gas was flown through the balance to create an inert atmosphere. Usually, nitrogen (N_2) is used as the purge gas. Prior to the reinforcement of the fibers with polymer matrices, the thermal stability of the natural fibers was measured. The thermogram of buriti fiber is shown in Figure 3.1. It can be noticed that the natural fiber undergone decomposition at various temperatures. The constituents of natural fiber such as cellulose, hemicellulose, pectin, lignin, moisture, etc. are volatiles and decomposes at elevated temperatures. The initial weight loss at 50–100 °C is due to the release of moisture from the fiber. The weight loss between 200 and 350 °C is due to the decomposition of hemicellulose, and the main degradation in the temperature range of 320–400 °C is due to the decomposition of cellulose [32].

Radzi et al. [33] studied the thermal behavior of roselle fiber (RF)/sugar palm fiber (SPF)-reinforced thermoplastic polyurethane (TPU) composites. The authors

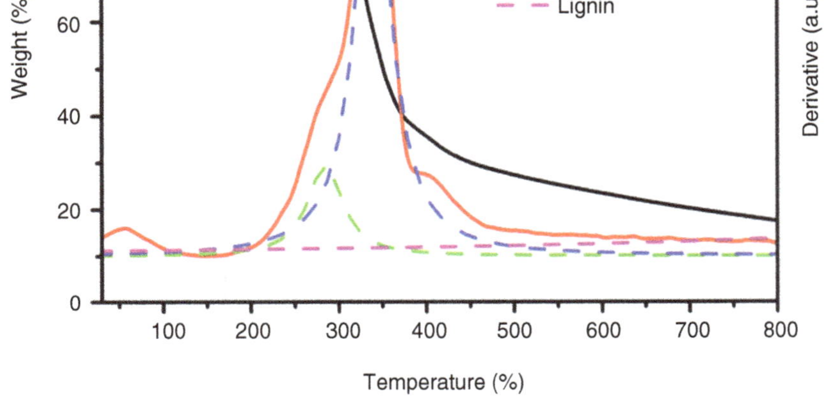

Figure 3.1 Thermogram of buriti fiber [32]. Source: Reproduced with thanks from John Wiley and Sons, License number: 4944760781666.

identified different stages of the degradation of plant fibers from the TGA profile as shown in Figure 3.2.

Atiqah et al. [34] studied the thermal properties of SPF/glass fiber (GF)-based TPU composites. Figure 3.3 shows the differential thermal analysis (DTA) of SPF/GF-based TPU composites. The TPU/GF composites showed two major peaks between 300–390 °C and 400–480 °C at the DTA curve typical for TPU. On the other hand, the hybrid composites with SPF showed two additional peaks, one at 50–150 °C due to the evaporation of the moisture and the other at 230–330 °C

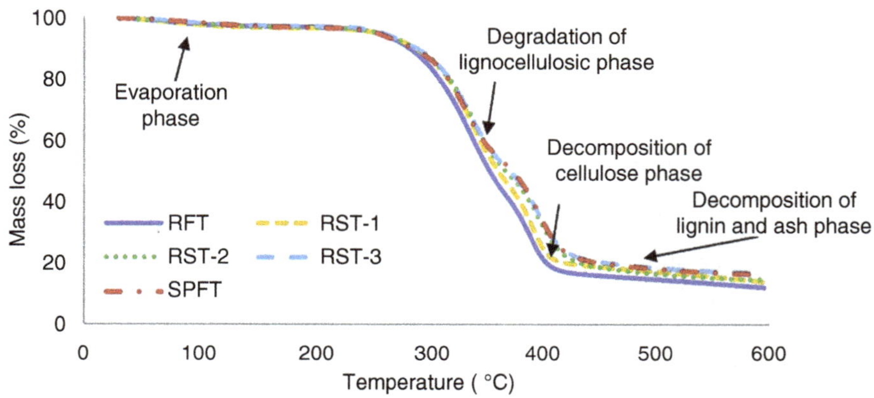

Figure 3.2 TGA of roselle fiber (RF)/sugar palm fiber (SPF)-based TPU composites. Source: Radzi et al. [33].

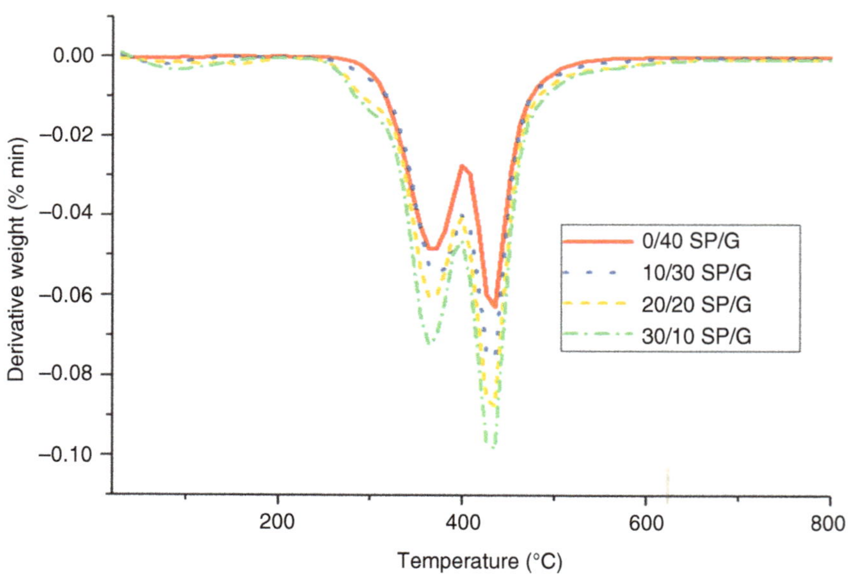

Figure 3.3 DTA of sugar palm (SP)/glass (G) fiber-based thermoplastic polyurethane (TPU) composites [34]. Source: Atiqah et al. [34]. Reproduced with thanks from Elsevier, License number: 4944800703160.

due to the decomposition of hemicellulose in the SPF. In another study, Atiqah et al. [35] fabricated hybrid TPU composites reinforced with the treated SPF and GF. The composites reinforced with treated fiber showed lower water absorption properties and higher thermal stability. The best thermal properties were observed for 6% alkaline-treated and 2% silane-treated fiber composites.

Boopalan et al. [36] studied the thermal properties of epoxy hybrid composites containing raw jute and banana fiber. The composites with the varying weight ratio of banana and jute fiber (100/0, 75/25, 50/50, 25/75, and 0/100) were prepared. Among the composites studied, an even combination of banana and jute fiber showed the best thermal properties. Figure 3.4 shows the differences in thermal stability of varying weight ratio of banana and jute fiber-reinforced epoxy composites.

Dhakal et al. [37] studied the thermal stability of flax/carbon fiber-based hybrid epoxy composites. Two types of flax fibers such as unidirectional (UD) and cross-ply (CP) were used for the fabrication of composites. On the other hand, unidirectional carbon fiber was commonly used. The flax fiber-reinforced epoxy composite showed the lowest thermal stability among the composites prepared. The hybridization of flax fiber with carbon fiber improved the onset temperature, main decomposition peak temperature, and char residue. The results suggest that hybridizing natural fiber with synthetic fiber was a good method to improve the thermal stability of the composites. Kumar and Reddy [38] studied the thermal behavior of 5% NaOH-treated and NaOH-untreated sisal/glass fiber-reinforced epoxy composites. The authors observed improved thermal stability and char residue for the epoxy hybrid composites when treated fibers were used. Srinivasan et al. [39] studied the thermal stability of glass/flax epoxy composites, glass/banana epoxy composites, and glass/flax/banana epoxy composites. The authors reported the highest thermal

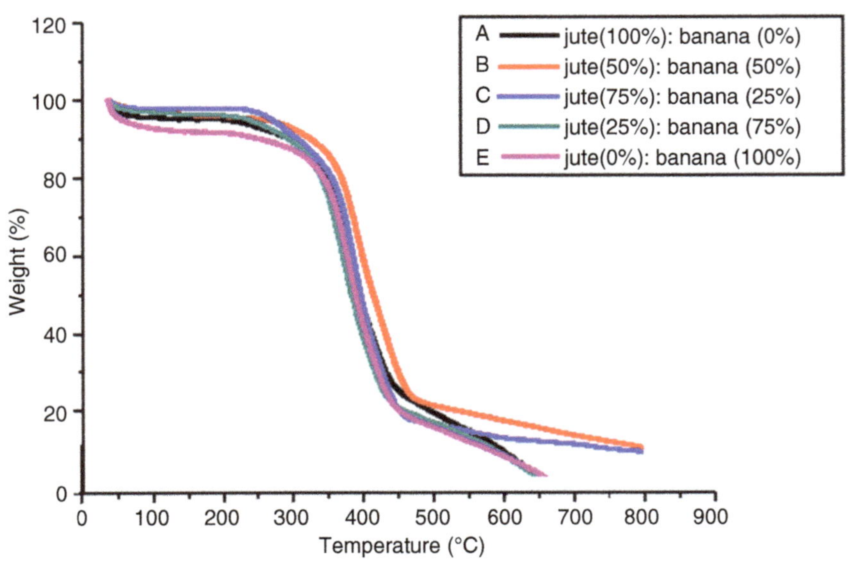

Figure 3.4 TGA thermogram of jute/banana fiber-reinforced epoxy hybrid composites [36]. Source: Boopalan et al. [36]. Reproduced with thanks from Elsevier, License number: 4944811464586.

stability and limiting oxygen index for the hybrid glass/flax/banana fiber-reinforced epoxy composites.

Jesuarockiam et al. [40] performed TGA of epoxy-based natural fiber composites. The fibers such as *Cocos nucifera* sheath, Kevlar, and graphene nanoplatelets were used to fabricate the composites. The authors observed three stages of thermal decomposition for the composite prepared. The initial degradation at 100 °C was due to the evaporation of moisture. The main degradation at 300 °C was due to the degradation of the epoxy matrix. Also, the final degradation above 500 °C was due to the degradation of Kevlar and graphene nanoplatelets. The authors proposed a schematic showing the thermal degradation of hemicellulose, cellulose, and lignin along with epoxy, Kevlar, and graphene nanoplatelets in the composites as shown in Figure 3.5.

Gupta et al. [41] studied the heat deflection temperature (HDT) and thermal stability of bio-poly(trimethylene terephthalate) (bio-PTT)-based hybrid composites containing lignin nanoparticle (LNP) and vapor grown carbon fiber (VGCF). TGA and differential thermogravimetric (DTG) curves of bio-PTT and hybrid nanocomposites are shown in Figure 3.6. Bio-PTT reported an HDT of ~55 °C, and this value increased to ~120 °C with the addition of 1.5% of LNP into bio-PTT. The HDT of bio-PTT further increases to ~180 °C by incorporating 1.5% of LNP and 7% VGCF. TGA results indicates a good improvement in the initial degradation temperature, the temperature at 50% degradation, and final degradation temperature with the incorporation of 1.5% of LNP and 7% VGCF in bio-PTT. The improvement in HDT and thermal stability of the hybrid composites was due to the better dispersion and interaction of the fillers with the polymer matrix.

Ngaowthong et al. [42] studied the thermal behavior of the sisal fiber-reinforced PLA composites. The thermal stability of the PLA composites was reduced with the

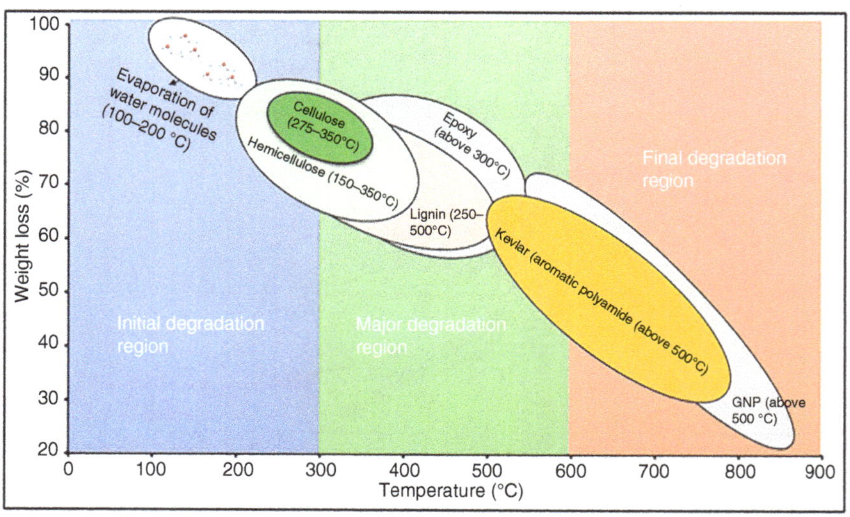

Figure 3.5 Schematic showing the thermal degradation of various components in the epoxy hybrid composite. Source: Jesuarockiam et al. [40].

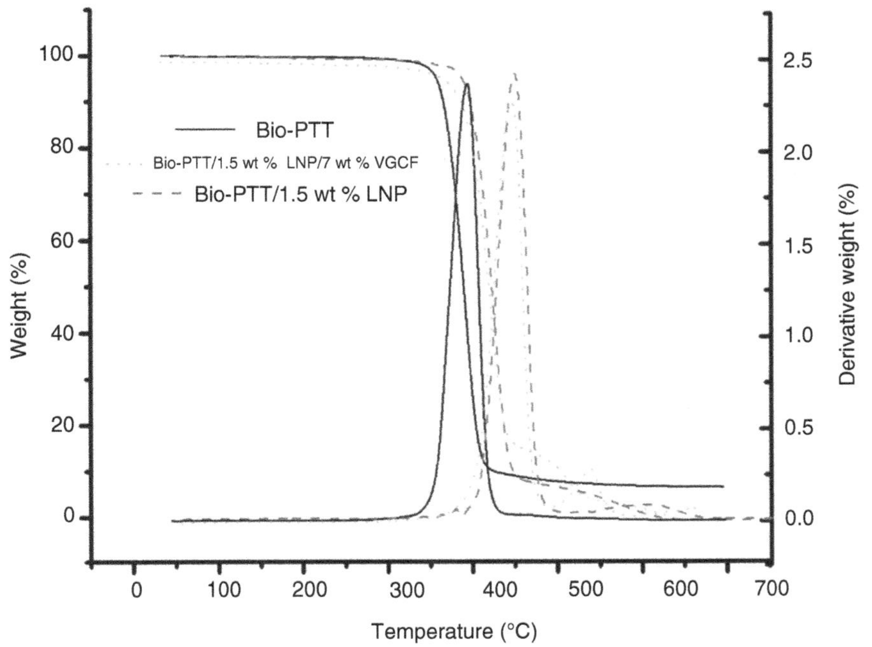

Figure 3.6 TGA and DTG curves of bio-PTT and bio-PTT-based hybrid nanocomposites. Source: Gupta et al. [41]. Reproduced with permission from The Royal Society of Chemistry, License number: 1144133-1.

incorporation of sisal fibers. It was ascribed to the lower thermal stability of the sisal fiber than the pure PLA matrix. The effect of recycling on the thermal stability of the composites was also studied. The use of recycled thermoplastics reduced the thermal stability of the composites, especially with fiber content above 20%. Varghese et al. [3] observed a reduction in thermal stability of PHBV matrix with the addition of *Ceiba pentandra* fiber. The hydrolysis of PHBV caused the drop in thermal stability due to the presence of the OH groups in the natural fiber. Akderya et al. [43] observed that the decomposition of the PLA composites occurred at a lower temperature with the incorporation of chicken feather fiber (CFF). This is due to the lower thermal stability of the CFF than that of PLA. The aging effect on the thermal decomposition of the PLA composites was also studied. The results showed that the thermal stability of pure PLA and its composites was marginally reduced after the aging studies.

Yorseng et al. [44] studied the thermal stability of sisal (S) and kenaf (K) woven fabrics-based hybrid bioepoxy composites before and after the accelerated weathering test. Five different samples were prepared, such as neat bioepoxy, KKK, KSK, SKS, and SSS. Irrespective of (i) the fiber fabric, (ii) fabric sequence, and (iii) accelerated weathering, the thermal stability of the epoxy composites showed identical results. Later, Yorseng et al. [45] studied the thermal stability of non-biodegradable epoxy composites, with the incorporation of sisal (S) and kenaf (K) woven fabrics. The thermal stability of the composites shows no change irrespective of (i) the type of fabric, (ii) fabric sequence, and (iii) accelerated weathering test.

Senthilkumar et al. [46, 47] prepared hybrid sisal/hemp bioepoxy composite and studied the thermal stability of the composites before and after the accelerated weathering test. The incorporation of hemp and sisal fiber mat in the bioepoxy system enhanced the thermal stability of the composites. The enhancement in the thermal stability of the composites was due to the high thermal stability of the sisal and hemp fiber mat compared with bioepoxy resin. After the accelerated weathering test, the thermal stability of the composites was reduced marginally due to the degradation of the matrix and/or fibers caused by temperature, ultraviolet (UV) radiation, and moisture absorption during the accelerated weathering test.

Samal et al. [48] fabricated polypropylene (PP) composites with banana fiber and banana + glass fiber and composites modified with maleic anhydride-*grafted* polypropylene (MAPP). The thermal properties of the banana fiber-reinforced PP was increased through the treatment with MAPP. The control PP shows a single degradation temperature due to the decomposition of the PP while the banana fiber-reinforced composites showed multiple degradation due to the decomposition of hemicellulose and cellulose. Interestingly the main DTG peak of PP shifted toward higher temperature for the hybrid banana + glass fiber + MAPP composites. This is due to three factors such as (i) the chain extension of PP chains in the presence of MAPP, (ii) the anhydride groups of MAPP forming a covalent bond with OH groups of banana fibers and SiO groups of glass fibers, and (iii) high thermal stability of the glass fiber. The covalent bond causes better interfacial adhesion between the fibers and polymers and hence better thermal resistance. Jarukumjorn and Suppakarn [49] studied thermal properties of PP, PP/sisal, PP/sisal/PP-g-MA, and hybrid PP composites with sisal + glass fiber + PP-g-MA. The authors observed an improvement in heat distortion temperature, the temperature at 5% degradation, and temperature at 50% degradation with the incorporation of PP-g-MA and glass fiber into PP/sisal composites. The highest thermal stability was observed at composites with the highest percentage of glass fiber (PP/10sisal/20GF/PP-g-MA) due to the high thermal stability of the glass fiber. Chanprapanon et al. [50] fabricated sisal fiber/organoclay-reinforced PP composites and studied the thermal stability of the composites. The incorporation of clay improved the thermal stability of the PP/sisal fiber composites. Here, the clay nanoparticles act as an insulator preventing the thermal degradation of the PP. Recently, Atagur et al. [51] studied the changes in the thermal stability of *Carpinus betulus* fiber-filled PP matrix. The initial decomposition temperature was reduced, due to the lower onset temperature of the natural fiber. On the other hand, T_{max} and percent char residue increased with the addition of *C. betulus* fiber, which may be due to the high lignin content in the plant fiber.

3.2.2 Differential Scanning Calorimetry (DSC)

Differential scanning calorimetry (DSC) was used to study the thermal transitions in natural fibers or natural fiber-reinforced polymer composites. It works on a simple principle by measuring the difference in the heat flow rate between the sample and the reference that permits the determination of various endothermic and exothermic processes taking place in the sample during thermal heating. The thermal parameters such as glass transition temperature (T_g), cold crystallization temperature (T_{cc}), melting temperature (T_m), enthalpy of melting (ΔH_m), crystallization temperature (T_c), enthalpy of crystallization (ΔH_c), and percent crystallinity are measurable from the DSC thermogram [52, 53].

Boumediri et al. [54] performed the DSC thermograms of vascular bundles and fiber strands of date palm tree, which is shown in Figure 3.7a,b. A series of exothermic and endothermic peaks were observed for both vascular bundles and fiber strands. The endothermic peak at ~93 °C of vascular bundles and ~89 °C of fiber strands showed the evaporation of moisture absorbed in the fibers. The exothermic peak at ~297 °C of vascular bundles and ~284 °C of fiber strands exhibited the degradation of hemicellulose of the fibers, while the peak at ~367 °C of vascular bundles and ~356 °C for fiber strands corresponds to cellulose degradation. On the other hand, the exothermic peak at ~432 °C of vascular bundles and ~419 °C of fiber strands corresponds to the degradation of lignin.

Tengsuthiwat et al. [55] fabricated PLA-based hybrid composites containing sisal fiber and cuprous oxide (Cu_2O) particles and studied the thermal transitions of the composites using the DSC. It was observed that the incorporation of sisal fiber and Cu_2O filler improves the percent crystallinity of the PLA because both the filler and fiber provide new sites for the crystallization of the polymer. Roy and Rhim [56] studied the DSC of curcumin-modified PLA composites. The incorporation of curcumin improved the crystallinity of the PLA due to the heterogeneous nucleation caused by the presence of curcumin. Akderya et al. [43] carried out the DSC analysis of before and after aged CFF-reinforced PLA composites. The concentration of the CFF used for the fabrication of the composites was 2%, 5%, and 10%. The T_g of PLA was marginally reduced with the incorporation of CFF, which may be due to the plasticizing action of CFF. The T_{cc} was not observed in the neat PLA sample. However, T_{cc} was observed in the composite samples, suggesting that the crystallization rate was decreased in the presence of CFF. The T_m of PLA showed no change with the incorporation of the CFF. On the other hand, the T_g of the aged samples (neat PLA and composites) increased by ~10 °C, possibly due to the restriction in the mobility of the PLA chains caused by the embrittlement of polymer during the aging process. The T_{cc} and T_m showed only marginal variations but within the experimental limits.

Kamau-Devers et al. [57] studied the hydrothermal aging of wood flour-reinforced PLA composites. Before starting the experiment, the composite samples were fully desiccated or fully submerged for 3 days at various temperatures. The DSC analysis reported that the T_g, T_{cc}, and T_m values of the submerged samples are lower than that of the fully desiccated samples. It may be attributed to the degradation of the samples in fully submerged conditions. Yetiş et al. [58] fabricated the acetylated microfibrillated cellulose with high cellulose modified PLA composites. DSC thermograms of PLA/acetylated microfibrillated cellulose composites and their photographs are shown in Figure 3.8a,b. The T_g, T_{cc}, and T_m of PLA exhibit marginal variations with the incorporation of the acetylated microfibrillated cellulose. On the other hand, the authors reported a good improvement in crystallinity with the incorporation of acetylated microfibrillated cellulose, because the filler acts as the

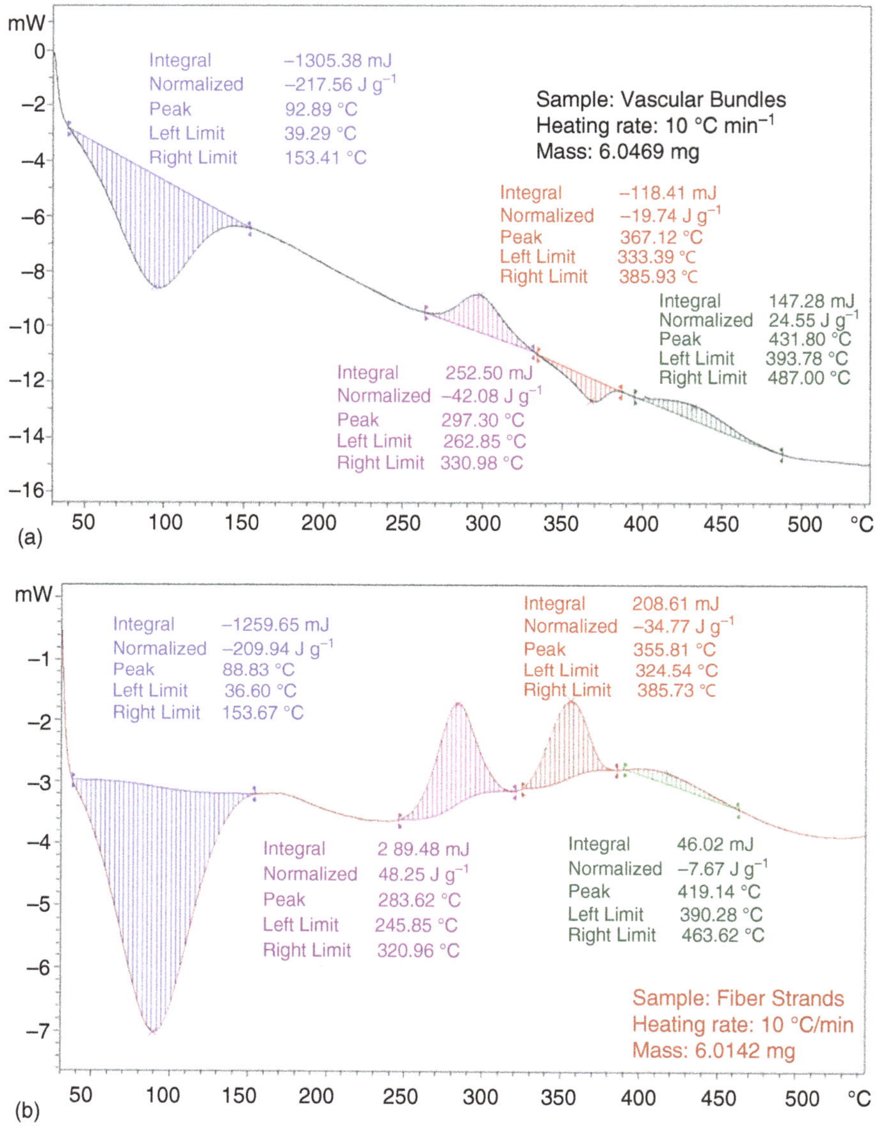

Figure 3.7 DSC thermogram of (a) vascular bundles and (b) fiber strands of date palm tree [54]. Source: Boumediri et al. [54]. Reproduced with thanks from Elsevier, License number: 4944870534040.

nucleating agent for the crystallization of the PLA matrix. The photographs of the composites shown in Figure 3.8b shows good transparency irrespective of the filler content, showing uniform distribution of the filler in the polymer matrix.

Almaadeed et al. [59] studied the changes in the T_m, T_c, and X_c of the date palm wood flour + glass fiber + recycled polypropylene (RPP) composites. The composites showed no change in T_m and T_c. However, the X_c was reduced with the incorporation of wood flour. This is because wood flour prevents the crystallization

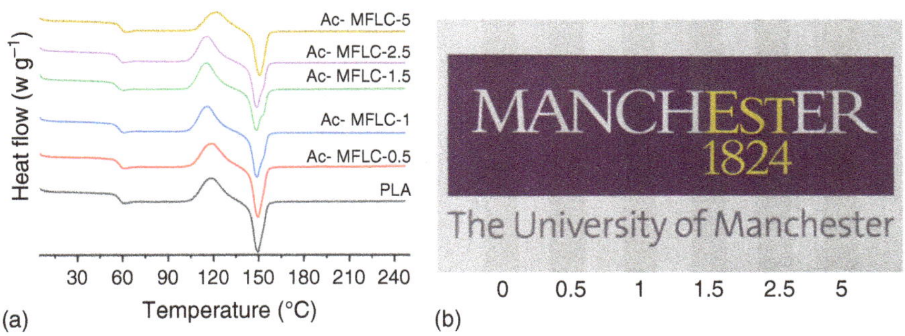

Figure 3.8 (a) DSC thermogram of PLA/acetylated microfibrillated cellulose composites and (b) their photographs [58]. Source: Yetiş et al. [58]. Reproduced with thanks from Elsevier, License number: 4945021161771.

of RPP. On the other hand, the introduction of glass fiber improved the percent crystallinity of the RPP, because glass fiber acts as a nucleating agent. Sarasini et al. [31] fabricated high-density polyethylene (HDPE)-based hybrid composites containing basalt and hemp fiber, with and without maleic anhydride-modified HDPE. The effect of fibers on the thermal properties of HDPE composites was studied. The melting temperature of the composites was reduced while crystallinity was increased due to the heterogeneous nucleation caused by the presence of fibers. On the other hand, the introduction of the coupling agent reduces the crystallinity of the composites. The overall crystallinity of the composites depends on the distribution of fibers and its nucleating ability. Atagur et al. [51] studied the changes in T_m, T_c, ΔH_m, ΔH_c, and X_c of *C. betulus* fiber-filled PP matrix. The authors observed an increase in T_c, ΔH_m, ΔHc, and percent crystallinity with the incorporation of up to 15 wt% of *C. betulus* fiber. These results show the nucleating action of this fiber.

3.2.3 Dynamic Mechanical Analysis (DMA)

In dynamic mechanical analysis (DMA), a sinusoidal load is applied to the sample with respect to temperature, and the resultant strain is measured. DMA provides a detailed overview of the variations in the storage modulus, loss modulus, and damping factor (tan δ) with respect to temperature or frequency. The storage modulus, loss modulus, and tan δ of the composites depend on the parameters such as the dispersion of the fiber, functional groups present in the fiber, interfacial interaction, modulus of the fibers, aspect ratio of the fibers, stacking sequence, and wetting of the fibers. The distribution of fibers in the polymer matrix and their interfacial adhesion can be understood using several mathematical equations [45, 60–63]. Several attempts have been made to fabricate advanced biocomposites with various natural fibers, combination of natural fibers and synthetic fibers, natural fibers with nanofillers, natural fibers with synthetic fibers and nanofillers, and chemical treatment of natural fibers. In this section, an overview of the viscoelastic properties of various natural fiber-based composites is discussed.

Manral et al. [64] studied the DMA of flax/jute-reinforced PLA hybrid composites. The weight percentage of the fiber was maintained at 30 for the fabrication of

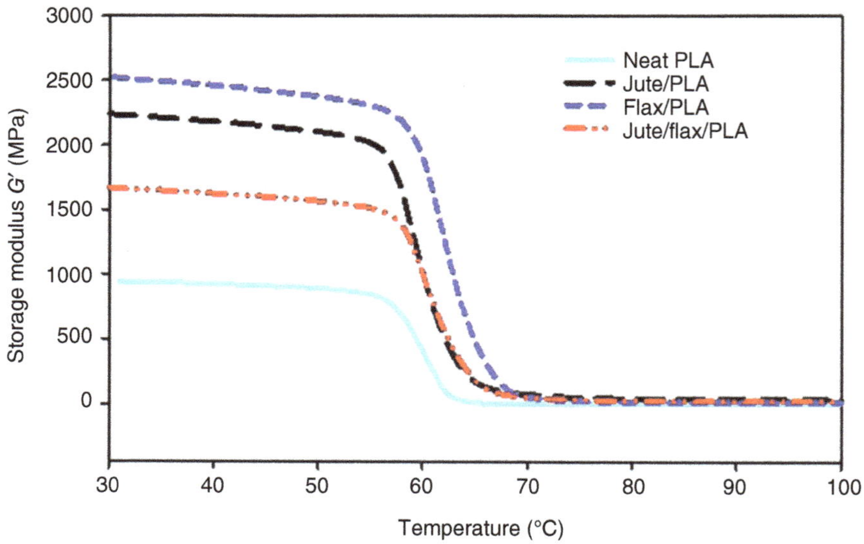

Figure 3.9 The storage modulus of flax, jute, and flax + jute in PLA composites [64]. Source: Manral et al. [64]. Reproduced with thanks from Elsevier, License number: 4945030162653.

the composites. The incorporation of flax, jute, and flax + jute in PLA improved the storage modulus significantly with good improvement in T_g. Figure 3.9 shows the storage modulus of flax, jute, and flax + jute in PLA composites. The highest modulus and T_g were observed for flax fiber-reinforced PLA composites. Similarly, the loss modulus was highest for flax fiber-reinforced PLA composites, while the damping is found to be the lowest. This result suggests good reinforcement of the flax fiber with the PLA matrix.

Balakrishnan et al. [65] fabricated thermoplastic starch composites based on cellulose nanofiber (CNF) and cellulose nanocrystals (CNC). Both CNF and CNC were isolated from pineapple leaf fiber. The concentration of the filler used is 1 and 3 wt%. The storage modulus of the composites was increased with the incorporation of the nanofillers. This result showed that the nanofiller improved the stiffness of the composites due to the reinforcement effect of the filler with the polymer matrix. An and Ma [66] studied the DMA of poly(3-hydroxybutyrate-*co*-4-hydroxybutyrate) (P34HB) composites of reinforced wood fiber in the presence of the different amount of maleic anhydride. The study showed that the applied frequency has an influence on the viscoelastic properties of the composites. The storage modulus and $\tan\delta$ are maximum at high frequency, because during this, the polymer has less time for molecular rearrangement [61]. On the other hand, the change in weight percentage of MAPP in P34HB/wood flour composites showed fluctuations in storage modulus while the T_g remains constant.

Yorseng et al. [44, 45] studied the storage modulus of kenaf woven mat and sisal woven mat-reinforced single-fiber and/or hybrid epoxy and bioepoxy composites. Changing the fiber sequence had no effect on the storage modulus of the composites in the glassy state. The aging studies of the composites were also carried out to study the variations of storage modulus of the composites with respect to accelerated weathering. Interestingly, the composites retained the storage modulus even after the weathering test. Sathyaseelan et al. [67] studied the storage modulus of areca (A) and kenaf (K) fiber mat-reinforced hybrid epoxy composites. Four samples with different stacking sequences such as AKAKA, KAKAK, KKKKK, and AAAAA were prepared. The authors observed a significantly high storage modulus and T_g for KKKKK-based composites. This is because of the stiffness of kenaf and good bonding strength between the fiber and epoxy. Manral and Bajpai [68] studied the effect of orientation of kenaf fiber mat on the viscoelastic properties of PLA. The photographs of kenaf fiber mat in different orientations and the storage modulus curve of the composites are shown in Figure 3.10a,b. Among the composites studied, the randomly oriented natural fiber mat-based composites give the highest storage modulus and T_g, because this mat provides a better reinforcement effect with the polymer matrix.

Fiore et al. [69] studied the effect of aging studies on the dynamic mechanical properties of untreated and treated fiber (flax and jute fabrics)-reinforced single-fiber and hybrid epoxy composites. Five layers of fabrics were utilized for the preparation of composites. For untreated fiber composites, the storage modulus gradually reduced with aging time. On the other hand, sodium bicarbonate-treated flax fiber retained the storage modulus even after two months of aging in the salt-fog environment. Similarly, hybrid composites with flax outer layer aged two months retained the storage modulus. However, the treatment showed no effect on the storage modulus of the jute fiber-reinforced composites after the aging test. In another work, Mazlan et al. [70] studied the storage modulus, loss modulus, and $\tan\delta$ of treated and untreated kenaf woven fabric and glass fiber-reinforced epoxy hybrid composites in the presence of varying concentrations of clay. The authors fabricated composites such as untreated kenaf + epoxy single-fiber composite, untreated kenaf + glass + clay-based epoxy hybrid composites, treated kenaf + epoxy single-fiber composite, and treated kenaf + glass + clay-based epoxy hybrid composites. Irrespective of the treatment, the hybrid composites with 1% clay showed high storage modulus and loss modulus compared with single-fiber composite. This is due to the synergistic effect caused by the high modulus of the glass fiber and the uniform distribution of the clay particles. On the other hand, the $\tan\delta$ peak height is reduced for the hybrid composites due to the reinforcement effect of the fibers and nanofiller. Interestingly, the T_g was marginally reduced for the hybrid composites because the arrangement of fillers will be disturbed at higher temperatures and hence catalyzes the mobility of the polymer chains at lower temperatures.

Tajvidi et al. [71] studied the effect of incorporating of various short fibers on the storage, loss modulus, and $\tan\delta$ behavior of the PP composites with respect to temperature. Different natural fibers such as wood flour, kenaf fiber, rice hulls, and newsprint were used as additives for PP. The compatibilizer, maleated PP, was used in improving the compatibility between the fiber and polymer. Irrespective of the type of fibers, the introduction of fibers improved the storage modulus and loss modulus of the PP, while the $\tan\delta$ was reduced. This result showed that the fibers improved the stiffness of the composites due to the reinforcement effect of the natural fibers with the polymer matrix. However, the T_g was slightly reduced possibly due

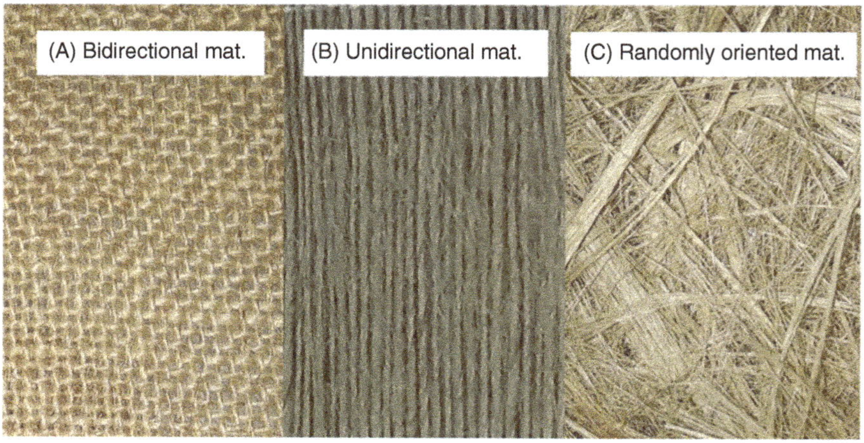

(a)

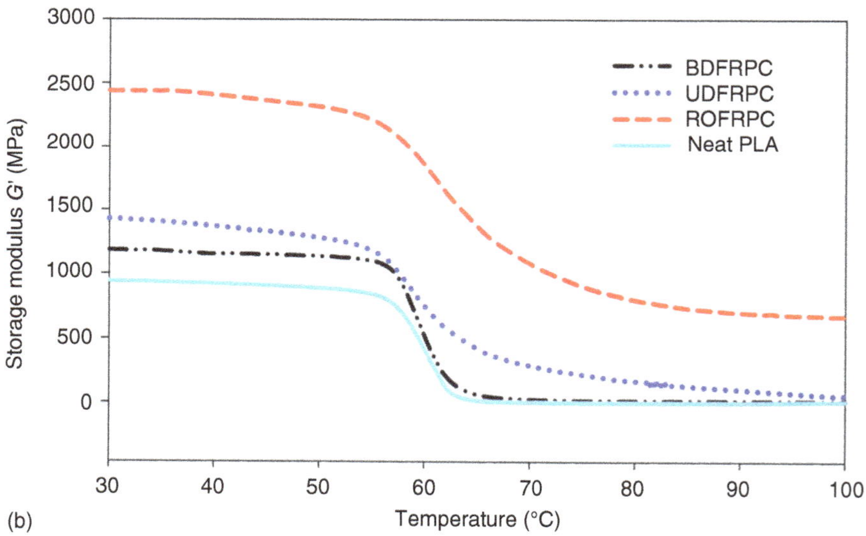

(b)

Figure 3.10 (a) Photographs of the different orientation of kenaf fiber mat and (b) storage modulus of the PLA composites. Source: Manral and Bajpai [68]. Reproduced with thanks from John Wiley and Sons, License number: 4945030647545.

to the presence of maleated PP. Zuhudi et al. [72] studied the dynamic mechanical properties of hybridized bamboo fabric and glass fiber-reinforced PP composites. They observed good improvement in the storage modulus of the PP with the incorporation of bamboo fibers due to the high modulus of the bamboo fibers and the reinforcement effect imparted by the bamboo fiber. However, a further improvement in the storage modulus of the single bamboo fiber-reinforced PP composites was observed with the replacement of the bamboo fabrics with glass fibers. This is due to the hybrid effect caused by the high modulus of the glass fiber compared with bamboo fabric.

Devi et al. [60] studied the effect of incorporation of hybrid fibers on the storage, loss modulus, and tan δ behavior of the unsaturated polyester composites. Glass (G) and pineapple (P) fibers were used for the fabrication of the composites. Three different sample sequences were prepared with GPG, PGP, and intimately mixed composites by varying the glass fiber content. The total fiber content used was 40 wt%. A significant increase in storage modulus and loss modulus was observed with the introduction of glass fiber while the tan δ was reduced. This result showed that the introduction of glass fibers improved the stiffness of the composites due to the reinforcement effect with the polymer matrix. Gupta et al. [73] studied the DMA of long jute and sisal fiber-reinforced hybrid unsaturated polyester resin composites. Five different composites were prepared by varying fiber content, where it was maintained at 30 wt%. The hybrid composites with jute fiber 25 parts and sisal fiber 75 parts showed the highest storage modulus, lowest damping, and highest T_g. The 50 : 50 (jute/sisal) composition also showed reasonably good stiffness compared with single-fiber-reinforced composites. The results showed the possibility of hybridizing natural fibers for improved thermomechanical properties.

3.3 Conclusion

Natural fiber-reinforced composites possess many advantages such as (i) low density, (ii) biodegradability, and (iii) recyclability. Thus, natural fiber composites have been considered as an alternative to synthetic fiber-based composites. Though the natural fibers are (i) hydrophilic, (ii) less resistant to wet environmental conditions, and (iii) highly susceptible to moisture, they are widely used in many applications, including automotive, aerospace, and infrastructure, whereby these composites are limited to semi-structural and non-structural applications. In this context, hybrid fiber composites have been developed. Thus, the result of hybridized composites exhibits a balance of (i) mechanical properties, (ii) thermal stability, (iii) aggressive environmental conditions, and (iv) cost reduction. In this chapter, the thermal properties of natural fiber-based hybrid composites were discussed. Many researchers reported that the thermal stability of hybridized composites provided enhanced properties; also, the hybrid composites overcome the drawbacks of single fiber-reinforced composites. Besides, the hybridization helps to improve the physical property and dimensional stability of the composites. The performance of the hybrid composites could also be influenced by chemical treatments, type of matrix, bonding strength between the fiber and matrix, and weight percentages of fibers used.

Acknowledgment

Jyotishkumar Parameswaranpillai would like to thank King Mongkut's University of Technology North Bangkok, Thailand, for their research funding (2020–2021) with grant numbers KMUTNB-64-KNOW03 and KMUTNB BasicR-64-16.

References

1 Thakur, V.K., Thakur, M.K., and Gupta, R.K. (2014). Review: Raw natural fiber – based polymer composites. *International Journal of Polymer Analysis and Characterization* 2013: 256–271. https://doi.org/10.1080/1023666X.2014.880016.

2 Varghese, S.A., Siengchin, S., and Parameswaranpillai, J. (2020). Essential oils as antimicrobial agents in biopolymer-based food packaging – a comprehensive review. *Food Bioscience* https://doi.org/10.1016/j.fbio.2020.100785.

3 Varghese, S.A., Pulikkalparambil, H., Rangappa, S.M. et al. (2020). Novel biodegradable polymer films based on poly(3-hydroxybutyrate-*co*-3-hydroxyvalerate) and *Ceiba pentandra* natural fibers for packaging applications. *Food Packaging and Shelf Life* 25: 100538.

4 Vijayan, P.P., Bhanu, A.A., Archana, S.R. et al. (2020). Development of chicken feather fiber filled epoxy protective coating for metals. *Materials Today: Proceedings* 41: 468–472.

5 Vinod, A., Siengchin, S., and Parameswaranpillai, J. (2020). Renewable and sustainable biobased materials: an assessment on biofibers, biofilms, biopolymers and biocomposites. *Journal of Cleaner Production* 258: 120978.

6 Patel, J.P. and Parsania, P.H. (2018). Characterization, testing, and reinforcing materials of biodegradable composites, Chapter 3. In: *Biodegradable and Biocompatible Polymer Composites* (ed. N.G. Shimpi), 55–79. Elsevier Ltd. https://doi.org/10.1016/B978-0-08-100970-3.00003-1.

7 Saba, N., Tahir, P., and Jawaid, M. (2014). A review on potentiality of nano filler/natural fiber filled polymer hybrid. *Composites*: 2247–2273. https://doi.org/10.3390/polym6082247.

8 Lalit, R., Mayank, P., and Ankur, K. (2018). Natural fibers and biopolymers characterization: a future potential composite material. *Strojnicky Casopis* 68 (1): 33–50. https://doi.org/10.2478/scjme-2018-0004.

9 Rodriguez, S., Margem, F.M., Monteiro, S.N., and Calado, V. (2012). Thermogravimetric behavior of natural fibers reinforced polymer composites – an overview. *Materials Science and Engineering: A* 557: 17–28. https://doi.org/10.1016/j.msea.2012.05.109.

10 Carvalho, H., Salman, H., and Leite, M. (2018). Natural fibre composites and their applications: a review. *Journal of Composites Science*: 1–20. https://doi.org/10.3390/jcs2040066.

11 Bhattacharyya, D., Subasinghe, A., and Kim, N.K. (2015). Natural fibers: their composites and flammability characterizations, Chapter 4. In: *Multi-Functionality of Polymer Composites* (eds. K. Friedrich and U. Breuer) https://doi.org/10.1016/B978-0-323-26434-1.00004-0, 102–143. Elsevier Inc.

12 Peng, X., Fan, M., Hartley, J., and Al-zubaidy, M. (2011). Properties of natural fiber composites made by pultrusion process. *Journal of Composites Materials* https://doi.org/10.1177/0021998311410474.

13 Fuqua, M.A., Huo, S., and Ulven, C.A. (2012). Natural fiber reinforced composites. *Polymer Reviews* 52: 259–320.

14 Radoor, S., Karayil, J., Rangappa, S.M. et al. (2020). A review on the extraction of pineapple, sisal and abaca fibers and their use as reinforcement in polymer matrix. *Express Polymer Letters* 14 (4): 309–335.

15 Arrakhiz, F.Z., El Achaby, M., Malha, M. et al. (2013). Mechanical and thermal properties of natural fibers reinforced polymer composites: doum/low density polyethylene. *Materials & Design* 43: 200–205.

16 Ganapathy, T., Sathiskumar, R., Sanjay, M.R. et al. (2019). Effect of graphene powder on banyan aerial root fibers reinforced epoxy composites. *Journal of Natural Fibers* 18: 1–8.

17 Pickering, K.L., Efendy, M.G.A., and Le, T.M. (2016). A review of recent developments in natural fibre composites and their mechanical performance. *Composites Part A: Applied Science and Manufacturing* 83: 98–112. https://doi.org/10.1016/j.compositesa.2015.08.038.

18 Sanjay, M.R., Siengchin, S., Parameswaranpillai, J. et al. (2019). A comprehensive review of techniques for natural fibers as reinforcement in composites: preparation, processing and characterization. *Carbohydrate Polymers* 207: 108–121.

19 Bakar, N., Chin, S.C., Siregar, J.P., and Ngien, S.K. (2020). A review on physical, mechanical, thermal properties and chemical composition of plant fibers. *IOP Conference Series: Materials Science and Engineering* https://doi.org/10.1088/1757-899X/736/5/052017.

20 Younesi-Kordkheili, H. and Pizzi, A. (2020). Ionic liquid-modified lignin as a bio-coupling agent for natural fiber-recycled polypropylene composites. *Composites Part B: Engineering* 181: 107587.

21 Taneli, V., Das, O., and Tomppo, L. (2017). A review on new bio-based constituents for natural fiber-polymer composite. *Journal of Cleaner Production* 149: 582–596. https://doi.org/10.1016/j.jclepro.2017.02.132.

22 Eselini, N., Tirkes, S., Akar, A.O., and Tayfun, U. (2020). Production and characterization of poly(lactic acid)-based biocomposites filled with basalt fiber and flax fiber hybrid. *Journal of Elastomers & Plastics* 52 (8): 701–716.

23 Rajesh, M., Jayakrishna, K., Sultan, M.T.H. et al. (2020). The hydroscopic effect on dynamic and thermal properties of woven jute, banana, and intra-ply hybrid natural fiber composites. *Journal of Materials Research and Technology* 9 (5): 10305–10315.

24 Ramesh, P., Prasad, B.D., and Narayana, K.L. (2021). Influence of montmorillonite clay content on thermal, mechanical, water absorption and biodegradability properties of treated kenaf fiber/PLA-hybrid biocomposites. *Silicon* 13: 109–118.

25 Sekar, S., Suresh Kumar, S., Vigneshwaran, S., and Velmurugan, G. (2020). Evaluation of mechanical and water absorption behavior of natural fiber-reinforced hybrid biocomposites. *Journal of Natural Fibers*: 1–11. https://doi.org/10.1080/15440478.2020.1788487.

26 Pérez-Fonseca, A.A., Robledo-Ortíz, J.R., Moscoso-Sánchez, F.J. et al. (2014). Injection molded self-hybrid composites based on polypropylene and natural fibers. *Polymer Composites* 35 (9): 1798–1806.

27 Nunna, S., Chandra, P.R., Shrivastava, S., and Jalan, A.K. (2012). A review on mechanical behavior of natural fiber based hybrid composites. *Journal of Reinforced Plastics and Composites* 31 (11): 759–769. https://doi.org/10.1177/0731684412444325.

28 Cheng, M., Zhong, Y., Kureemun, U. et al. (2020). Environmental durability of carbon/flax fiber hybrid composites. *Composite Structures* 234: 111719.

29 Nurazzi, N.M., Khalina, A., Sapuan, S.M. et al. (2020). Thermal properties of treated sugar palm yarn/glass fiber reinforced unsaturated polyester hybrid composites. *Journal of Materials Research and Technology* 9 (2): 1606–1618.

30 Rodriguez-Castellanos, W. and Rodrigue, D. (2016). Production and characterization of production of hybrid polymer composites based on natural fibers. In: *Composites from Renewable and Sustainable Materials* (ed. M. Poletto). InTech https://doi.org/10.5772/64995.

31 Sarasini, F., Tirillò, J., Sergi, C. et al. (2018). Effect of basalt fibre hybridisation and sizing removal on mechanical and thermal properties of hemp fibre reinforced HDPE composites. *Composite Structures* 188: 394–406.

32 Poletto, M., Júnior, H.L.O., and Zattera, A.J. (2015). Thermal decomposition of natural fibers: kinetics and degradation mechanisms. In: *Reactions and Mechanisms in Thermal Analysis of Advanced Materials* (eds. A. Tiwari and B. Raj), 515–545. John Wiley & Sons.

33 Radzi, A.M., Sapuan, S.M., Jawaid, M., and Mansor, M.R. (2019). Water absorption, thickness swelling and thermal properties of roselle/sugar palm fibre reinforced thermoplastic polyurethane hybrid composites. *Journal of Materials Research and Technology* 8 (5): 3988–3994.

34 Atiqah, A., Jawaid, M., Sapuan, S.M. et al. (2018). Thermal properties of sugar palm/glass fiber reinforced thermoplastic polyurethane hybrid composites. *Composite Structures* 202: 954–958.

35 Atiqah, A., Jawaid, M., Sapuan, S.M. et al. (2019). Physical and thermal properties of treated sugar palm/glass fibre reinforced thermoplastic polyurethane hybrid composites. *Journal of Materials Research and Technology* 8 (5): 3726–3732.

36 Boopalan, M., Niranjanaa, M., and Umapathy, M.J. (2013). Study on the mechanical properties and thermal properties of jute and banana fiber reinforced epoxy hybrid composites. *Composites Part B: Engineering* 51: 54–57.

37 Dhakal, H.N., Zhang, Z.Y., Guthrie, R. et al. (2013). Development of flax/carbon fibre hybrid composites for enhanced properties. *Carbohydrate Polymers* 96 (1): 1–8.

38 Kumar, M.A. and Reddy, G.R. (2014). Mechanical & thermal properties of epoxy based hybrid composites reinforced with jute/sansevieria cylindrica fibres. *International letters of chemistry, physics and astronomy* 38: 191–197.

39 Srinivasan, V.S., Boopathy, S.R., Sangeetha, D., and Ramnath, B.V. (2014). Evaluation of mechanical and thermal properties of banana–flax based natural fibre composite. *Materials & Design* 60: 620–627.

40 Jesuarockiam, N., Jawaid, M., Zainudin, E.S. et al. (2019). Enhanced thermal and dynamic mechanical properties of synthetic/natural hybrid composites with graphene nanoplateletes. *Polymers* 11 (7): 1085.

41 Gupta, A.K., Mohanty, S., and Nayak, S.K. (2015). Influence of addition of vapor grown carbon fibers on mechanical, thermal and biodegradation properties of lignin nanoparticle filled bio-poly(trimethylene terephthalate) hybrid nanocomposites. *RSC Advances* 5 (69): 56028–56036.

42 Ngaowthong, C., Borůvka, M., Běhálek, L. et al. (2019). Recycling of sisal fiber reinforced polypropylene and polylactic acid composites: thermo-mechanical properties, morphology, and water absorption behavior. *Waste Management* 97: 71–81.

43 Akderya, T., Özmen, U., and Baba, B.O. (2020). Investigation of long-term ageing effect on the thermal properties of chicken feather fibre/poly(lactic acid) biocomposites. *Journal of Polymer Research* 27: 162.

44 Yorseng, K., Rangappa, S.M., Pulikkalparambil, H. et al. (2020). Accelerated weathering studies of kenaf/sisal fiber fabric reinforced fully biobased hybrid bioepoxy composites for semi-structural applications: morphology, thermo-mechanical, water absorption behavior and surface hydrophobicity. *Construction and Building Materials* 235: 117464.

45 Yorseng, K., Mavinkere Rangappa, S., Parameswaranpillai, J., and Siengchin, S. (2020). Influence of accelerated weathering on the mechanical, fracture morphology, thermal stability, contact angle, and water absorption properties of natural fiber fabric-based epoxy hybrid composites. *Polymers* 12 (10): 2254.

46 Senthilkumar, K., Thiagamani, S.M.K., Muthukumar, C. et al. (2019). Effects of stacking sequences on static, dynamic mechanical and thermal properties of completely biodegradable green epoxy hybrid composites. *Materials Research Express* 6 (10): 105351.

47 Senthilkumar, K., Ungtrakul, T., Chandrasekar, M. et al. (2021). Performance of sisal/hemp bio-based epoxy composites under accelerated weathering. *Journal of Polymers and the Environment* 29: 1–13.

48 Samal, S.K., Mohanty, S., and Nayak, S.K. (2009). Banana/glass fiber-reinforced polypropylene hybrid composites: fabrication and performance evaluation. *Polymer – Plastics Technology and Engineering* 48 (4): 397–414. https://doi.org/10.1080/03602550902725407.

49 Jarukumjorn, K. and Suppakarn, N. (2009). Effect of glass fiber hybridization on properties of sisal fiber–polypropylene composites. *Composites Part B: Engineering* 40 (7): 623–627. https://doi.org/10.1016/j.compositesb.2009.04.007.

50 Chanprapanon, W., Suppakarn, N., and Jarukumjorn, K. (2012). Flame retardancy, thermal stability, and mechanical properties of sisal fiber/organoclay/polypropylene composites. In: *Advanced Materials Research*, vol. 410 (eds. A.K.T. Lau, T.S. Srivatsan, D. Bhattacharyya, et al.), 47–50. Trans Tech Publications Ltd.

51 Atagur, M., Seki, Y., Pasaoglu, Y. et al. (2020). Mechanical and thermal properties of *Carpinas betulus* fiber filled polypropylene composites. *Polymer Composites* 41 (5): 1925–1935.

52 Krishnasamy, S., Thiagamani, S.M.K., Kumar, C.M. et al. (2019). Recent advances in thermal properties of hybrid cellulosic fiber reinforced polymer composites. *International Journal of Biological Macromolecules* 141: 1–13. https://doi.org/10.1016/j.ijbiomac.2019.08.231.

53 Techawinyutham, L., Siengchin, S., Dangtungee, R., and Parameswaranpillai, J. (2019). Influence of accelerated weathering on the thermo-mechanical, antibacterial, and rheological properties of polylactic acid incorporated with porous silica-containing varying amount of capsicum oleoresin. *Composites Part B: Engineering* 175: 107108.

54 Boumediri, H., Bezazi, A., Del Pino, G.G. et al. (2019). Extraction and characterization of vascular bundle and fiber strand from date palm rachis as potential bio-reinforcement in composite. *Carbohydrate Polymers* 222: 114997.

55 Tengsuthiwat, J., Yorseng, K., Siengchin, S., and Parameswaranpillai, J. (2019). Thermomechanical, water absorption, ultraviolet resistance and laser-assisted electroless plating behavior of Cu_2O and melamine–formaldehyde-coated sisal fiber-modified poly(lactic acid) composites. *Polymer Composites* 40 (8): 3264–3274.

56 Roy, S. and Rhim, J.W. (2020). Preparation of bioactive functional poly(lactic acid)/curcumin composite film for food packaging application. *International Journal of Biological Macromolecules* 162: 1780–1789.

57 Kamau-Devers, K., Kortum, Z., and Miller, S.A. (2019). Hydrothermal aging of bio-based poly(lactic acid) (PLA) wood polymer composites: studies on sorption behavior, morphology, and heat conductance. *Construction and Building Materials* 214: 290–302.

58 Yetiş, F., Liu, X., Sampson, W.W., and Gong, R.H. (2020). Acetylation of lignin containing microfibrillated cellulose and its reinforcing effect for polylactic acid. *European Polymer Journal*: 109803.

59 AlMaadeed, M.A., Kahraman, R., Khanam, P.N., and Madi, N. (2012). Date palm wood flour/glass fibre reinforced hybrid composites of recycled polypropylene: mechanical and thermal properties. *Materials & Design* 42: 289–294.

60 Devi, L.U., Bhagawan, S.S., and Thomas, S. (2010). Dynamic mechanical analysis of pineapple leaf/glass hybrid fiber reinforced polyester composites. *Polymer Composites* 31 (6): 956–965.

61 Hameed, N., Sreekumar, P.A., Francis, B. et al. (2007). Morphology, dynamic mechanical and thermal studies on poly(styrene-*co*-acrylonitrile) modified epoxy resin/glass fibre composites. *Composites Part A: Applied Science and Manufacturing* 38 (12): 2422–2432.

62 Idicula, M., Malhotra, S.K., Joseph, K., and Thomas, S. (2005). Dynamic mechanical analysis of randomly oriented intimately mixed short banana/sisal hybrid fibre reinforced polyester composites. *Composites Science and Technology* 65 (7-8): 1077–1087.

63 Pothan, L.A., Oommen, Z., and Thomas, S. (2003). Dynamic mechanical analysis of banana fiber reinforced polyester composites. *Composites Science and Technology* 63 (2): 283–293.

64 Manral, A., Ahmad, F., and Chaudhary, V. (2020). Static and dynamic mechanical properties of PLA bio-composite with hybrid reinforcement of flax and jute. *Materials Today: Proceedings* 25: 577–580.

65 Balakrishnan, P., Gopi, S., Geethamma, V.G., Kalarikkal, N. and Thomas, S., 2018. Cellulose nanofiber vs nanocrystals from pineapple leaf fiber: a comparative studies on reinforcing efficiency on starch nanocomposites. *Macromolecular Symposia*. Vol. 380, No. 1, p. 1800102.

66 An, S. and Ma, X. (2017). Properties and structure of poly(3-hydroxybutyrate-*co*-4-hydroxybutyrate)/wood fiber biodegradable composites modified with maleic anhydride. *Industrial Crops and Products* 109: 882–888.

67 Sathyaseelan, P., Sellamuthu, P., and Palanimuthu, L. (2021). Dynamic mechanical analysis of areca/kenaf fiber reinforced epoxy hybrid composites fabricated in different stacking sequences. *Materials Today: Proceedings* 39: 1202–1205.

68 Manral, A. and Bajpai, P.K. (2020). Static and dynamic mechanical analysis of geometrically different kenaf/PLA green composite laminates. *Polymer Composites* 41 (2): 691–706.

69 Fiore, V., Sanfilippo, C., and Calabrese, L. (2020). Dynamic mechanical behavior analysis of flax/jute fiber-reinforced composites under salt-fog spray environment. *Polymers* 12 (3): 716.

70 Mazlan, N., Chai Hua, T., Sultan, M.T.H., and Abdan, K. (2020). Thermogravimetric and dynamic mechanical analysis of woven glass/kenaf/epoxy hybrid nanocomposite filled with clay. *Advances in Materials and Processing Technologies*: 1–14. https://doi.org/10.1080/2374068X.2020.1755114.

71 Tajvidi, M., Falk, R.H., and Hermanson, J.C. (2006). Effect of natural fibers on thermal and mechanical properties of natural fiber polypropylene composites studied by dynamic mechanical analysis. *Journal of Applied Polymer Science* 101 (6): 4341–4349.

72 Zuhudi, N.Z.M., Lin, R.J., and Jayaraman, K. (2016). Flammability, thermal and dynamic mechanical properties of bamboo–glass hybrid composites. *Journal of Thermoplastic Composite Materials* 29 (9): 1210–1228.

73 Gupta, M.K., Choudhary, N., and Agrawal, V. (2018). Static and dynamic mechanical analysis of hybrid composite reinforced with jute and sisal fibres. *Journal of the Chinese Advanced Materials Society* 6 (4): 666–678.

4

Thermal Properties of Sugar Palm Fiber-Based Hybrid Composites

R.A. Ilyas[1,2], S.M. Sapuan[3,4], A. Atiqah[5], M.R.M. Asyraf[6], N.M. Nurazzi[7], Mohd N. F. Norrrahim[8], Mohd A. Jenol[9], M.M. Harussani[3,4], R. Ibrahim[10], M.S.N. Atikah[11], and Chandrasekar Muthukumar[12]

[1] *Universiti Teknologi Malaysia, School of Chemical and Energy Engineering, Faculty of Engineering, Johor Bahru, 81310, Johor, Malaysia*
[2] *Universiti Teknologi Malaysia, Centre for Advanced Composite Materials (CACM), Johor Bahru, 81310, Johor, Malaysia*
[3] *Universiti Putra Malaysia, Advanced Engineering Materials and Composites Research Centre (AEMC), Faculty of Engineering, Department of Mechanical and Manufacturing Engineering, Serdang, 43400, Selangor, Malaysia*
[4] *Universiti Putra Malaysia, Institute of Tropical Forestry and Forest Products (INTROP), Laboratory of Biocomposite Technology, Serdang, 43400, Selangor, Malaysia*
[5] *Universiti Kebangsaan Malaysia, Institute of Microengineering and Nanoelectronics, Bangi, 43600, Selangor, Malaysia*
[6] *Universiti Putra Malaysia, Department of Aerospace Engineering, Serdang, 43400, Selangor, Malaysia*
[7] *National Defence University of Malaysia, Centre for Defence Foundation Studies, Kem Sungai Besi, Kuala Lumpur 57000, Malaysia*
[8] *National Defence University of Malaysia, Research Centre for Chemical Defence, Kem Sungai Besi, Kuala Lumpur 57000, Malaysia*
[9] *Universiti Putra Malaysia, Faculty of Biotechnology and Biomolecular Sciences, Department of Bioprocess Technology, Serdang, 43400, Selangor, Malaysia*
[10] *Forest Research Institute Malaysia (FRIM), Innovation & Commercialization Division, Kepong, 52109, Selangor Darul Ehsan, Malaysia*
[11] *Universiti Putra Malaysia, Department of Chemical and Environmental Engineering, Serdang, 43400, Selangor, Malaysia*
[12] *Hindustan Institute of Technology & Science, School of Aeronautical Sciences, Chennai, 603103, Tamil Nadu, India*

4.1 Introduction

Currently, polymer-based products are vital and are on high demand in numerous industrial applications, including cross arm in transmission line tower [1], wind turbine [2], automotive [3–7], boat [8], packaging [9–17], antimicrobial packaging [18, 19], and household products [20]. The polymer-based products have significant mechanical performance in conducting and resisting tension and compressive force [21, 22]. Owing to the global demands on sustainable energy and green technology, the current development in polymer sector is shifting toward reducing the dependency on synthetic-based products. At the moment, the most feasible and promising practice to cope with ecologically friendly polymeric composite is

Natural Fiber-Reinforced Composites: Thermal Properties and Applications, First Edition.
Edited by Senthilkumar Krishnasamy, Senthil Muthu Kumar Thiagamani,
Chandrasekar Muthukumar, Rajini Nagarajan and Suchart Siengchin.

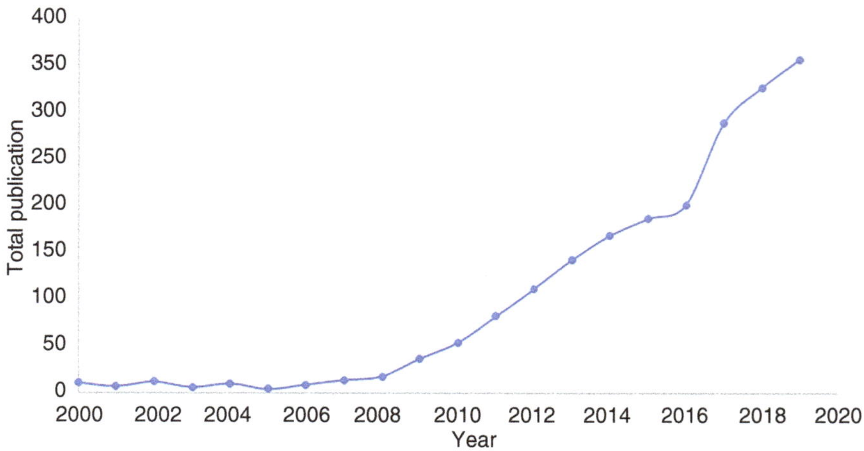

Figure 4.1 Number of published outputs, based on hybrid composites (between 2000 and 2019), found in the lens.org by searching the words "applications of fiber hybrid composites."

the implementation of natural fiber as its reinforcement [23–27]. The introduction of green-based products, such as natural fiber reinforcement composites aids to cut down the demands on synthetic-based composites, especially coming from depleting petroleum resources [28]. Moreover, the application of natural fibers in polymer-based products helps to increase the demand on agricultural waste products, subsequently providing prospective source of income for farmers [29–32]. Besides that, natural fiber also exhibits lightweight and low cost relative to synthetic counterpart. However, there are some drawbacks with the application of polymer composites reinforced with natural fiber, including lack of thermal and mechanical performances, such as quasi-static, creep, and fatigue [33–38]. Within the scope, various approaches have been conducted to tackle this problem, such as hybridization of two different fibers [39], treating the fiber using chemical modification, and inclusion of filler particles in composite laminate [40]. One of the viable techniques used by many researchers was hybridization of two different fibers, since it is able to overcome the disadvantages of one another. Figure 4.1 denotes the total publications based on hybrid composites from year 2000 to 2019, searched from lens .org website by searching the "applications of fiber hybrid composites" keywords. It would seem that publications relating to applications for hybrid fiber composites have received a lot of attention from 2008 and kept increasing ever since. Currently, carbon fiber-based hybrid composite appears to have a big future in many sectors, especially in automotive and aerospace. However, limitations related to the wider application of carbon fibers is their fluctuating price and availability in comparison with other structural reinforcements [41].

Sugar palm fibers (SPFs) are harvested from the sugar palm tree, an indigenous species of Malaysia and several other Asian countries. This fiber is produced from various parts of the tree including its bunch, frond, trunk, and also the "gentian ijuk," placed around the trunk [42–44], as shown in Figure 4.2. SPF are used to manufacture traditional products, such as coating brush, bottle brush, mats, and rope [46]. Many researchers have studied the potential of SPF originating from different sections of the tree as alternative enhancement materials in the polymer composites [42, 47–56]. The fiber has been applied as reinforcement since they are naturally available in the form of woven structure, and extraction is simple without cutting the tree. In past decades, the reinforcement of SPFs with various types of polymers had been implemented. The main polymers used are corn starch, cassava starch (CS), sago starch, sugar palm starch, vinyl ester, phenolic, polyurethane, polyurethane, high-impact polystyrene, phenol formaldehyde, polyester resin, unsaturated polyester, and epoxy (Table 4.1). SPF-reinforced composites have been discovered to have poor strength and modulus; thus, the drawbacks restrict their usefulness in applications requiring higher strength and durability. SPF exhibit excellent water resistance; hence, their hybridization with other fibers such as cassava showed to be a perfect pairing for their material properties enhancements, especially on thermal performance [77]. The behaviors of hybrid fiber composites are highly dependent on fiber intermingling, individual fiber content, fiber orientation, compatibility of fiber to matrix, and hybrid fiber arrangement [80]. Recently, studies on the SPF hybridization with other fibers has attracted many researches in discovering the augmentation of the thermal properties of the fiber. In this case, the effects from natural fibers hybridization are being considered as one of the main paramount issues in dealing with hybrid composites [76, 81, 82].

Material characterization is one of the vital parts in understanding the properties of natural fibers and their composites to various conditions, including elevated loads, moisture, and temperature.

Thermal gravimetric analysis (TGA) is a thermal evaluation approach that implements mass change (in %). It is used to assess the thermal decomposition in elevated temperature and thermal stability of the natural fiber before and after any modifications [83, 84]. These depend on the three quantities: temperature, mass, and changes of mass with temperature. The TGA has emerged as a fast, easy, less expensive, and alternative approach to regulate the composition of lignocellulosic biomass. By using this method, the α-cellulose and hemicellulose contents can be determined accurately, as compared with the wet chemical methods [45]. This is very beneficial and useful due to its high-throughput natural (at line) and is ideal for accessing polymers such as lignin. However, this method is commonly known as destructive test in temperature ramping that leads to changes in polymer analyses [85]. For instance, Figure 4.3 shows the thermogram of the treated and untreated SPF [52]. The thermal degradation of treated and untreated fibers was found to have occurred in five phases: the moisture content evaporation that happened at a temperature of 45–123 °C within the fiber, deterioration of the hemicellulose lignocellulosic components (220–315 °C), cellulose degradation (300–370 °C), lignin degradation (160–900 °C), and formation of ash at 1723 °C [52].

Apart from that, derivative thermogravimetry (DTG) is another thermal analysis carried out using differential thermal analysis (DTA)–TGA thermal analyzer such as

Figure 4.2 Photographs of (a) the sugar palm tree, (b) sugar palm fibers natural woven structure, (c) sugar palm short fibers. FESEM images of sugar palm fibers: (d) longitudinal section, (e) cross section, and (f) primary and secondary cell wall and middle lamella. Source: Adapted with permission from Ilyas et al. [45].

Shimadzu DTG-60H [86]. Usually, this thermal analysis measures the rate of material weight changes upon heating dependably on temperature. Moreover, it is also being applied to simplify the data acquisition on weight vs. temperature in thermogram [87]. For instance, a study led by Ibrahim et al. [76] showed that each event related with notable peak observed from the DTG curve that corresponds to weight loss displayed in the thermograph. In this case, the initial weight loss happened at below 100 °C because of evaporation and dehydroxylation processes of water. Higher heating rates was experienced by the samples with higher moisture content and therefore larger weight loss, and it is coherent with the results of moisture content.

Dynamic mechanical analysis is a testing approach using analytical instrument to measure the physical behavior of solids and polymer melts to assess their thermal moduli and damping properties. The program involves the measurement of force, stress, strain, frequency, and temperature to describe the rheology of solids along with the temperature response [88, 89]. In specific, the test implements the oscillating force in terms of stress to evaluate the elastic response undergoing viscoelastic behavior. Later, the test would aid stress to discover glass transition temperature or (T_g) and to determine the transition for other molecular motion. During the test, sinusoidal stress is applied to estimate thermostrain and complex modulus of the polymeric material [90, 91]. Meanwhile, the measurement of the heat energy loss along the viscous region of the plastic is called loss modulus.

Table 4.1 Reinforced sugar palm fiber with different polymer composites.

Fiber	Matrix	References
Sugar palm fiber	Epoxy	[57]
	Unsaturated polyester	[58]
		[59]
		[60, 61]
	Polyester	[62]
		[63]
		[64]
	Phenol formaldehyde	[43]
	High impact polystyrene	[65]
	Polyurethane	[66]
	Phenolic	[67]
	Sugar palm starch	[68]
	Vinyl ester	[69]
		[70]
		[71]
		[69]
	Polyurethane	[72]
		[73]
	Sago starch	[74, 75]
	Corn starch	[76]
	Cassava starch	[77]
Sugar palm cellulose	Sugar palm starch	[78]
Sugar palm nanocrystalline cellulose	Sugar palm starch	[53]
Sugar palm nanofibrillated cellulose	Sugar palm starch	[51]
		[49, 79]

For the instrumentation, a dynamic mechanical analysis (DMA) is composed of displacement sensor (linear variable differential transformer), which measures the voltage shift as a consequence of an instrument probe moving through a magnetic core [92]. In the machine, the sample laminates are clamped to allow placing the sample in a fixed position in the heat chamber by a guidance system.

The objective of this chapter is to review and discuss the thermal properties of SPF-based hybrid polymer composites, including TGA, DTG, and dynamic mechanical analysis (DMA).

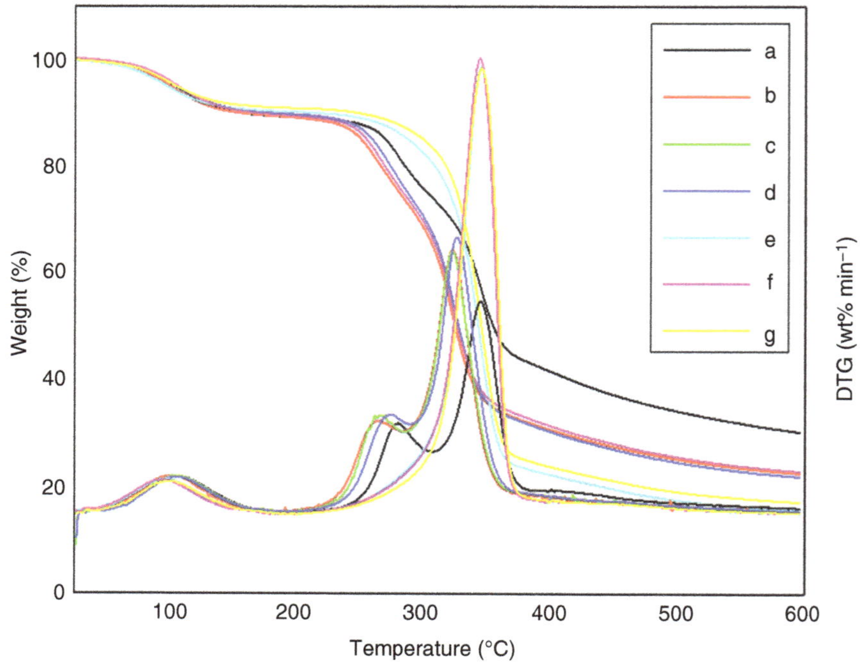

Figure 4.3 TGA and DTG curves of the samples: (a) raw SPF, bleached fiber; (b) SPBF05, (c) SPBF06, and (d) SPBF07, alkali-treated fiber; (e) SPC05; (f) SPC06; and (g) SPC07. Sugar palm bleached fiber 5 hours of treatment, SPBF05. Sugar palm bleached fiber 6 hours of treatment, SPBF06. Sugar palm bleached fiber 7 hours of treatment, SPBF07. Source: Ilyas et al. [52].

4.2 Thermal Analysis of Sugar Palm Fiber-Based Hybrid Composites

4.2.1 TGA Properties of Sugar Palm Composites

Referring to the research on thermal properties of SPFs, *Arenga pinnata*, from various heights of sugar palm tree, 1, 3, 5, 7, 9, 11, 13, and 15 m [93], there are different thermal properties of different height of the SPFs. Moreover, there have been four stages of fibers decomposition. The decomposition sequence began with the decomposition of moisture, hemicelluloses, cellulose, and lignin, respectively, whereas the ash remained. Moreover, the thermal decomposition of these fibers were identified in ranges of 45–123, 210–300, 300–400, 160–900, and 1723 °C, respectively. From the curves of the TGA and DTG, it was found that the 1 m fiber displayed greater thermal stability than the 3–15 m fibers. Here, various thermal stabilities recorded for each fiber were contributed by the different chemical compositions, i.e. fiber containing high ash content indicated higher thermal stability.

Based on the study of TGA of SPF, the first phase of decomposition observed was the evaporation of the absorbed moisture in fibers within temperature range of 45–123 °C. The second stage was decomposition of hemicelluloses. The elevated

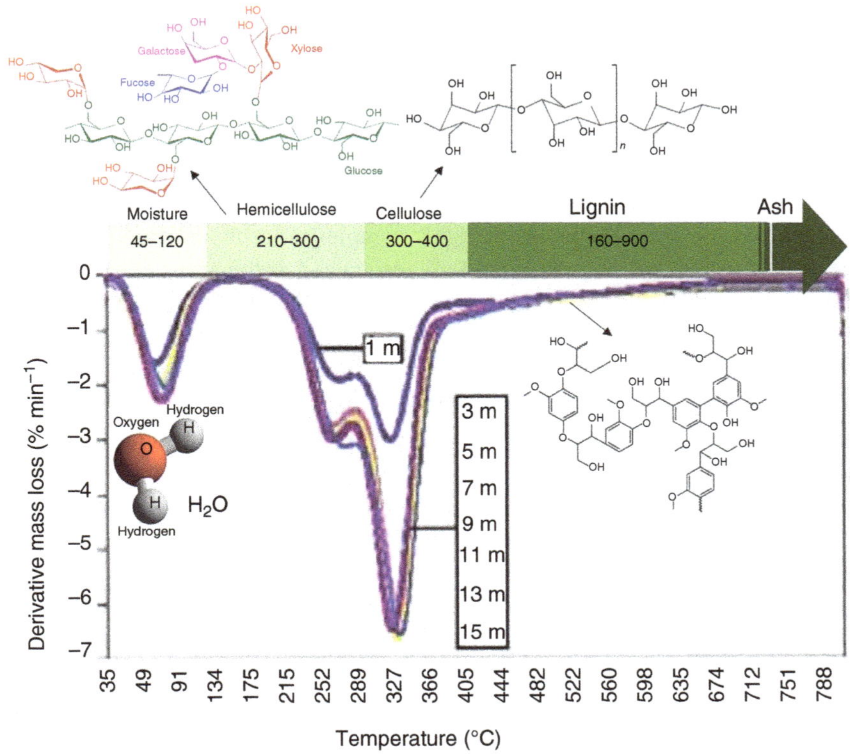

Figure 4.4 DTG thermogram and summary of decomposition of sugar palm fibers attained from sugar palm tree at different heights.

temperature leads to the changes in the hemicelluloses components chemically, as it was caused by the breakdown of the cellular. It was reported that the temperature needed to decay the hemicelluloses of the fibers (1–15 m) was in the range of 228–312 °C. Nevertheless, the authors reported that in terms of mass loss rate at 271 °C, 1 m of fiber has lower mass loss rate, approximately 1.91 wt% min^{-1}, relative to the 3–15 m of fibers with 2.95 wt% min^{-1}. It was attributed by the lesser content of hemicelluloses (4.71%) compared with the fiber of 3–15 m (7.47%). Compared with other major compositions of natural fiber, there are many explanations for early degradation of hemicelluloses relative to other compositions. Hemicelluloses consist of numerous saccharides, such as glucose, mannose, xylose, etc., which are random, with amorphous structure plus easily detached from the main stem. Furthermore, even at low temperatures, they easily degraded into volatiles such as CO, CO_2, and other hydrocarbons [94]. Therefore, hemicelluloses have low thermal stability rather than cellulose that constructed with closely associated number of chains through networks of hydrogen bonding to form the microfibril. Hence, the microfibril formation acts as the main reinforcing compound in the cell wall ([95, 96]). Figure 4.4 shows the DTG thermograph and summary of decomposition of SPF collected from various heights.

The moment the hemicelluloses had fully decomposed, the third stage of decomposition of SPF will take place. Due to the high-crystalline nature of cellulose relative to hemicellulose, it has quite high thermal stability, which is normally at high temperature, 315 °C. The fourth stage of decomposition of SPF is lignin, which is harder to be deteriorated rather than the cellulose and hemicellulose. Kim et al. [97] reported that the decomposition of lignin began at 160 °C. Yet, the lignin slowly decomposes and prolongs its temperature up to 900 °C to be decomposed completely. Thus, owing to its highly complex amorphous structure, lignin is considered as a tough materials and thus recognized as the core compound that offers rigidity to the plants. As recorded from the previous study, lignin also acts as a cementing material that fills up the spaces among the cellulose and the hemicelluloses [98]. The final stage is the decomposition of inorganic material in the SPFs, also known as ash. This indicates the presence of inorganic materials in the fiber, for example, silica or silicon dioxide, SiO_2, has high degradation point, at temperatures greater than 1723 °C.

In another study done by Norizan et al. [99] on the reinforced SPF thermal properties with unsaturated polyester composites and from the research, three main phases of composites weight loss within the specified temperature ranges were discovered. Based on the curve, the first stage of weight loss was indicated by a small prominent step that occurred at the range of temperature from 30 to 110 °C. Hence, it was associated toward the moisture content loss of the fiber. Most of the fiber deterioration occurred at temperature range between 150 and 380 °C during the second stage. The degradation of three major natural fiber components such as hemicellulose, cellulose, and lignin occurred at this crucial phase. For neat unsaturated polyester, a single degradation step is seen by statistical chain rupture. It was discovered that styrene is the main product that is extracted at temperature range between 360 and 400 °C. In total, the composites' thermal stability decreased in correlation with the onset and maximum temperatures, as for the residue percentage kept increased with the increasing fiber loadings. Based on Table 4.2, the summarized outcome displayed the decreasing of the decomposition onset temperature of the unsaturated polyester due to the addition of SPFs into the composite. It is due to the substitution of certain parts of the unsaturated polyester matrix with SPFs with lower thermal stability; therefore, the polymer matrix thermal resilience also decreased. Next, the thermal stability found to decrease with increasing loading of SPFs, which also attributed to the fiber–fiber interaction. In addition, this interaction was stronger with higher loadings of SPFs than with fiber-unsaturated polyester interactions, which could also mitigate the temperature of decomposition due to the unrestricted mobility.

4.2.2 TGA Properties of Sugar Palm Hybrid Composites

Hybridization with synthetic fibers like carbon fiber, glass fiber, aramid fiber, and Kevlar that are discovered to contribute in enhancing the mechanical, physical, and even thermal properties of the composites [86]. A comprehensive review conducted by Atiqah et al. [100] discovered that the loading of glass fibers to polyurethane-reinforced SPF exhibited stronger thermal stability as shown by TGA.

Table 4.2 Characteristic temperature of sugar palm-unsaturated polyester composites at elevated weight loss.

Fiber loading (wt%)	T_{onset} (°C)	T_{max} (°C)	T_{end} (°C)	$T_{10\%}$ (°C)	Residue (%)
0	340.53	380.42	593.00	282.90	0.42
10	307.79	366.74	595.59	253.85	4.03
20	301.48	370.30	595.22	265.62	6.75
30	301.50	370.96	596.01	263.86	8.28
40	295.91	366.84	595.36	246.48	9.07
50	294.88	343.04	595.59	248.05	10.78

The composites exhibited three phases of degradation. The first stage was contributed by the sugar palm degradation, followed by the deterioration of the glass fiber and thermoplastic polyurethane (TPU). Moreover, the temperature at the maximum weight loss rate (T_{max}) rose proportionally with the addition of SPF contributing to the improvement of the thermal stability of the composites. Furthermore, lower temperature of decomposition was recorded for the higher fiber loading of SPF, 30/10 of SP/G ratio. Thus, concluded that increasing amount of fiber loading decreases the temperature of decomposition of the hybrid composites. Besides, the authors also have stated that the consistent changes in the final stage of the degradation behavior of hybrid SPFglass and glass-reinforced polyurethane hybrid composites. However, hybrid composites with 30 wt% of fiber demonstrated greater weight loss with lower char residue, compared with the glass fiber-reinforced composites. Thus, it indicates a greater composition of SiO_2 in SPF, which was primarily responsible for the production of thermal residues of char. Next, the chemically treated SPF hybrids possessed better thermal properties than the untreated composites [101]. In overall, the hybrid composite with combined alkaline–silane (6% alkaline–2% silane) degraded at lower temperature, while the alkaline-treated and silane-treated composites degraded at higher temperature.

Researchers have discovered the effect of hybridization and alkaline treatment on the reinforced SPF polymer composite. According to Nurazzi et al. [60], treated sugar palm yarn reinforced with glass fiber composites relatively affect its thermal properties. The fiber was treated for 60 minutes with 1% sodium hydroxide (NaOH) solution with matrix reinforcement ratio of 70/30, 60/40, and 50/50 wt%, accordingly. Results from TGA showed that the residue percentage declined as the loading of glass fiber raised. In the end, the hybridization of treated SPF with glass fiber improved the efficiency of its thermal properties that could be utilized for structural purposes. A slight decline in the structural destabilization could well be observed from samples with 30 and 40 wt% of the untreated and treated SPFs. It happened as the balanced ratio between the SPFs and the glass fibers that have higher thermal stability in the hybrid composites. The equilibrium ratio allowed the SPF to consume more heat at early phase of the temperature rise compared with the glass fiber. Furthermore, the treated SPF composites are more thermally stable than the

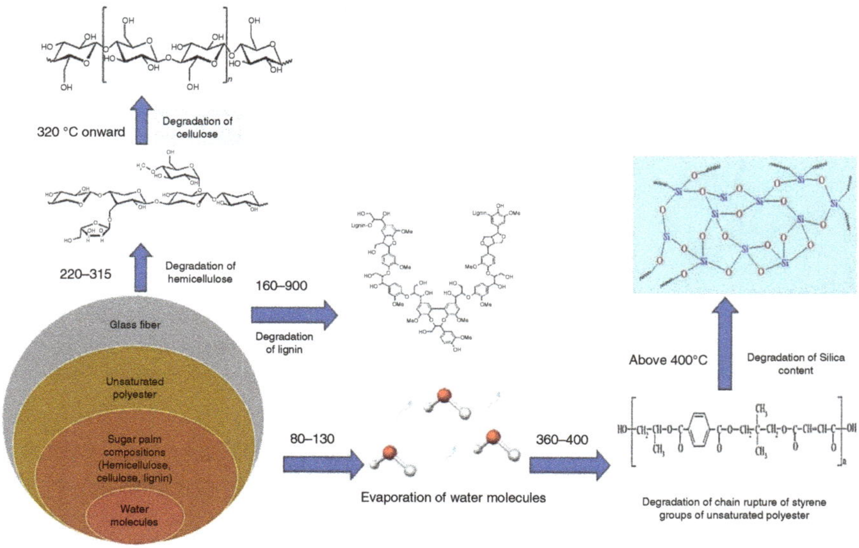

Figure 4.5 Decomposition route of hybrid sugar palm/glass fiber/unsaturated polyester composites.

untreated one [102]. The alkaline treatment eliminated the fiber surface wax; hence, it caused reduced content of hemicellulose and lignin. Owing to the higher crystallinity and thermal stability of cellulose, compared with amorphous hemicellulose, the enhancement of the crystalline structure may lead to the increasing amount of hydrogen binding instead of hydroxyl structure, which happened due to the elimination of the hydroxyl group. Hence, improved hydrogen bonding in the composite structure leads to better thermal energy among the bond. These took place after the alkaline treatment [103]. Figure 4.5 shows the decomposition route of hybrid sugar palm/glass fiber/unsaturated polyester composites [102].

Study conducted by Jumaidin et al. [80] had showed the influences of the SPF loading toward the thermal properties of thermoplastic sugar palm starch agar (TPSA) reinforced with seaweed composites. TPSA acts as a polymer matrix in which the hybridized seaweed/SPF filled at weight ratio of 25 : 75, 50 : 50, and 75 : 25 was fabricated. From the study, it was found that the composites' thermal stability was improved, specified by the larger onset temperature of 25 : 75 seaweed/SPF composites degradation at 259 °C compared with the individual seaweed composites at temperature 253 °C. The results showed the early decomposition phase that occurred below 100 °C associated with the process of water dehydration and compound with lower molecular weight. While at 100–200 °C, the occurred decomposition phase was attributed to water and glycerol evaporation [104]. The decomposition of agar occurred at a temperature over 270 °C [105]. In the meantime, at around 300 °C, the starch carbon chains were degraded and depolymerized, and thus, it contributed to the maximum decomposition stage [106]. Moreover, the carbohydrate and protein decomposition originated from seaweed occurred at temperature range between 180 and 450 °C [107]. Moreover, based on Ross et al. [108], the decomposition of cellulose and hemicellulose in SPF occurred at temperature between 200 and 270 °C, and the lignin breakdown and the cellulose final decomposition occurred at 270–370 °C. Hence, the decomposition that occurred at temperature over 500 °C is consistent with the deterioration of carbonate in seaweed, resulting in the char formation. Hybridization of SPF with seaweed hybrid composites enhanced the thermal stability of the polymer composites. Moreover, higher loading of SPF in composites has been found to upsurge their onset and maximum temperature of degradation. It may be due to the greater amount of lignocellulosic components in SPF.

Furthermore, the study on the stimuli of alkaline treatment toward thermal properties of TPU reinforced with hybrids of roselle fiber/SPF composites, whereas the thermal stability of the composites were improved [82]. Prior to the characterizations, the roselle fiber/SPF hybrid composites were treated with various concentrations of NaOH such as 3%, 6%, and 9% and then fabricated via melt mixing and compression molding techniques. In addition, these composites decomposed thermally within four stages, including moisture evaporation, lignocellulosic compound decomposition, cellulose and lignin degradation, and ash formation. From the results, the treated hybrid composites showed higher thermal stability than the untreated hybrid composites [109]. Thus, this phenomenon was contributed by the alkalization that removes all the impurities, wax, and hemicellulose within the composites. Besides that, TGA thermogram results displayed increasing amount of char residue percentage relative to the untreated one by approximately 14.98%.

Another hybridization approach of sugar palm hybrids with cassava bagasse is by using casting approach where cassava starch is the matrix while fructose acts as the plasticizer, as reported by Ahmed Edhirei et al. [77]. SPF were added to the cassava starch/cassava bagasse composite film with 6% cassava bagasse at different loadings of 2%, 4%, 6%, and 8% w/w dry starch. From the thermal properties observations, the fiber loading led to the decrease of degradation temperature of the hybrid. Thus, it is due to the low degradation temperature of the natural fibers itself. Hence, the incorporation of SPFs resulted in the changes in properties of the starch film, affecting the performance of the films.

4.3 Dynamic Mechanical Properties of Sugar Palm Fiber-Based Hybrid Composites

4.3.1 Sugar Palm/Glass Fiber

Hybrid particulate with sugar palm and glass as fibers was mixed using melt mixing/compounding and examined by DMA at frequency of 1 Hz. The sugar palm-reinforced composites resulted in slightly lower storage modulus (E') than sugar palm/glass composites. Moreover, storage modulus (E') values of sugar palm/glass hybrid composites were 551.22 MPa higher than glass composites. The optimum fiber concentration at 10/30 wt% of sugar palm/glass fiber is due to the hybrid effect by the presence of stiffer glass fiber [100]. Similarly, sugar palm- and

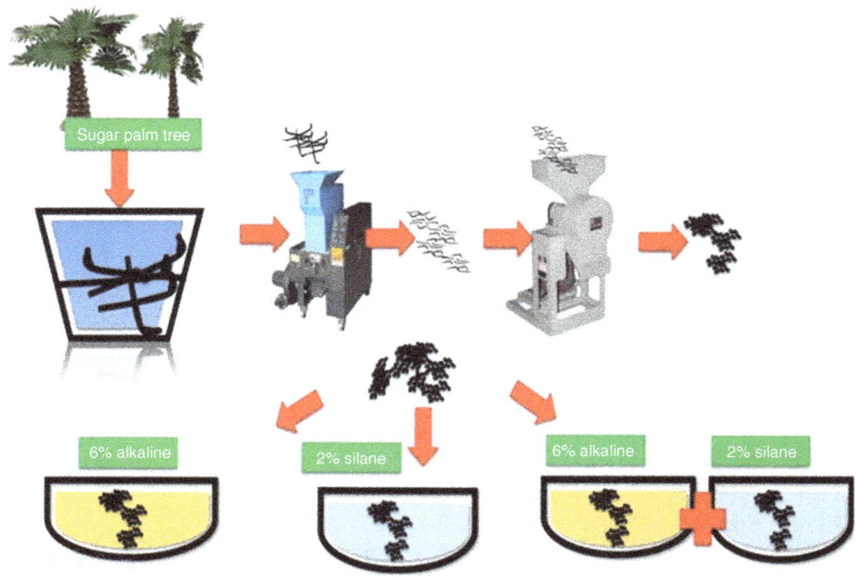

Figure 4.6 Fiber treatment for sugar palm fiber in the particulate form. Source: Atiqah et al. [110].

glass fiber-reinforced TPUs revealed that the properties such as stiffness and E' of the hybrid composites would depend on the fiber treatment (such as 6% alkaline, 2% silane, and combination of 6% NaOH and 2% silane treatment). The storage modulus E' and loss modulus (E'') for treated combined alkaline and silane showed improved thermal mechanical properties as compared with individual treatment and untreated of SPF [110]. Figure 4.6 shows the fiber treatment process for sugar palm/glass-reinforced TPU hybrid composites.

Safri et al. [111] used benzoyl treatment for sugar palm with glass fiber and showed the highest loss and storage modulus than untreated sugar palm-reinforced epoxy composites. The optimum SPF concentration up to 30 wt% showed the enhancement of the hybrid composites mechanical properties that was affected by the temperature. Also, the highest damping factor with 0.6807 at 71.55 °C showing benzoyl treatment enhanced the interaction between sugar palm and glass fibers. Moreover, the positive shift of T_g for benzoyl-treated fibers indicated the excellent interaction of fiber and matrix in the hybrid composites. The Cole–Cole plot gave the homogeneity and heterogeneity of individual composites [112]. The relationship between storage and modulus also gave the validity of a time–temperature superposition. Moreover, none of the curves showed heterogeneity, indicating the hybrid composites were finely dispersed.

Hybrid composites made of sugar palm yarn (untreated and 1% NaOH treated, as shown in Figures 4.7 and 4.8) with glass fiber were studied for dynamic mechanical properties over a predominant range of temperature [61]. The E' of the alkaline treated with 50/50 wt% hybrid samples was found to increase due to the good fiber matrix interaction. Also, the increased E" and the stiffness were observed for

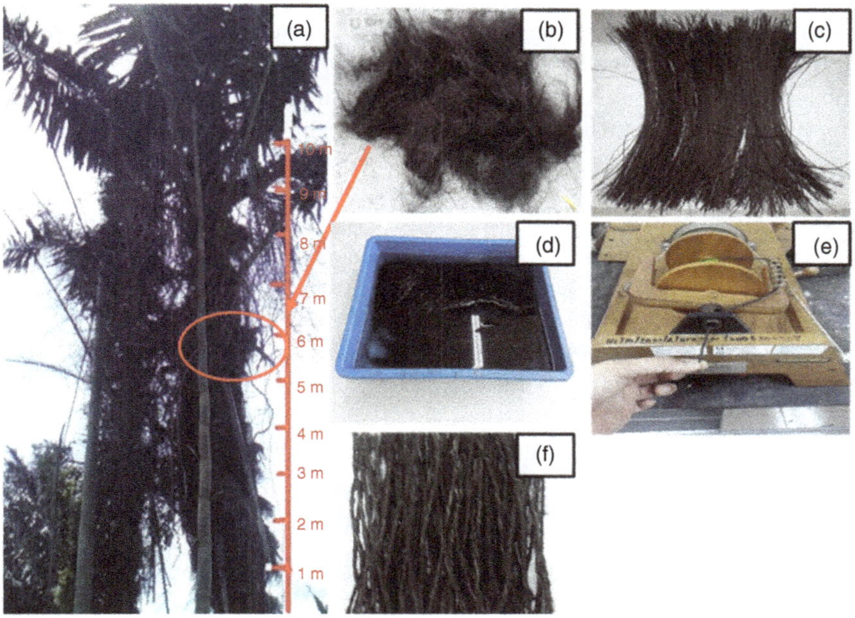

Figure 4.7 Alkaline-treated sugar palm/glass fiber yarn reinforced polyester hybrid composites. Source: Nurazzi et al. [61].

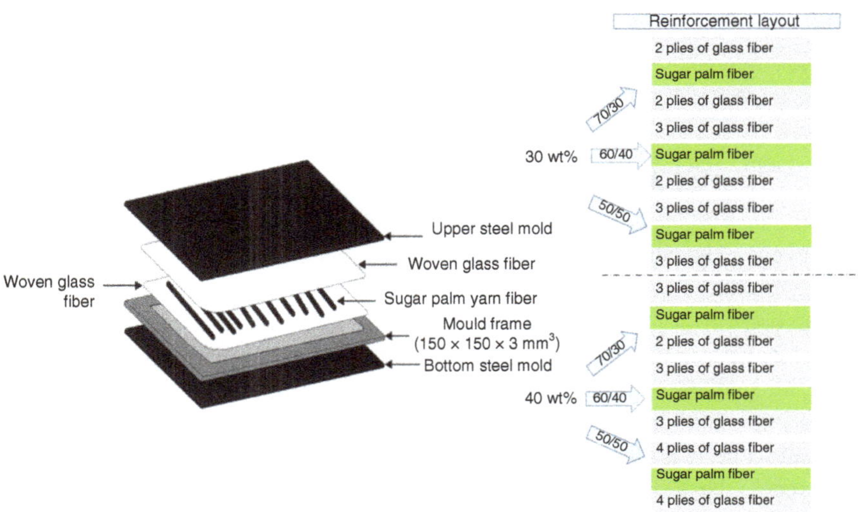

Figure 4.8 Layup segmentation and reinforcement layout schematic diagram. Source: Nurazzi et al. [61].

composites with fiber treated with 1% NaOH that was caused by the reduction of hydrophilicity of the SPF. This was due to the removal of 16% of hemicellulose content that occurred during the post-alkaline treatment. The shifting T_g from the peak of E'' was due to the reduced movement of unsaturated polyester resin that exhibited the excellent interfacial adhesion of fiber and matrix. The lower damping factor ($\tan \delta$) exhibited treated 50/50 wt% of sugar palm yarn/glass fiber-reinforced polyester composites [61].

4.3.2 Sugar Palm/Cassava Bagasse

Cassava bagasse-reinforced cassava starch (CS) composite, cassava bagasse (CB)/ sugar palm hybrid-reinforced CS composites, and neat CS composite film demonstrated the drop in the DMA properties. The addition of SPF at 8 wt% increased the E' value at 1.49 GPa as compared with CS composites having 0.457 GPa at −75°. Also, the E'' showed the broad peak for CS–CB/SPF hybrid that could be due to the increase in flexibility of cassava starch caused by the addition of SPF [95].

4.3.3 Sugar Palm/Flax

Variations in the fiber content of sugar palm and flax fibers have shown improved E' for all fiber loading studied (0/30, 30/0, and 15/15 wt%). SPF/flax with two different stacking sequences, either the flax or SPF, was arranged intermediate of the layer of hybrid composites. The loss modulus E'' for the flax/sugar palm fiber (F/SPF) hybrid composite was 970 MPa and for sugar palm fiber/flax/sugar palm fiber (SPF/F/SPF) composite was 240 MPa. The incorporation of flax and sugar palm can lower the damping factor that represented elastic response characteristics permitting broader applications. The Cole–Cole plot also displayed perfect semicircular plots, indicating the improved fiber matrix compatibility that led to effective interfacial interactions of hybrid composites. Table 4.3 shows the summary of works on DMA of sugar palm hybrid composites.

4.4 Potential Applications

In the twenty-first century, the demand to create better performance and environmentally friendly fiber hybrid composites is increasing. This could be achieved by the development of the fiber-based hybrid composites as a sustainable alternative material for numerous applications, such as aerospace, automotive, packaging, electronics, biomedical, transmission, renewable energy, and other applications [113]. Scientists are continuously discovering several new developments on fiber hybrid composites that can give more benefits to the society. Natural fiber hybrid composites could be a great alternative to carbon fiber hybrid composites, which worked to improve the mechanical properties of the composites and reduce moisture absorption, production costs, environmental impact, and carbon and energy footprint of the materials [114]. However, poor thermal properties of natural fibers hinder their utilizations and improvements. The thermal properties such as TGA, thermomechanical analysis (TMA), DMA, and differential scanning

Table 4.3 Summary of previous work on DMA of sugar palm hybrid composites.

Hybrid	Fiber treatment	Polymer matrix	Fabrication process	Testing parameter	DMA properties	T_g at E'' peak (°C)	Peak height of tan δ	References
Sugar palm/glass fiber (particle)	Untreated	Thermoplastic polyurethane (TPU)	Melt mixing followed by hot pressing	Temp: −100 to 100 °C Heating rate: 10 °C min^{-1} Oscillation frequency: 1.0 Hz	E', E'' increase and lower tan δ with the addition of 10–30 wt% sugar palm fiber	16.17–17.27	0.2113–0.2288	[110]
Sugar palm/glass fiber (particle)	6% NaOH, 2%, combined silane 6% NaOH + 2% silane for 3 h	TPU	Melt mixing followed by hot pressing	Temp: −100 to 100 °C Heating rate: 10 °C min^{-1} Oscillation frequency: 1.0 Hz	E', E'' increase and lower tan δ with the treated of sugar palm fiber	(−26.71)–(−26.71)	0.1944–0.2626	[100]
Sugar palm/cassava bagasse (particle)	Untreated	Cassava starch	Casting technique	Temp: −75 to 75 °C Heating rate: 1.5 °C min^{-1} Oscillation frequency: 1.0 Hz	E', E'' increase with the addition of sugar palm fiber. No trend for tan δ	−30	0.1–0.7	[95]
Sugar palm/glass (yarn)	1% NaOH for 1 h	Unsaturated polyester resin	Hot pressing	Temp: 30–150 °C Heating rate: 5 °C min^{-1} Oscillation frequency: 1.0 Hz	E', E'' increase and lower tan δ with the treated and addition of sugar palm at 50/50 wt% fiber and alkaline treated	73.11	0.3550	[61]

Table 4.3 (Continued)

Hybrid	Fiber treatment	Polymer matrix	Fabrication process	Testing parameter	DMA properties	T_g at E'' peak (°C)	Peak height of tan δ	References
Sugar pam/glass (chopped random fiber)	Benzoyl	Epoxy	Hand layup	Temp: 25–200 °C Heating rate: 5 °C min^{-1} Oscillation frequency: 1.0 Hz	E', E″ increase and lower tan δ with the treated and addition of 30 v/v sugar palm fiber	77.1	0.6807	[111]
Sugar palm/flax (unidirectional long fiber)	Untreated	Epoxy	Hand layup	Temp: 30–145 °C Heating rate: 5 °C min^{-1} Oscillation frequency: 1.0 Hz	E', E″ increase and lower tan δ with the treated and addition of flax fiber	68.18	0.41	[39]

Table 4.4 Thermal properties and application highlights of selected studies on natural fiber hybrid composites.

Material	Highlight on thermal/ thermomechanical properties	Suggested application	References
Sisal/basalt/epoxy	The hybrid highlighted shown better DMA as compared to other hybrid composites	Aeronautic	[116]
Glass–PEEK (polyether ether ketone) hybrid	Stable properties of real permittivity and loss tangent of glass–PEEK laminates at low frequencies	Aeronautic	[117]
Bamboo and flax fiber (BFBF) mat-reinforced epoxy hybrid	Improved thermal properties of the mat reinforcement content	Automotive	[118]
cork/cellulose pulp/hemp fibers	The developed board has a good thermal conductivity with those of other materials used for similar applications	Boards and mortars for nonstructural elements of construction material	[119]
Oil palm fiber/clay-reinforced HDPE	The DMA of complex modulus was increased with addition of filler. The hybrid composites thermal stability was increased with inclusion of clay, more thermally stable than oil palm fiber	Packaging plastics	[120]
Corn husk (CH)–cornstarch (CS)/sugar palm fiber (SPF) hybrid	Degradation temperature (298–300 °C) slightly lower than CH–CS composite	*Not mentioned*	[76]
Treated sugar palm yarn/glass fiber-reinforced unsaturated polyester hybrid	The hybrid composite possessed the largest storage and loss modulus, as well as the lowest damping factor	*Not mentioned*	[60]

calorimetry (DSC) of natural fibers hybrid composites are among important criteria that influence their applicability in advanced applications, especially for automotive and aerospace [115]. The developed natural fiber hybrid composite for aeronautic and automotive products is vulnerable to fire and hence is desirable to make them fire resistant. Table 4.4 summarizes selected studies on the thermal properties and their potential applications. This section has the potential to guide all composite designers, manufacturers, and users on the right selection of composite materials based on the thermal properties for several applications.

The aircraft industry is the most important sector that utilizes the fiber hybrid composites since it relies on higher stress components, especially on safety and weight. According to Ezugwu [121], this industry has been known to be a pacemaker for innovations and development of new material frameworks. This has important thermal properties for the fiber to debilitate their limitation, since the matrix material used with carbon and glass fiber could maintain its temperatures beyond 426–482 °C [122]. In spite of this, Mochane et al. [114] suggested that carbon fibers have offered excellent properties as composite material to be used in high-end applications, including aeronautic, wind turbines, and large-scale energy storage since it renders high thermal and electrical conductivities [114]. There are several studies conducted on the development of natural hybrid composite for aeronautic applications. Alexander and Churchill [116] have developed a new hybrid composite by combining basalt fiber and natural fibers, including sisal and glass fibers. The hybrid composite (sisal/basalt/epoxy composites) was found to have better DMA properties (storage modulus and loss modulus) than other combinations. This study also suggested that the bond strength formed between sisal and basalt fiber has great potential to be utilized in aircraft structural applications. In another finding, Shekar et al. [117] demonstrated the fiber–fiber hybrid composite by combining the extruded polyether ether ketone (PEEK) and glass fiber using compression molding technique for the application of aerospace industry. The result showed that the fabricated hybrid composite with 17 layers of fabric exhibited better modulus and flexural strength than 15 and 20 layers of fabric. In addition, at low frequencies, this hybrid showed stable real permittivity and loss tangent behavior due to the dipole line-up direction in the field and consistent behavior at the temperature range of 50–300 °C.

In the automotive industry, the demand for lightweight vehicles has tremendously increased as it helped in low-carbon emission and reduces fuel consumption. In fact, 25% reduction in the vehicle's weight could minimize about 250 million barrels of fuels and 220 billion in energy cost, which subsequently reduce in CO_2 emission [123, 124]. Prabhakaran et al. [125] suggested that the structural component, such as automotive bumper, has a vital contribution in vehicle crashworthiness, which leads to the protection of the occupant during collisions. In another report, Kim et al. [126] reported that the vehicle's bumper system made up of three main components, (i) an energy absorber, (ii) beam, and (iii) vis fascia, as illustrated in Figure 4.9.

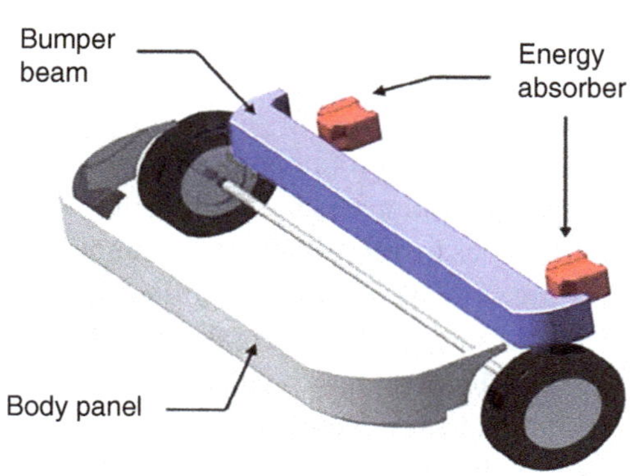

Figure 4.9 The schematic diagram of bumper system. Source: Kim et al. [126].

As demonstrated by Shukla and Srivastava [118], the bamboo and flax (BFBF) mat-reinforced epoxy hybrid composite was analyzed for its application in automotive industry. This composite exhibited higher first and second stages of thermal degradation, in the range of 215–390 and 431–860 °C, respectively. The author also reported that this natural hybrid composite had increased glass transition temperature compared with plain epoxy. This is due to reinforcement of bamboo and flax mat.

The application of natural fiber hybrid composites in marine systems also received much attention in the last decades. It is potentially usable to replace the several components in the ship, including wind and tidal turbine blades, ship hulls, and propeller blades. Rubino et al. [127] had reviewed the applications of fiber hybrid composite in marine. The authors discovered that environmental issues are driving the production of highly recyclable materials like natural cellulose fibers and bio-based polymers that could be utilized in marine applications. Figure 4.10 shows the utilization of the composite materials on the military ships and even submarine body parts. Fiber hybrid composite for the marine applications also requires resins (thermosetting and thermoplastics) that have good thermal properties. Moreover, the composite should also have higher processability especially during the manufacturing of composite parts [127]. The resins are easy to be laminated via room

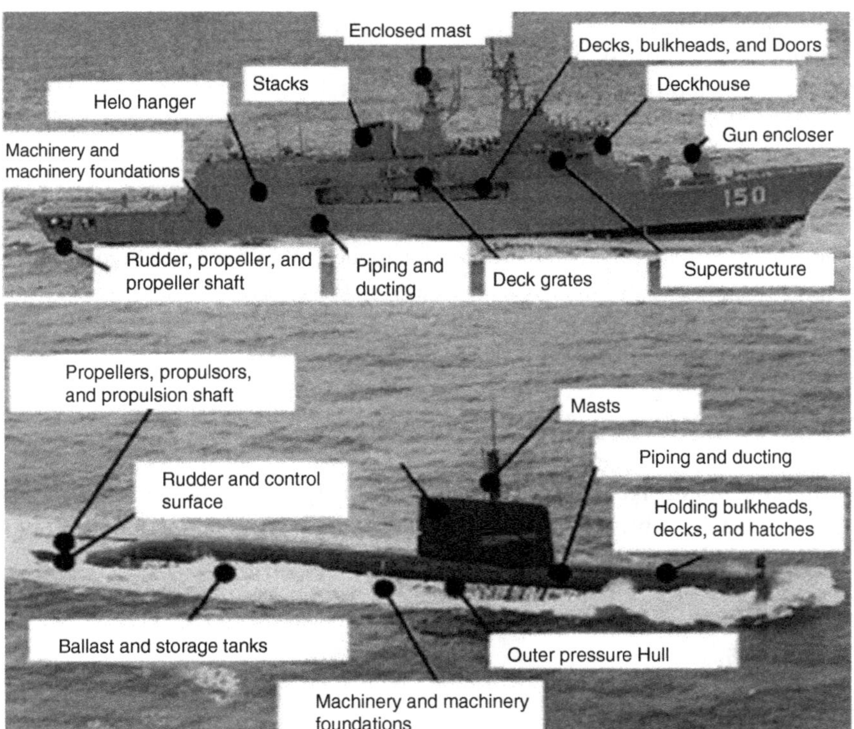

Figure 4.10 Military applications of fiber composites to naval ships and submarines. Source: Rubino et al. [127].

temperature curing and even compatible with the reinforcing materials. In offshore oil and gas field, thermosets are commonly used and even have more drawbacks and hindrances than thermoplastics. Whereas thermosets are comparatively low costs and use facile process, yet, its shelf life is considerably lower. Additionally, thermosets also deliver greater mechanical and thermal properties than thermoplastics.

Besides that, some hybrid composites have inferior thermal conductivity and stability; therefore, it can be used for other applications such as packaging, where outstanding heat dissipation and low thermal expansion are not necessary. Essabir et al. [120] had discovered the potential of natural fiber hybrid composites for packaging application. The oil palm fibers and clay particles have been modified and then introduced to strengthen high-density polyethylene (HDPE). Based on the result obtained, the complex modulus of the dynamic mechanical characterization was increased with filler addition. Moreover, as compared to oil palm fiber, the thermal stability of the hybrid composites was enhanced further with clay inclusion. Hybrid composites demonstrated high thermal, shear, and rheological attributes since those are the key characteristics needed for plastics applications.

4.5 Conclusions

The improvement of SPF-reinforced polymer composites using two-fiber hybrid together in a composite laminate has been studied comprehensively to be applied in automotive and structural applications. Since the SPF is hybridized with other fiber, the thermal properties specifically thermal stability could be altered and vary for each of the combination fibers in a composite laminate. The enhancement of thermal stability of the hybrid SPF composites is due to the balanced ratio of the low thermal stability SPFs and the glass fibers. Thus, it is attributed by the heat absorption by the SPF during the earlier increasing temperature. Hence, the application of SPF hybrid composites can be applied for either household products or heavy scale industrial applications due to its high mechanical and improved thermal stability properties.

Acknowledgment

The authors would like to express their grateful thanks to the Universiti Putra Malaysia (UPM), Universiti Kebangsaan Malaysia (UKM), and National Defence University of Malaysia (NDUM).

References

1 Asyraf, M.R.M., Ishak, M.R., Sapuan, S.M. et al. (2019). Creep test rig for full-scale composite crossarm: simulation modelling and analysis. In: *Seminar Enau Kebangsaan*, 34–38. Persatuan Pembangunan dan Industri Enau Malaysia (PPIEM).

2 Johari, A.N., Ishak, M.R., Leman, Z. et al. (2019). Fabrication and cut-in speed enhancement of savonius vertical axis wind turbine (SVAWT) with hinged blade using fiberglass composites. In: *Seminar Enau Kebangsaan*, 978–983. Persatuan Pembangunan dan Industri Enau Malaysia (PPIEM)

3 Afzaluddin, A., Jawaid, M., Salit, M.S., and Ishak, M.R. (2019). Physical and mechanical properties of sugar palm/glass fiber reinforced thermoplastic polyurethane hybrid composites. *Journal of Materials Research and Technology* 8 (1): 950–959. https://doi.org/10.1016/j.jmrt.2018.04.024.

4 Jaafar, C.N.A., Rizal, M.A.M., and Zainol, I. (2018). Effect of kenaf alkalization treatment on morphological and mechanical properties of epoxy / silica / kenaf composite. *International Journal of Engineering and Technology* 7: 258–263. https://doi.org/10.14419/ijet.v7i4.35.22743.

5 Jaafar, C.N.A., Zainol, I., and Rizal, M.A.M. (2018). Preparation and characterisation of epoxy/silica/kenaf composite using hand lay-up method. In: *27th Scientific Conference of the Microscopy Society Malaysia (27th SCMSM 2018)*, 2–6. Melaka, Malaysia: Universiti Putra Malaysia (UPM).

6 Mastura, M.T., Sapuan, S.M., Mansor, M.R., and Nuraini, A.A. (2017). Conceptual design of a natural fibre-reinforced composite automotive anti-roll bar using a hybrid approach. *International Journal of Advanced Manufacturing Technology* 91 (5–8): 2031–2048. https://doi.org/10.1007/s00170-016-9882-8.

7 Norizan, M.N., Abdan, K., and Ilyas, R.A. (2019). Effect of water absorption on treated sugar palm yarn fibre/glass fibre hybrid composites. In: *Prosiding Seminar Enau Kebangsaan 2019*, 78–81. Bahau, Negeri Sembilan, Malaysia: Institute of Tropical Forest and Forest Products (INTROP), Universiti Putra Malaysia.

8 Ishak, M.R., Leman, Z., Sapuan, S.M. et al. (2013). Impregnation modification of sugar palm fibres with phenol formaldehyde and unsaturated polyester. *Fibers and Polymers* 14 (2): 250–257. https://doi.org/10.1007/s12221-013-0250-0.

9 Abral, H., Ariksa, J., Mahardika, M. et al. (2019). Highly transparent and antimicrobial PVA based bionanocomposites reinforced by ginger nanofiber. *Polymer Testing*: 106186. https://doi.org/10.1016/j.polymertesting.2019.106186.

10 Abral, H., Ariksa, J., Mahardika, M. et al. (2020). Transparent and antimicrobial cellulose film from ginger nanofiber. *Food Hydrocolloids* 98: 105266. https://doi.org/10.1016/j.foodhyd.2019.105266.

11 Abral, H., Atmajaya, A., Mahardika, M. et al. (2020). Effect of ultrasonication duration of polyvinyl alcohol (PVA) gel on characterizations of PVA film. *Journal of Materials Research and Technology* 9 (2): 2477–2486. https://doi.org/10.1016/j.jmrt.2019.12.078.

12 Abral, H., Basri, A., Muhammad, F. et al. (2019). A simple method for improving the properties of the sago starch films prepared by using ultrasonication treatment. *Food Hydrocolloids* 93: 276–283. https://doi.org/10.1016/j.foodhyd.2019.02.012.

13 Ayu, R.S., Khalina, A., Harmaen, A.S. et al. (2020). Characterization study of empty fruit bunch (EFB) fibers reinforcement in poly(butylene) succinate (PBS)/starch/glycerol composite sheet. *Polymers* 12 (7): 1571. https://doi.org/10.3390/polym12071571.

14 Ilyas, R.A., Sapuan, S.M., Ishak, M.R., and Zainudin, E.S. (2018). Water transport properties of bio-nanocomposites reinforced by sugar palm (*Arenga pinnata*) nanofibrillated cellulose. *Journal of Advanced Research in Fluid Mechanics and Thermal Sciences* 51 (2): 234–246.

15 Ilyas, R.A., Sapuan, S.M., Atikah, M.S.N. et al. (2020). Effect of hydrolysis time on the morphological, physical, chemical, and thermal behavior of sugar palm nanocrystalline cellulose (*Arenga pinnata* (Wurmb.) Merr). *Textile Research Journal*, 004051752093239. https://doi.org/10.1177/0040517520932393.

16 Ilyas, R.A., Sapuan, S.M., Ibrahim, R. et al. (2020). Thermal, biodegradability and water barrier properties of bio-nanocomposites based on plasticised sugar palm starch and nanofibrillated celluloses from sugar palm fibres. *Journal of Biobased Materials and Bioenergy* 14 (2): 234–248. https://doi.org/10.1166/jbmb.2020.1951.

17 Ilyas, R.A., Sapuan, S.M., Ibrahim, R. et al. (2019). Sugar palm (*Arenga pinnata* (Wurmb.) Merr) cellulosic fibre hierarchy: a comprehensive approach from macro to nano scale. *Journal of Materials Research and Technology* 8 (3): 2753–2766. https://doi.org/10.1016/j.jmrt.2019.04.011.

18 Rozilah, A., Jaafar, C.N.A., Sapuan, S.M. et al. (2020). The effects of silver nanoparticles compositions on the mechanical, physiochemical, antibacterial, and morphology properties of sugar palm starch biocomposites for antibacterial coating. *Polymers* 12 (11): 2605. https://doi.org/10.3390/polym12112605.

19 Syafiq, R., Sapuan, S.M., Zuhri, M.Y.M. et al. (2020). Antimicrobial activities of starch-based biopolymers and biocomposites incorporated with plant essential oils: a review. *Polymers* 12 (10) https://doi.org/10.3390/polym12102403.

20 Mazani, N., Sapuan, S.M., Sanyang, M.L. et al. (2019). Design and fabrication of a shoe shelf from kenaf fiber reinforced unsaturated polyester composites. In: *Lignocellulose for Future Bioeconomy* (eds. H. Ariffin, S.M. Sapuan, and Mohd A. Hassan), 315–332. Elsevier. https://doi.org/10.1016/B978-0-12-816354-2.00017-7.

21 Ashraf, W., Ishak, M.R., Zuhri, M.Y.M. et al. (2019). Investigation of different facesheet materials on compression properties of honeycomb sandwich composite. In: *Seminar Enau Kebangsaan*, 129–132. Bahau, Negeri Sembilan, Malaysia: Persatuan Pembangunan dan Industri Enau Malaysia (PPIEM).

22 Sapuan, S.M., Aulia, H.S., Ilyas, R.A. et al. (2020). Mechanical properties of longitudinal basalt/woven-glass-fiber-reinforced unsaturated polyester-resin hybrid composites. *Polymers* 12 (10) https://doi.org/10.3390/polym12102211.

23 Jumaidin, R., Ilyas, R.A., Saiful, M. et al. (2019). Water transport and physical properties of sugarcane bagasse fibre reinforced thermoplastic potato starch biocomposite. *Journal of Advanced Research in Fluid Mechanics and Thermal Sciences* 61 (2): 273–281.

24 Jumaidin, R., Khiruddin, M.A.A., Asyul Sutan Saidi, Z. et al. (2020). Effect of cogon grass fibre on the thermal, mechanical and biodegradation properties of thermoplastic cassava starch biocomposite. *International Journal of Biological Macromolecules* 146: 746–755. https://doi.org/10.1016/j.ijbiomac.2019.11.011.

25 Jumaidin, R., Saidi, Z.A.S., Ilyas, R.A. et al. (2019). Characteristics of cogon grass fibre reinforced thermoplastic cassava starch biocomposite: water absorption and physical properties. *Journal of Advanced Research in Fluid Mechanics and Thermal Sciences* 62 (1): 43–52.

26 Shahroze, R.M., Chandrasekar, M., Senthilkumar, K. et al. (2019). A review on the various fibre treatment techniques used for the fibre surface modification of the sugar palm fibres. In: *Seminar Enau Kebangsaan*, 48–52. Bahau, Negeri Sembilan, Malaysia: Persatuan Pembangunan dan Industri Enau Malaysia (PPIEM).

27 Syafri, E., Sudirman, M., Yulianti, E. et al. (2019). Effect of sonication time on the thermal stability, moisture absorption, and biodegradation of water hyacinth (*Eichhornia crassipes*) nanocellulose-filled bengkuang (*Pachyrhizus erosus*) starch biocomposites. *Journal of Materials Research and Technology* 8 (6): 6223–6231. https://doi.org/10.1016/j.jmrt.2019.10.016.

28 Ilyas, R.A. and Sapuan, S.M. (2020). The preparation methods and processing of natural fibre bio-polymer composites. *Current Organic Synthesis* 16 (8): 1068–1070. https://doi.org/10.2174/157017941608200120105616.

29 Asyraf, M.R.M., Rafidah, M., Ishak, M.R. et al. (2020). Integration of TRIZ, Morphological Chart and ANP method for development of FRP composite portable fire extinguisher. *Polymer Composites*: 1–6. https://doi.org/10.1002/pc.25587.

30 Ilyas, R.A., Sapuan, S.M., Norrrahim, M.N.F. et al. (2021). Nanocellulose/starch biopolymer nanocomposites: processing, manufacturing, and applications. In: *Advanced Processing, Properties, and Application of Strach and Other Bio-based Polymer*, 1e (ed. F.M. Al-Oqla). Amsterdam, Netherland: Elsevier Inc. https://doi.org/10.1016/B978-0-12-189661-8.00006-8.

31 Nazrin, A., Sapuan, S.M., Zuhri, M.Y.M. et al. (2020). Nanocellulose reinforced thermoplastic starch (TPS), polylactic acid (PLA), and polybutylene succinate (PBS) for food packaging applications. *Frontiers in Chemistry* 8 (213): 1–12. https://doi.org/10.3389/fchem.2020.00213.

32 Norizan, M.N., Harussani, M.M., Demon, S.Z.N. et al. (2020). Carbon nanotubes: functionalisation and their application in chemical sensors. *RSC Advances* 10: 43704–43732. https://doi.org/10.1039/D0RA09438B.

33 Asyraf, M.R.M., Ishak, M.R., Razman, M.R., and Chandrasekar, M. (2019). Fundamentals of creep, testing methods and development of test rig for the full-scale crossarm: a review. *Jurnal Teknologi* 81 (4): 155–164.

34 Asyraf, M.R.M., Ishak, M.R., Sapuan, S.M., and Yidris, N. (2019). Conceptual design of creep testing rig for full-scale cross arm using TRIZ-Morphological chart-analytic network process technique. *Journal of Materials Research and Technology* 8 (6): 5647–5658. https://doi.org/10.1016/j.jmrt.2019.09.033.

35 Asyraf, M.R.M., Ishak, M.R., Sapuan, S.M., and Yidris, N. (2020). Conceptual design of multi-operation outdoor flexural creep test rig using hybrid concurrent engineering approach. *Journal of Materials Research and Technology* 9 (2): 2357–2368. https://doi.org/10.1016/j.jmrt.2019.12.067.

36 Asyraf, M.R.M., Ishak, M.R., Sapuan, S.M. et al. (2020). Woods and composites cantilever beam: a comprehensive review of experimental and numerical creep methodologies. *Journal of Materials Research and Technology* https://doi.org/10.1016/j.jmrt.2020.01.013.

37 Asyraf, M.R.M., Ishak, M.R., Sapuan, S.M. et al. (2020a). Evaluation of design and simulation of creep test rig for full-scale cross arm structure. *Advances in Civil Engineering* https://doi.org/10.1155/2019/6980918.

38 Asyraf, M.R.M., Ishak, M.R., Sapuan, S.M. et al. (2020c). Creep test rig for cantilever beam: Fundamentals, prospects and present views. *Journal of Mechanical Engineering Science.*

39 Chandrasekar, M., Shahroze, R.M., Ishak, M.R. et al. (2019). Flax and sugar palm reinforced epoxy composites: effect of hybridization on physical, mechanical, morphological and dynamic mechanical properties. *Materials Research Express* 6 (10): 105331. https://doi.org/10.1088/2053-1591/ab382c.

40 Shahroze, R.M., Ishak, M.R., Salit, M.S. et al. (2018). Effect of organo-modified nanoclay on the mechanical properties of sugar palm fiber-reinforced polyester composites. *Bioresources* 13 (4): 7430–7444. https://doi.org/10.15376/biores.13.4.7430-7444.

41 Michal, D. (2015). Application of natural fibres in hybrid composite materials. *Materials Science Forum* 818: 311–315. https://doi.org/10.4028/www.scientific.net/MSF.818.311.

42 Ilyas, R.A., Sapuan, S.M., Ishak, M.R. et al. (2018). Characterization of sugar palm nanocellulose and its potential for reinforcement with a starch-based composite. In: *Sugar Palm Biofibers, Biopolymers, and Biocomposites*, 1e, 189–220. Boca Raton, FL: CRC Press/Taylor & Francis Group https://doi.org/10.1201/9780429443923-10.

43 Ishak, M.R., Sapuan, S.M., Leman, Z. et al. (2013). Sugar palm (*Arenga pinnata*): its fibres, polymers and composites. *Carbohydrate Polymers* 91 (2): 699–710. https://doi.org/10.1016/j.carbpol.2012.07.073.

44 Sapuan, S.M., Ilyas, R.A., Ishak, M.R. et al. (2018). Development of sugar palm–based products: a community project. In: *Sugar Palm Biofibers, Biopolymers, and Biocomposites*, 1e, 245–266. Boca Raton, FL: CRC Press/Taylor & Francis Group https://doi.org/10.1201/9780429443923-12.

45 Ilyas, R.A., Sapuan, S.M., and Ishak, M.R. (2018). Isolation and characterization of nanocrystalline cellulose from sugar palm fibres (*Arenga pinnata*). *Carbohydrate Polymers* 181: 1038–1051. https://doi.org/10.1016/j.carbpol.2017.11.045.

46 Sapuan, S.M., Ishak, M.R., Leman, Z. et al. (2017). *Pokok Enau: Potensi dan pembangunan produk*, 1e (ed. S.M. Sapuan). Serdang, Selangor: Penerbit Universiti Putra Malaysia.

47 Atikah, M.S.N., Ilyas, R.A., Sapuan, S.M. et al. (2019). Degradation and physical properties of sugar palm starch / sugar palm nanofibrillated cellulose bionanocomposite. *Polimery* 64 (10): 27–36. https://doi.org/10.14314/polimery.2019.10.5.

48 Bachtiar, D., Sapuan, S.M.M., and Hamdan, M.M.M. (2008). The effect of alkaline treatment on tensile properties of sugar palm fibre reinforced epoxy composites. *Materials & Design* 29 (7): 1285–1290. https://doi.org/10.1016/j.matdes.2007.09.006.

49 Hazrol, M.D., Sapuan, S.M., Ilyas, R.A. et al. (2020). Electrical properties of sugar palm nanocrystalline cellulose reinforced sugar palm starch nanocomposites. *Polimery* 65 (05): 363–370. https://doi.org/10.14314/polimery.2020.5.4.

50 Ilyas, R.A., Sapuan, S.M., Atiqah, A. et al. (2019). Sugar palm (*Arenga pinnata* [Wurmb.] Merr) starch films containing sugar palm nanofibrillated cellulose as reinforcement: water barrier properties. *Polymer Composites*: 1–9. https://doi.org/10.1002/pc.25379.

51 Ilyas, R.A., Sapuan, S.M., Ibrahim, R. et al. (2019). Effect of sugar palm nanofibrillated cellulose concentrations on morphological, mechanical and physical properties of biodegradable films based on agro-waste sugar palm (*Arenga pinnata* (Wurmb.) Merr) starch. *Journal of Materials Research and Technology* 8 (5): 4819–4830. https://doi.org/10.1016/j.jmrt.2019.08.028.

52 Ilyas, R.A., Sapuan, S.M., Ishak, M.R., and Zainudin, E.S. (2017). Effect of delignification on the physical, thermal, chemical, and structural properties of sugar palm fibre. *Bioresources* 12 (4): 8734–8754. https://doi.org/10.15376/biores.12.4.8734-8754.

53 Ilyas, R.A., Sapuan, S.M., Ishak, M.R., and Zainudin, E.S. (2018a). Development and characterization of sugar palm nanocrystalline cellulose reinforced sugar palm starch bionanocomposites. *Carbohydrate Polymers* 202: 186–202. https://doi.org/10.1016/j.carbpol.2018.09.002.

54 Ilyas, R.A., Sapuan, S.M., Ishak, M.R., and Zainudin, E.S. (2018b). Sugar palm nanocrystalline cellulose reinforced sugar palm starch composite: degradation and water-barrier properties. *IOP Conference Series: Materials Science and Engineering* 368 https://doi.org/10.1088/1757-899X/368/1/012006.

55 Ilyas, R.A., Sapuan, S.M., Ishak, M.R., and Zainudin, E.S. (2019). Sugar palm nanofibrillated cellulose (Arenga pinnata (Wurmb.) Merr): effect of cycles on their yield, physic-chemical, morphological and thermal behavior. *International Journal of Biological Macromolecules* 123: 379–388. https://doi.org/10.1016/j.ijbiomac.2018.11.124.

56 Maisara, A.M.N., Ilyas, R.A., Sapuan, S.M. et al. (2019). Effect of fibre length and sea water treatment on mechanical properties of sugar palm fibre reinforced unsaturated polyester composites. *International Journal of Recent Technology and Engineering* 8 (2S4): 510–514. https://doi.org/10.35940/ijrte.b1100.0782s419.

57 Leman, Z., Sapuan, S.M., Azwan, M. et al. (2008). The effect of environmental treatments on fiber surface properties and tensile strength of sugar palm fiber-reinforced epoxy composites. *Polymer - Plastics Technology and Engineering* 47 (6): 606–612. https://doi.org/10.1080/03602550802059451.

58 Sahari, J., Sapuan, S.M., Ismarrubie, Z.N., and Rahman, M.Z.a. (2011). Investigation on bending strength and stiffness of sugar palm fibre from different parts reinforced unsaturated polyester composites. *Key Engineering Materials*

471–472 (October 2015): 502–506. https://doi.org/10.4028/www.scientific.net/KEM.471-472.502.

59 Misri, S., Leman, Z., Sapuan, S.M., and Ishak, M.R. (2010). Mechanical properties and fabrication of small boat using woven glass/sugar palm fibres reinforced unsaturated polyester hybrid composite. *IOP Conference Series: Materials Science and Engineering* 11: 012015. https://doi.org/10.1088/1757-899x/11/1/012015.

60 Nurazzi, N.M., Khalina, A., Sapuan, S.M. et al. (2019). Thermal properties of treated sugar palm yarn/glass fiber reinforced unsaturated polyester hybrid composites. *Journal of Materials Research and Technology* https://doi.org/10.1016/j.jmrt.2019.11.086.

61 Nurazzi, N.M., Khalina, A., Sapuan, S.M. et al. (2020). Thermal properties of treated sugar palm yarn/glass fiber reinforced unsaturated polyester hybrid composites. *Journal of Materials Research and Technology* 9 (2): 1606–1618. https://doi.org/10.1016/j.jmrt.2019.11.086.

62 Ticoalu, A., Aravinthan, T., and Cardona, F. (2010). Experimental investigation into gomuti fibres/polyester composites. In: *21st Australasian Conference on the Mechanics of Structures and Materials (ACMSM 21)*, 451–456. Melbourne, Australia: CRC Press/Balkema.

63 Oumer, A.N. and Bachtiar, D. (2014). Modeling and experimental validation of tensile properties of sugar palm fiber reinforced high impact polystyrene composites. *Fibers and Polymers* 15 (2): 334–339. https://doi.org/10.1007/s12221-014-0334-5.

64 Irfa'i, M.A., Dzulkiflih, D., Wulandari, D. et al. (2019). Study of the orientation of Palm Sugar E-Glass hybrid fiber - Polyester composite for Running Prostheses Application. In: *2019 International Biomedical Instrumentation and Technology Conference (IBITeC)*, 108–111. US and Canada IEEE https://doi.org/10.1109/IBITeC46597.2019.9091720.

65 Bachtiar, D., Salit, M.S., Zainudin, E. et al. (2011). Effects of alkaline treatment and a compatibilizing agent on tensile properties of sugar palm fibrereinforced high impact polystyrene composites. *Bioresources* 6 (4): 4815–4823.

66 Mohammed, A.A., Bachtiar, D., Rejab, M.R.M., and Siregar, J.P. (2018). Effect of microwave treatment on tensile properties of sugar palm fibre reinforced thermoplastic polyurethane composites. *Defence Technology* 14 (4): 287–290. https://doi.org/10.1016/j.dt.2018.05.008.

67 Agrebi, F., Ghorbel, N., Rashid, B. et al. (2018). Influence of treatments on the dielectric properties of sugar palm fiber reinforced phenolic composites. *Journal of Molecular Liquids* 263: 342–348. https://doi.org/10.1016/j.molliq.2018.04.130.

68 Sahari, J., Sapuan, S.M., Zainudin, E.S., and Maleque, M.A. (2013). Thermo-mechanical behaviors of thermoplastic starch derived from sugar palm tree (*Arenga pinnata*). *Carbohydrate Polymers* 92 (2): 1711–1716. https://doi.org/10.1016/j.carbpol.2012.11.031.

69 Razali, N., Sapuan, S.M., and Razali, N. (2018). Mechanical properties and morphological analysis of roselle/sugar palm fiber reinforced vinyl ester hybrid composites. In: *Natural Fibre Reinforced Vinyl Ester and Vinyl Polymer*

Composites (eds. S.M. Sapuan, H. Ismail, and E.S. Zainudin), 169–180. Elsevier https://doi.org/10.1016/B978-0-08-102160-6.00008-1.

70 Huzaifah, M.R.M., Sapuan, S.M., Leman, Z. et al. (2018). Effect of soil burial on water absorption of sugar palm fibre reinforced vinyl ester composites. In: *6th Postgraduate Seminar on Natural Fiber Reinforced Polymer Composites 2018* (eds. S.M. Sapuan, Z. Leman, and E.S. Zainudin), 52–54. Serdang, Selangor: Institute of Tropical Forestry and Forest Products.

71 Ammar, I.M., Huzaifah, M.R.M., Sapuan, S.M. et al. (2018). Development of sugar palm fiber reinforced vinyl ester composites. In: *Natural Fibre Reinforced Vinyl Ester and Vinyl Polymer Composites* (eds. S.M. Sapuan, H. Ismail, and E.S. Zainudin), 211–224. Elsevier https://doi.org/10.1016/B978-0-08-102160-6.00011-1.

72 Atiqah, A., Mastura, M.T., Jawaid, M. et al. (2020). Interfaces in sugar palm fibres reinforced composites. *Interfaces in Particle and Fibre Reinforced Composites* https://doi.org/10.1016/b978-0-08-102665-6.00008-x.

73 Radzi, A.M., Sapuan, S.M., Jawaid, M., and Mansor, M.R. (2017). Influence of fibre contents on mechanical and thermal properties of roselle fibre reinforced polyurethane composites. *Fibers and Polymers* 18 (7): 1353–1358. https://doi.org/10.1007/s12221-017-7311-8.

74 Halimatul, M.J., Sapuan, S.M., Jawaid, M. et al. (2019a). Effect of sago starch and plasticizer content on the properties of thermoplastic films: Mechanical testing and cyclic soaking-drying. *Polimery* 64 (6): 32–41. https://doi.org/10.14314/polimery.2019.6.5.

75 Halimatul, M.J., Sapuan, S.M., Jawaid, M. et al. (2019b). Water absorption and water solubility properties of sago starch biopolymer composite films filled with sugar palm particles. *Polimery* 64 (9): 27–35. https://doi.org/10.14314/polimery.2019.9.4.

76 Ibrahim, M.I.J., Sapuan, S.M., Zainudin, E.S., and Zuhri, M.Y.M. (2020). Preparation and characterization of cornhusk/sugar palm fiber reinforced cornstarch-based hybrid composites. *Journal of Materials Research and Technology* 9 (1): 200–211. https://doi.org/10.1016/j.jmrt.2019.10.045.

77 Edhirej, A., Sapuan, S.M., Jawaid, M., and Zahari, N.I. (2017a). Cassava/sugar palm fiber reinforced cassava starch hybrid composites: physical, thermal and structural properties. *International Journal of Biological Macromolecules* 101: 75–83. https://doi.org/10.1016/j.ijbiomac.2017.03.045.

78 Sanyang, M.L., Sapuan, S.M., Jawaid, M. et al. (2016). Effect of sugar palm-derived cellulose reinforcement on the mechanical and water barrier properties of sugar palm starch biocomposite films. *Bioresources* 11 (2): 4134–4145. https://doi.org/10.15376/biores.11.2.4134-4145.

79 Hazrol, M.D., Sapuan, S.M., Zuhri, M.Y.M., and Ilyas, R.A. (2019). Electrical properties of sugar palm nanocellulose fibre reinforced sugar palm starch biopolymer composite. In: *Prosiding Seminar Enau Kebangsaan 2019*, 57–62. Bahau, Negeri Sembilan, Malaysia: Institute of Tropical Forest and Forest Products (INTROP)/Universiti Putra Malaysia.

80 Jumaidin, R., Sapuan, S.M., Jawaid, M. et al. (2017). Thermal, mechanical, and physical properties of seaweed/sugar palm fibre reinforced thermoplastic sugar palm starch/agar hybrid composites. *International Journal of Biological Macromolecules* 97 (1): 606–615. https://doi.org/10.1016/j.ijbiomac.2017.01.079.

81 Azammi, A.M.N., Ilyas, R.A., Sapuan, S.M. et al. (2019). Characterization studies of biopolymeric matrix and cellulose fibres based composites related to functionalized fibre-matrix interface. In: *Interfaces in Particle and Fibre Reinforced Composites- From Macro to Nano Scales*, 1e, 1–68. London: Woodhead Publishing https://doi.org/10.1016/B978-0-08-102665-6.

82 Radzi, A.M., Sapuan, S.M., Jawaid, M., and Mansor, M.R. (2019a). Effect of alkaline treatment on mechanical, physical and thermal properties of roselle/sugar palm fiber reinforced thermoplastic polyurethane hybrid composites. *Fibers and Polymers* 20 (4): 847–855. https://doi.org/10.1007/s12221-019-1061-8.

83 Kumar, T.S.M., Chandrasekar, M., Senthilkumar, K. et al. (2020). Characterization, thermal and antimicrobial properties of hybrid cellulose nanocomposite films with in-situ generated copper nanoparticles in *Tamarindus indica* nut powder. *Journal of Polymers and the Environment*: 1–10. https://doi.org/10.1007/s10924-020-01939-w.

84 Mahardika, M., Abral, H., Kasim, A. et al. (2018). Production of nanocellulose from pineapple leaf fibers via high-shear homogenization and ultrasonication. *Fibers* 6 (2): 28. https://doi.org/10.3390/fib6020028.

85 Gomes, J., Batra, J., Chopda, V.R. et al. (2018). Monitoring and control of bioethanol production from lignocellulosic biomass. In: *Waste Biorefinery: Potential and Perspectives*, 727–749. Elsevier. https://doi.org/10.1016/B978-0-444-63992-9.00025-2.

86 Aisyah, H.A., Paridah, M.T., Sapuan, S.M. et al. (2019). Thermal properties of woven kenaf/carbon fibre-reinforced epoxy hybrid composite panels. *International Journal of Polymer Science* 2019: 1–8. https://doi.org/10.1155/2019/5258621.

87 Dhakal, H.N. and Zhang, Z. (2015). The use of hemp fibres as reinforcements in composites. In: *Biofiber Reinforcements in Composite Materials* (eds. O. Faruk and M. Sain), 86–103. Woodhead Publishing. https://doi.org/10.1533/9781782421276.1.86.

88 Guedes, R.M., Sá, A., and Faria, H. (2007). Influence of moisture absorption on creep of GRP composite pipes. *Polymer Testing* 26 (5): 595–605. https://doi.org/10.1016/j.polymertesting.2007.02.007.

89 Mohanty, S. and Nayak, S.K. (2010). Short bamboo fiber-reinforced HDPE composites: influence of fiber content and modification on strength of the composite. *Journal of Reinforced Plastics and Composites* 29 (14): 2199–2210. https://doi.org/10.1177/0731684409345618.

90 Ferry, J.D. (1991). Some reflections on the early development of polymer dynamics: viscoelasticity, dielectric dispersion, and self-diffusion. *Macromolecules* 24 (19): 5237–5245. https://doi.org/10.1021/ma00019a001.

91 Ferry, J.D. and Rice, S.A. (1962). Viscoelastic properties of polymers. *Physics Today* 15 (1): 76–78. https://doi.org/10.1063/1.3057989.

92 Faydi, Y., Lachat, R., Lesage, P., and Meyer, Y. (2015). Experimental characterization method of the gas diffusion layers compression modulus for high compressive loads and based on a dynamic mechanical analysis. *Journal of Fuel Cell Science and Technology* 12 (5) https://doi.org/10.1115/1.4031695.

93 Ishak, M.R., Sapuan, S.M., Leman, Z. et al. (2012). Characterization of sugar palm (*Arenga pinnata*) fibres. *Journal of Thermal Analysis and Calorimetry* 109 (2): 981–989. https://doi.org/10.1007/s10973-011-1785-1.

94 Zhao, C., Jiang, E., and Chen, A. (2017). Volatile production from pyrolysis of cellulose, hemicellulose and lignin. *Journal of the Energy Institute* 90 (6): 902–913. https://doi.org/10.1016/j.joei.2016.08.004.

95 Edhirej, A., Sapuan, S.M., Jawaid, M., and Zahari, N.I. (2017). Tensile, barrier, dynamic mechanical, and biodegradation properties of cassava/sugar palm fiber reinforced cassava starch hybrid composites. *Bioresources* 12 (4): 7145–7160. https://doi.org/10.15376/biores.12.4.7145-7160.

96 Razali, N., Salit, M.S., Jawaid, M. et al. (2015). A study on chemical composition, physical, tensile, morphological, and thermal properties of roselle fibre: effect of fibre maturity. *Bioresources* 10 (1) https://doi.org/10.15376/biores.10.1.1803-1824.

97 Kim, D.-Y., Nishiyama, Y., Wada, M. et al. (2001). Thermal decomposition of cellulose crystallites in wood. *Holzforschung* 55 (5): 521–524. https://doi.org/10.1515/HF.2001.084.

98 Hao, L.C., Sapuan, S.M., Hassan, M.R., and Sheltami, R.M. (2018). Natural fiber reinforced vinyl polymer composites. In: *Natural Fibre Reinforced Vinyl Ester and Vinyl Polymer Composites* (eds. S.M. Sapuan, H. Ismail, and E.S. Zainudin), 27–70. Elsevier Ltd. https://doi.org/10.1016/b978-0-08-102160-6.00002-0.

99 Norizan, M.N., Abdan, K., Salit, M.S., and Mohamed, R. (2017). Physical, mechanical and thermal properties of sugar palm yarn fibre loading on reinforced unsaturated polyester composite. *Journal of Physical Science* 28 (3): 115–136. https://doi.org/10.21315/jps2017.28.3.8.

100 Atiqah, A., Jawaid, M., Sapuan, S.M. et al. (2018). Thermal properties of sugar palm/glass fiber reinforced thermoplastic polyurethane hybrid composites. *Composite Structures* https://doi.org/10.1016/j.compstruct.2018.05.009.

101 Atiqah, A., Jawaid, M., Sapuan, S.M. et al. (2019). Physical and thermal properties of treated sugar palm/glass fibre reinforced thermoplastic polyurethane hybrid composites. *Journal of Materials Research and Technology* 8 (5): 3726–3732. https://doi.org/10.1016/j.jmrt.2019.06.032.

102 Mohd Nurazzi, N., Khalina, A., Sapuan, S.M., and Rahmah, M. (2018). Development of sugar palm yarn/glass fibre reinforced unsaturated polyester hybrid composites. *Materials Research Express* 5 (4): 045308. https://doi.org/10.1088/2053-1591/aabc27.

103 Chü, N.J. (1970). The conformation of the anhydrocellobiose units in cellulose I and II. *Journal of Applied Polymer Science* 14 (12): 3129–3136. https://doi.org/10.1002/app.1970.070141215.

104 Prachayawarakorn, J., Chaiwatyothin, S., Mueangta, S., and Hanchana, A. (2013). Effect of jute and kapok fibers on properties of thermoplastic cassava starch composites. *Materials and Design* 47: 309–315. https://doi.org/10.1016/j.matdes.2012.12.012.

105 Nadlene, R., Sapuan, S.M., Jawaid, M. et al. (2018). The effects of chemical treatment on the structural and thermal, physical, and mechanical and morphological properties of roselle fiber-reinforced vinyl ester composites. *Polymer Composites* 39 (1): 274–287. https://doi.org/10.1002/pc.23927.

106 Nascimento, T.A., Calado, V., and Carvalho, C.W.P. (2012). Development and characterization of flexible film based on starch and passion fruit mesocarp flour with nanoparticles. *Food Research International* 49 (1): 588–595. https://doi.org/10.1016/j.foodres.2012.07.051.

107 Sanchez-Silva, L., López-González, D., Garcia-Minguillan, A.M., and Valverde, J.L. (2013). Pyrolysis, combustion and gasification characteristics of *Nannochloropsis gaditana* microalgae. *Bioresource Technology* 130: 321–331. https://doi.org/10.1016/j.biortech.2012.12.002.

108 Ross, A.B., Jones, J.M., Kubacki, M.L., and Bridgeman, T. (2008). Classification of macroalgae as fuel and its thermochemical behaviour. *Bioresource Technology* 99 (14): 6494–6504. https://doi.org/10.1016/j.biortech.2007.11.036.

109 Radzi, A.M., Sapuan, S.M., Jawaid, M., and Mansor, M.R. (2019b). Water absorption, thickness swelling and thermal properties of roselle/sugar palm fibre reinforced thermoplastic polyurethane hybrid composites. *Journal of Materials Research and Technology* https://doi.org/10.1016/j.jmrt.2019.07.007.

110 Atiqah, A., Jawaid, M., Sapuan, S.M., and Ishak, M.R. (2019). Dynamic mechanical properties of sugar palm/glass fiber reinforced thermoplastic polyurethane hybrid composites. *Polymer Composites* 40 (4): 1329–1334. https://doi.org/10.1002/pc.24860.

111 Safri, S.N.A., Sultan, M.T.H., Jawaid, M., and Abdul Majid, M.S. (2019). Analysis of dynamic mechanical, low-velocity impact and compression after impact behaviour of benzoyl treated sugar palm/glass/epoxy composites. *Composite Structures* 226 (July): 111308. https://doi.org/10.1016/j.compstruct.2019.111308.

112 Mohammed, B.R., Leman, Z., Jawaid, M. et al. (2017). Dynamic mechanical analysis of treated and untreated sugar palm fibre-based phenolic composites. *Bioresources* 12 (2): 3448–3462.

113 Asyraf, M.R.M., Ishak, M.R., Sapuan, S.M. et al. (2020). Potential application of green composites for cross arm component in transmission tower: a brief review. *International Journal of Polymer Science* 2020: 1–15. https://doi.org/10.1155/2020/8878300.

114 Mochane, M.J., Mokhena, T.C., Mokhothu, T.H. et al. (2019). Recent progress on natural fiber hybrid composites for advanced applications: a review. *eXPRESS Polymer Letters* 13 (2): 159–198.

115 Krishnasamy, S., Muthu, S., Thiagamani, K. et al. (2019). Recent advances in thermal properties of hybrid cellulosic fiber reinforced polymer composites. *International Journal of Biological Macromolecules* 141: 1–13. https://doi.org/10.1016/j.ijbiomac.2019.08.231.

116 Alexander, J. and Elphej Churchill, S.J. (2017). Mechanical characterization of baslat based natural hybrid composites for aerospace applications. *IOP Conference Series: Materials Science and Engineering* 197 (1) https://doi.org/10.1088/1757-899X/197/1/012008.

117 Shekar, R.I., Rao, P.M.D., Siddaramaiah et al. (2010). Fibre-fibre hybrid composites for aerospace applications. *Advanced Materials Research* 123–125: 1231–1234. https://doi.org/10.4028/www.scientific.net/AMR.123-125.1231.

118 Shukla, R. and Srivastava, A.K. (2019). FTIR analysis of bamboo and flax (BFBF) mat reinforced epoxy hybrid composite. *International Journal of Scientific Research in Science, Engineering and Technology* 6 (July): 119–124.

119 Eires, R., Nunes, J. P., Fangueiro, R. et al. (2006). New eco-friendly hybrid composite materials for civil construction. *European Conference on Composite Materials*, 1–9. Portugal: University of Minho.

120 Essabir, H., Boujmal, R., Bensalah, M.O. et al. (2016). Mechanical and thermal properties of hybrid composites: oil-palm fiber/clay reinforced high density polyethylene. *Mechanics of Materials* https://doi.org/10.1016/j.mechmat.2016.04.008.

121 Ezugwu, E.O. (2005). Key improvements in the machining of difficult-to-cut aerospace superalloys. *International Journal of Machine Tools and Manufacture* 45 (12–13): 1353–1367. https://doi.org/10.1016/j.ijmachtools.2005.02.003.

122 Gururaja, M.N. and Rao, H. (2012). A review on recent applications and future prospectus of hybrid composites. *International Journal of Soft Computing and Engineering* 16: 2231–2307.

123 Bledzki, A.K., Franciszczak, P., Osman, Z., and Elbadawi, M. (2015). Polypropylene biocomposites reinforced with softwood, abaca, jute, and kenaf fibers. *Industrial Crops and Products* 70: 91–99. https://doi.org/10.1016/j.indcrop.2015.03.013.

124 Pandey, J.K., Ahn, S.H., Lee, C.S. et al. (2010). Recent advances in the application of natural fiber based composites. *Macromolecular Materials and Engineering* 295 (11): 975–989.

125 Prabhakaran, S., Chinnarasu, K., and Kumar, M.S. (2012). Design and fabrication of composite bumper for light passenger vehicles. *International Journal of Modern Engineering Research* 2 (4): 2552–2556.

126 Kim, D.H., Kim, H.G., and Kim, H.S. (2015). Design optimization and manufacture of hybrid glass/carbon fiber reinforced composite bumper beam for automobile vehicle. *Composite Structures* 131: 742–752. https://doi.org/10.1016/j.compstruct.2015.06.028.

127 Rubino, F., Nisticò, A., Tucci, F., and Carlone, P. (2020). Marine application of fiber reinforced composites: a review. *Journal of Marine Science and Engineering* 8 (1): 26. https://doi.org/10.3390/jmse8010026.

5

Thermal Properties of Sisal Fiber-Based Hybrid Composites

Tamil Moli Loganathan[1], Jesuarockiam Naveen[2], Koduri Naga Ganapathi Lakshmi Reshwanth[2], Kandasamy Jayakrishna[2], and Chandrasekar Muthukumar[3]

[1] *Universiti Putra Malaysia, Department of Aerospace Engineering, Faculty of Engineering, Serdang, 43400, Selangor, Malaysia*
[2] *Vellore Institute of Technology, School of Mechanical Engineering, Vellore, 632014, India*
[3] *Hindustan Institute of Technology & Science, School of Aeronautical Sciences, Padur, Kelambakkam, Chennai, 603103, India*

5.1 Introduction

Thermal stability of the polymeric composites refers to their ability in retaining mechanical strength and stiffness over a range of temperatures. Thermogravimetric analysis (TGA) is an effective technique to study the thermal decomposition characteristics and thermal stability of the polymeric materials, including thermoplastics, thermosets, elastomers, and fiber-reinforced polymeric composites. A thermogravimetric analyzer quantifies mass loss of the sample at various temperature ranges under a confined environment. Mass, temperature, and time are viewed as the base estimations in TGA. Also, temperature beyond which the material will undergo thermal decomposition can be identified. In general, most of the polymers soften or decompose before 200 °C. However, thermally stable polymers that can withstand temperatures of 300 °C in air and 500 °C in the presence of inert gases without changes in the morphology or strength reduction can also be investigated by TGA [1].

Thermal degradation of natural fiber-reinforced composites occurs in three different stages such as onset, major, and endset degradations. Initial degradation is mainly due to the evaporation of water molecules, and major degradation is due to the decomposition of natural constituents such as cellulose, hemicelluloses, lignin, etc. Final degradation occurs at higher temperatures, attributed to decomposition of the polymeric chain segments. Higher char residue at the end of the final degradation can be interpreted for lower degradation rate and superior thermal stability [2].

Natural Fiber-Reinforced Composites: Thermal Properties and Applications, First Edition.
Edited by Senthilkumar Krishnasamy, Senthil Muthu Kumar Thiagamani, Chandrasekar Muthukumar, Rajini Nagarajan and Suchart Siengchin.

(a) (b)

Figure 5.1 (a) Sisal plant and (b) sisal fiber.

Polymeric composite possesses viscoelastic properties. At elevated temperatures, deterioration in the mechanical strength and stiffness can be noticed. Viscoelastic characteristics of the polymeric composites can be assessed from the parameters such as storage modulus (E'), loss modulus (E''), and damping factor ($\tan \delta$) obtained from the dynamic mechanical analysis (DMA). Storage modulus is associated with the "stiffness," while loss modulus represents the heat energy dissipated during the heating. Tan δ is the ratio between (E'') loss modulus and storage modulus (E'). Lower value indicates better stiffness and elasticity that is a favorable characteristic for the polymeric composites, while higher $\tan \delta$ value indicates the nonelastic tendency of composites [3]

Lignocellulosic fibers such as jute, sisal, pineapple, and coir are potentially used in non-load-bearing applications in the automotive sector and other industries. Fishing net and ropes in the marine industry have long been made from the sisal fiber (SF). The name "sisal" means "cold waters" and is originated from a harbor town in Mexico. Initially, sisal plants were cultivated by "Maya Indians" before the arrival of Europeans. It is also found across countries such as India, Indonesia, Brazil, Haiti, and East Africa. Also, these countries cultivate a huge amount of sisal plants. Generally, SF is extracted through decortication process from its leaves (*Agave sisalana*) (Figure 5.1a).

SF is a coarse, strong, and hard fiber that takes only one half of the cost required by other textile fibers. It can be cultivated in hot and dry climate that is not suitable for other plants, and it can grow in almost all the type of soil except clay. Generally, after two years of plantation, the fibers can be extracted, and its product life span is around 12 years. A typical sisal plant is capable of producing 120–240 leaves. Each leaf contains moisture, fiber, cuticle, and other dry matters that constitute to 87.25%, 4%, 0.75%, and 8%, respectively. Fibers are aligned in the longitudinal direction of the leaves and are usually present in the surface of the leaves. Each leaf can weigh approximately 600 g and contains approximately 1000 fibers. This chapter focuses on the thermal properties of sisal fiber and its hybrid composites.

5.2 Thermal Properties of Sisal Fiber-Reinforced Polymeric Composites

Thermal properties such as the transition of temperature from end to end, load capability at specified temperature, material behavior, and dimensional stability at high temperatures are vital for the structural applications [4]. The natural fiber's predominant constituents such as cellulose, hemicellulose, and lignin and their thermal degradation are shown in Table 5.1. Decomposing at higher temperatures contributes to a significant deterioration in the properties of the composites. Therefore, the thermal behavior of sisal fibers and their composites is to be studied to determine their performance in the final product.

Gudayu et al. investigated thermal degradation characteristics of the raw and chemically treated (alkaline treated, acetylated, and combination of alkaline–acetylation treated) sisal fiber to evaluate their thermal stability [5]. From their TGA results, decomposition of up to 150 °C was found to be for moisture evaporation in the fibers involving intramolecular and intermolecular dehydration reactions [6].

Table 5.2 shows the decomposition pattern in percentage of raw and chemically treated sisal fiber. Based on the decomposition analysis, alkaline-treated

Table 5.1 Chemical composition (%) and thermal degradation of sisal fiber.

Chemical composition	Percentage (%) [5]	Thermal degradation (°C) [4]
Cellulose	67–78	300–360
Hemicellulose	10–14	200–300
Lignin	8–11	300–450
Pectin	10	270–330
Ash	0.6–1	—
Moisture content	10–22	50–100
Wax	2	—

Table 5.2 Decomposition pattern (%) of raw and chemically treated sisal fiber.

	% decomposed in the range of temperature		
Fiber	150–250 °C	150–300 °C	Up to 300 °C
Raw sisal fiber	1.92	13.39	22.0
Alkaline-treated sisal fiber	1.66	9.49	14.7
Acetylation-treated sisal fiber	9.36	26.25	30.7
Alkaline–acetylation-treated sisal fiber	3.00	11.5	16.8

fiber recorded minimum weight loss, while acetylation-treated fiber recorded highest weight loss. The thermal stability of the alkaline-treated fiber and alkaline–acetylation-treated sisal fiber improved by 49.6% and 30.9%, while acetylation-treated sisal fiber decreased by 39.5% in relative to raw sisal fiber. The alkaline-treated sisal fiber displayed better thermal stability as it promotes ionization of the hydroxyl groups by alkoxide [7]. The chemical treatment was found to cause changes in the chemical constituents. In addition, alkali treatment contributed to an increase of crystalline cellulose from amorphous cellulose content [8]. The acetylation-treated sisal fiber exhibited low thermal stability due to the cellulose degradation caused by fiber damage in acetylated solution containing sulfuric acid as a catalyst. Martin et al. recommended that the sisal fiber should be treated below 185 °C, on the basis that deterioration of holocellulose, hemicellulose, and cellulose begins at 186 °C [6].

Gañan et al. examined thermal decomposition properties of raw and treated sisal fiber (mercerization and silane)-reinforced epoxy composites. From TGA, three weight loss region can be determined: (i) below 100 °C is the weight loss assigned to moisture evaporation, (ii) between 210 and 300 °C is attributed to the decomposition of hemicellulose ad cellulose, and (iii) between 400 and 450 °C is associated with the decomposition of lignin [9]. A reduction of 25–52% in the moisture absorption of treated fiber relative to untreated fiber was observed. This is because alkaline and silane treatment reduced the hydrophilic tendency of the fiber. The decline in the moisture absorption achieved with silane treatment is attributed to the presence of silanol groups that act as barrier to moisture during the fiber treatment [9]. Thus, the thermal stability increased due to the new group of fiber's surface that restricts the mobility of segmental fiber [10]. Silane-treated sisal fibers reported a higher percentage of char that is associated with the degradation of grafted silane [11].

Effects of curing temperature on thermal properties of the sisal/glyoxal–phenolic composite were examined. According to their results, storage modulus increased while the lower loss modulus and tan ∂ decreased, indicating better fiber-matrix adhesion for the composite cured below 150 °C ([12]). Similarly, sisal/phenolic composite with fibers treated with the methylolated indulin and organosolv lignins displayed superior storage modulus [13]. The fiber treatment led to strong fiber-matrix interfacial bonding characteristic due to the chemical bonding between the hydroxymethyl groups in lignin and polymer phenolic rings in the resin.

5.3 Thermal Properties of Hybrid Sisal Fiber/Synthetic Fiber-Reinforced Polymeric Composites

Hybrid composites provide superior and combined properties compared with conventional monolithic composites. Different hybrid combinations are as follows:

(i) Two or more reinforcements in a single polymer matrix.
(ii) Single reinforcement in two or more matrices blended together.
(iii) Two or more reinforcement in one or more matrices.

The limitations of one of the constituents could be matched by the other constituent in a hybrid composite [14]. For example, hybrid fiber-reinforced polymer composites contain high modulus and low modulus fibers. The fiber with high modulus act as a load-carrying member, while the low modulus fiber improves the damage tolerance of the composites and reduces the materials cost. Synthetic fibers are man-made fibers derived from chemical resources or petroleum-based resources such as glass, carbon, and aramid fibers. Natural fibers extracted from the plants are composed of cellulose, hemicellulose, lignin, pectin, and wax. The percentage of individual constituent varies from one natural fiber to the other natural fiber.

Incorporating synthetic fibers into the natural fibers leads to the composite with superior strength and stiffness such that it could be suitable for high-performance applications. Replacing synthetic fibers with an ecofriendly lightweight material without affecting the performance has become an interesting research approach. Improvement in the thermal and dynamic properties of natural fiber-reinforced composites by adding various synthetic fibers has been demonstrated by researchers.

Jarukumjorn et al. studied the effect of hybridizing glass fiber on the thermal properties of sisal fiber/polypropylene composites. Alkali-treated sisal fibers were utilized in this research to remove the amorphous hemicelluloses. The length of both sisal and glass fiber was 4 mm. Brabender was used to mix the reinforcement, polymer, and compatibilizer (maleic anhydride). Three different hybrid sisal/glass fiber-reinforced polypropylene composites were fabricated through injection molding process by varying glass fiber content (10, 15, and 20 wt%). Thermal stability of the hybrid composites was analyzed through TGA. Decomposition temperatures were measured at 5% and 50% weight loss. This is mainly due to the different range in decomposition temperatures of sisal fiber and polypropylene. Increase in glass fiber content enhanced the decomposition temperatures and heat distortion temperature. At 5% weight loss, decomposition temperatures are 331, 337, and 364 °C by varying glass fiber by 10, 15, and 20 wt%. Similarly at 50% weight loss, decomposition temperatures are 460, 461, and 464 °C at 10, 15, and 20 wt%, respectively [1]. In addition to hybridization with the glass fibers, Nayak and Mohanty showed that the addition of maleic anhydride-grafted polypropylene (MAPP) can further enhance the crystallization and decomposition temperatures measured from the differential scanning calorimetry (DSC) test. Moreover, hybrid composite with MAPP exhibited higher storage modulus and loss modulus [15].

5.4 Thermal Properties of Sisal/Other Natural Fiber-Reinforced Polymeric Composites

Sumesh et al. [16] investigated the influence of adding biosynthesized alumina nanopowder (BSANP) up to 3% on the thermal properties of sisal/coir (SC) and sisal/banana (SB) and banana/coir (BC) composites. The addition of BSANP lowered the degradation rate and improved the degradation temperature of BC, SC, and SB composites as shown in Table 5.3.

Table 5.3 Effect of biosynthesized alumina nanopowder on the thermal stability of sisal/coir-reinforced composites [16].

Fibers	Parameters		Before adding BSANP	After adding BSANP
Sisal/coir-based hybrid composites	Degradation temperature (BC)	T-50	385.75 °C	397 °C
		T-70	468.54 °C	530.67 °C
		T-75	520.32 °C	580.14 °C
	Residual (%) (BC)		18.72	23.41
	Degradation temperature (SC)	T-50	377.98 °C	390.95 °C
		T-70	431.92 °C	477.01 °C
	Residual (%) (SC)		21.9	25.75

Source: Modified from Sumesh et al. [16].

Martin et al. utilized DSC to check the degradation of sisal/jute-based hybrid composites. As the lignin from fiber was exposed due to the temperature, thermal degradation of the holocellulose and cellulose occurred at lower temperatures in the controlled chamber. Under inert conditions, each of the peaks in the DSC curve represented the individual constituent of the sisal fiber [6].

Venkatram et al. examined the thermal degradation characteristics of the sisal/nanoclay-based composites. Initial weight loss occurred at 180 °C due to the evaporation of solvents from the polymer, followed by major weight loss at 300 °C and final residue at 510 °C for composites with nanoclay compared with 290 °C and final residue at 500 °C for composites without nanoclay. Similarly, the major weight loss also decreased from 82.5% to 79.25% due to the reinforcement of nanoclay in the sisal fiber-reinforced composites [17].

Sisal/propylene composites reinforced with the treated fibers displayed increased heat of fusion and crystallinity evident from the DSC [18]. Phiri et al. studied the effect of water glass (WG) coating on the thermal degradation properties of the sisal and kenaf fiber-based MAPP composites through DSC. It was observed that composites with WG-coated sisal fibers displayed marginally higher glass transition temperature (T_g). However, the heat of fusion declined due to the use of WG coating [19].

The addition of banana fibers on sisal/polyester-based composites was analyzed through the DMA. Increase of banana fibers in the overall volume fraction (V_f) led to better storage modulus and lower tan δ peak value. This is mainly due to the superior stiffness of the banana fibers over the sisal fibers. However, balanced properties under tensile, flexural, impact, and dynamic mechanical loads were obtained with the composite containing banana and sisal in 1 : 1 ratio at 0.40 V_f. At 0.40 V_f, the tan δ peak height was lower and broader. A positive shift in glass transition temperature at 0.4 V_f showed better interfacial adhesion [20].

5.5 Conclusions

Thermal properties of sisal and sisal fiber-based hybrid composites were reviewed. The following conclusions were drawn from the published works:

- Surface modification of sisal fiber through chemical treatments, the use of the coupling agent, and incorporation of the nanofillers greatly influences the thermal properties of sisal fiber-reinforced polymeric composites.
- Hybridizing sisal with man-made synthetic fibers will enhance thermal properties while it also helping to reduce the usage of synthetic fiber in polymeric composites. Hybridization combines the advantage of individual constituents while overcoming the limitations of individual constituents.
- Thermal properties of the sisal fiber-based polymer composites can be tailored with the addition of various natural fibers. Though it cannot match the increment in thermal properties obtained with hybridizing with synthetic fibers, it still helps to reduce the usage of synthetic fibers in the composite for structural applications requiring moderate strength.

References

1 Jarukumjorn, K. and Suppakarn, N. (2009). Effect of glass fiber hybridization on properties of sisal fiber–polypropylene composites. *Composites Part B: Engineering* 40 (7): 623–627.

2 Jesuarockiam, N., Jawaid, M., Zainudin, E.S. et al. (2019). Enhanced thermal and dynamic mechanical properties of synthetic/natural hybrid composites with graphene nanoplateletes. *Polymers* 11 (7): 1085.

3 Asim, M., Jawaid, M., Nasir, M., and Saba, N. (2018). Effect of fiber loadings and treatment on dynamic mechanical, thermal and flammability properties of pineapple leaf fiber and kenaf phenolic composites. *Journal of Renewable Materials* 6 (4): 383–393.

4 Asim, M., Paridah, M.T., Chandrasekar, M. et al. (2020). Thermal stability of natural fibers and their polymer composites. *Cellulose* 174: 175.

5 Gudayu, A.D., Steuernagel, L., Meiners, D., and Gideon, R. (2020). Effect of surface treatment on moisture absorption, thermal, and mechanical properties of sisal fiber. *Journal of Industrial Textiles*: 1528083720924774. https://doi.org/10.1177/1528083720924774.

6 Martin, A.R., Martins, M.A., da Silva, O.R., and Mattoso, L.H. (2010). Studies on the thermal properties of sisal fiber and its constituents. *Thermochimica Acta* 506 (1-2): 14–19.

7 Boran Torun, S., Pesman, E., and Donmez Cavdar, A. (2019). Effect of alkali treatment on composites made from recycled polyethylene and chestnut cupula. *Polymer Composites* 40 (11): 4442–4451.

8 Kathirselvam, M., Kumaravel, A., Arthanarieswaran, V., and Saravanakumar, S. (2019). Characterization of cellulose fibers in *Thespesia populnea* barks: influence of alkali treatment. *Carbohydrate Polymers* 217: 178–189.

9 Gañan, P., Garbizu, S., Llano-Ponte, R., and Mondragon, I. (2005). Surface modification of sisal fibers: effects on the mechanical and thermal properties of their epoxy composites. *Polymer Composites* 26 (2): 121–127.
10 Asim, M., Paridah, M., Saba, N. et al. (2018). Thermal, physical properties and flammability of silane treated kenaf/pineapple leaf fibres phenolic hybrid composites. *Composite Structures* 202: 1330–1338.
11 Orue, A., Jauregi, A., Peña-Rodriguez, C. et al. (2015). The effect of surface modifications on sisal fiber properties and sisal/poly(lactic acid) interface adhesion. *Composites Part B: Engineering* 73: 132–138.
12 Ramires, E.C., Megiatto, J.D. Jr., Gardrat, C. et al. (2010). Biobased composites from glyoxal–phenolic resins and sisal fibers. *Bioresource Technology* 101 (6): 1998–2006.
13 Megiatto, J.D. Jr., Silva, C.G., Rosa, D.S., and Frollini, E. (2008). Sisal chemically modified with lignins: correlation between fibers and phenolic composites properties. *Polymer Degradation and Stability* 93 (6): 1109–1121.
14 Amir, S.M., Sultan, M.T., Jawaid, M. et al. (2019). Effects of layering sequence and gamma radiation on mechanical properties and morphology of Kevlar/oil palm EFB/epoxy hybrid composites. *Journal of Materials Research and Technology* 8 (6): 5362–5373.
15 Nayak, S.K. and Mohanty, S. (2010). Sisal glass fiber reinforced PP hybrid composites: effect of MAPP on the dynamic mechanical and thermal properties. *Journal of Reinforced Plastics and Composites* 29 (10): 1551–1568.
16 Sumesh, K., Kanthavel, K., Ajithram, A., and Nandhini, P. (2019). Bioalumina nano powder extraction and its applications for sisal, coir and banana hybrid fiber composites: mechanical and thermal properties. *Journal of Polymers and the Environment* 27 (9): 2068–2077.
17 Venkatram, B., Kailasanathan, C., Seenikannan, P., and Paramasamy, S. (2016). Study on the evaluation of mechanical and thermal properties of natural sisal fiber/general polymer composites reinforced with nanoclay. *International Journal of Polymer Analysis and Characterization* 21 (7): 647–656.
18 Mohanty, S., Nayak, S., Verma, S., and Tripathy, S. (2004). Effect of MAPP as coupling agent on the performance of sisal–PP composites. *Journal of Reinforced Plastics and Composites* 23 (18): 2047–2063.
19 Phiri, G., Khoathane, M., and Sadiku, E. (2014). Effect of fibre loading on mechanical and thermal properties of sisal and kenaf fibre-reinforced injection moulded composites. *Journal of Reinforced Plastics and Composites* 33 (3): 283–293.
20 Idicula, M., Malhotra, S., Joseph, K., and Thomas, S. (2005). Dynamic mechanical analysis of randomly oriented intimately mixed short banana/sisal hybrid fibre reinforced polyester composites. *Composites Science and Technology* 65 (7-8): 1077–1087.

6

Thermal Properties of Flax Fiber Hybrid Composites

Carlo Santulli

Università di Camerino, School of Science and Technology, Geology Division, via Gentile III da Varano 7, Camerino, 62032, Italy

6.1 Introduction

Natural fiber composites (NFCs) are constituted by one or more fiber types of vegetable origin, which act as the reinforcement of a polymer matrix. This can be in turn thermosetting or thermoplastic and oil-derived or bio-based. A large number of natural fibers have been experimented for use in composites. Also, a number of literature reviews on NFCs do exist, starting from around two decades ago [1–3]. Some reviews concentrate on more specific aspects, such as the possibility of structural applications [4], effect of chemical treatment of fibers [5], mechanical properties [6], or more specifically tensile properties [7]. The performance was in other cases examined in combination with other factors, e.g. costs and lightweight [8], or in comparison with other classes of materials, such as fiberglass [9]. This is preparatory for the possible fiber selection, which for NFCs depends on a number of factors, including not only mechanical properties and density of the fibers themselves but also regularity of their geometry and certainly, as for all natural materials, considerations on their local availability [10].

In practice, among natural fibers used in composites, the most popular are those that have been applied in the textile sector, because this indicates also the possibility to obtain regular yarns, suitable for achieving, at least in principle, a strong interface with the polymer matrix. Of course, the quality of adhesion is essential for the use of the composite, which needs to behave as a compact material, avoiding as much as possible the pullout of fibers from the matrix during loading. In this view, the use of natural fibers in composites needs to be designed to provide a sufficiently strong and reliable interface to the level that is normally offered by traditional synthetic fibers, such as glass, whose performance in composites is more comparable with natural fiber ones [11].

In particular, some of the most used vegetable fibers are hemp, jute, sisal, and particularly flax, in which case the quasi-regular hexagonal geometry of the extracted fiber makes it especially adaptable to the production of textiles with different areal

Natural Fiber-Reinforced Composites: Thermal Properties and Applications, First Edition.
Edited by Senthilkumar Krishnasamy, Senthil Muthu Kumar Thiagamani,
Chandrasekar Muthukumar, Rajini Nagarajan and Suchart Siengchin.

weights and structures. The regularity of the structure of flax from the nanoscale and the not very variable dimension of the lumens has been often reported, such as in [12]. Not all flax fibers are obtained from cultivations aimed at the textile use; in other cases, the crop destination is in contrast the production of linseed oil. The latter fibers have received limited attention for manufacturing composite materials, though they may appear promising as a waste from other production, despite the limited length of fibers obtained [13].

As mentioned above, polymer matrices can have a thermoplastic or thermosetting behavior, therefore involving or not the possibility of softening it, which in practice indicates different ways of processing. More specifically, the use of thermoplastic matrices does require elucidating whether the fibers are able to withstand the temperature needed for processing itself. In practical terms, the degradation temperature of natural fibers needs to be higher than the processing temperature for the polymer matrix. Also, it is important that the thermal decomposition kinetics of this process is known, so to suggest how high the activation energy can be [14]. This suggests the interest for thermal analysis of natural fibers before their use in composites.

After these general considerations, it is worth investigating what the differences can be between the thermal behavior of a traditional polymer composite material, most of which are reinforced with glass, carbon, Kevlar or basalt fibers, or other less diffused polymeric or ceramic ones, and that of NFCs reinforced with lignocellulosic fibers. In attempting to explain the possible diversity of thermal behavior, some preliminary reflections are necessary. First and foremost, lignocellulosic fibers are poor thermal conducting materials compared with carbon fibers and some ceramic fibers, such as boron nitride, obviously with respect to metals. In practice, only the substitution of glass fibers with lignocellulosic fibers would keep the bulk thermal conductivity of the composite to the same levels: some values are reported in Table 6.1, keeping constant the use of epoxy resin as the composite matrix, so to allow comparison.

However, their behavior is not orthotropic, i.e. not constant along the same direction, because of the deviation of lignocellulosic fibers from a cylindrical geometry

Table 6.1 Thermal conductivity of epoxy resin composites with different reinforcements.

Materials	Thermal conductivity ($W\,m^{-1}K^{-1}$)	References
Pure epoxy resin	0.2	[15]
K1100 carbon/epoxy	300	[16]
Epoxy/boron nitride	0.25–1	[15]
Glass/epoxy	0.16–0.26	[16]
Abaca/epoxy	0.22–0.34	[17]
Pine dust/epoxy	0.18–0.22	[18]
Kevlar/epoxy	0.08–0.15	[19]

and also for the internal presence of a large hollow of variable dimension, referred to as *lumen*. The influence of the lumen size on transverse thermal conductivity has been investigated and modeled according to an electrical–thermal analogy by [20]. Other factors are also important, if a possible model of thermal conductivity for a composite has to be proposed, such as in [21]. These are fiber length, fiber aspect ratio, relative modulus of fiber and matrix, and therefore thermal expansion mismatch between the two.

6.2 Techniques for Thermal Analysis of Natural Fiber Composites

The first studies that have been performed to clarify the thermal expansion behavior of the different components (cellulose, hemicellulose, and lignin) of lignocellulosic materials were carried out using dilatometric analysis (DLA) [22]. The objective of DLA is to measure the thermal expansion of the material in a range between ambient temperature and approximately 1000 °C. This application proved to offer a sequence of degradation among the three components not completely reliable, due to the variety of alteration process occurring, the most common being depolymerization and hydrolysis [23].

Therefore, the information on thermal analysis of composites is most frequently obtained by performing thermogravimetric analysis (TGA) scans, in which small samples of the material are subjected to a heating cycle. This is terminating in the case of composite at a temperature between 700 and 900 °C. Thermogravimetric curve and the relevant differential thermogravimetric (DTG) one allow the establishment of the degradation onset and temperature range of the different components of lignocellulosic fibers, as reported in Figure 6.1. This is carried out over a range extending from ambient temperature up to a temperature above which degradation brings the material to be completely combusted, so to leave only a pure carbon residue. Sample preparation was reported to have some influence on the obtained results. In particular, different heat rates (5, 20, and 50 °C min^{-1}) of fine, medium, and coarse flax fiber samples obtained by cutting, milling, or freeze grinding resulted in different conditions of the fiber ends and in a variation of the peaks corresponding to the onset of fiber degradation. This suggested that medium fibers at the intermediate heat rate provided more reliable data [25]. Conversely, keeping the heat rate constant, it was also demonstrated that the fineness of the fibers after scutching could be determined by the temperature range of the two degradation peaks, the main one usually located at 340–360 °C and the secondary one at 420–460 °C, with the coarser fibers showing a higher value for the peaks [26].

TGA can be carried out in air or in nitrogen, offering distinctly different results and in the latter case excluding the effect of oxidation and therefore simulating the behavior of the material in an air-free (hence also moisture-free) ambient. In the case of natural fibers, TGA offers also the possibility to give information on the effect of chemical treatment, such as alkali, acetylation, benzoil, silanization, etc. and physical treatment, such as plasma or ultraviolet irradiation. In general terms,

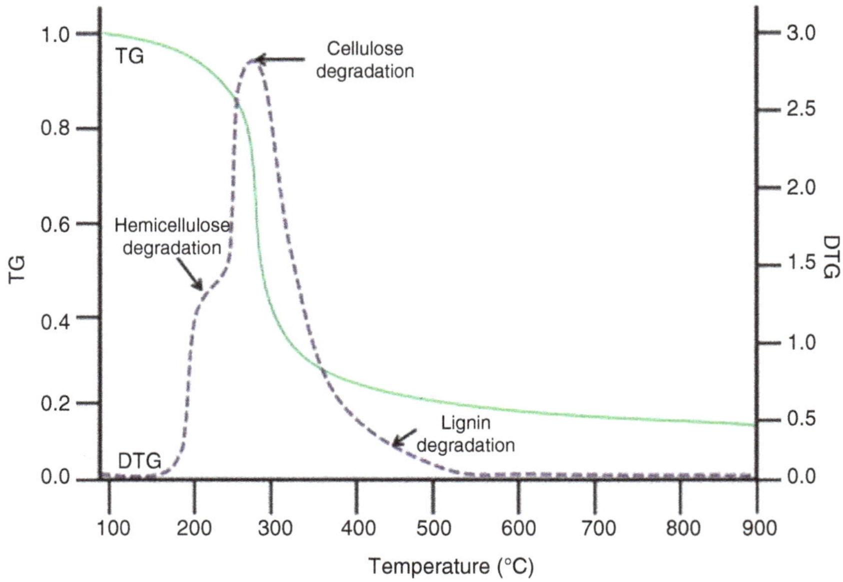

Figure 6.1 Typical thermal degradation patterns for lignocellulosic fibers. Source: Azwa et al. [24].

the objectives of treatment are to modify the moisture absorption process, clean, roughen, and chemically modify the fiber surface. In more practical terms, they are required to improve the mechanical and technological properties of the fibers, on their degradation with temperature [27]. Some treatments, such as acylation, do require specific reaction times with catalysts (in the specific case sulfuric acid) to reduce the weight loss, concentrating on a specific target temperature (e.g. 150 °C), related to the relevant intended processing of the polymer matrix [28]. TGA proved also effective in elucidating effects of specific treatments, e.g. laccase, aimed at improving the workability of other flax products, such as cellulose pulp, acting as mediators for bleaching [29]. Once again, these studies are also able to offer reliable measurements of the crystalline and amorphous fraction of the relevant by-product, be it flax fiber or pulp [30].

Differential scanning calorimetry (DSC) can be useful to see the change of behavior with respect to heat transmission with temperature; therefore, the presence of transitions, such as the one from glassy state into rubbery state, in polymer matrices can be obviously also due to the addition of fillers, such as natural fibers. This is very important to optimize processing in view of an enhanced thermal stability during the production [31]. It is also noteworthy that a higher thermal stability bears a relation with reducing flammability and the likeliness to reach a high limiting oxygen index (LOI) [32].

A combination of TGA and DSC is also able to inform on the comparative effect of different treatment methods, such as scouring and bio-scouring, on flax roves, while considering that all treatments in general tend to improve the thermal stability of the fibers [33]. The two techniques together prove the thermal stability of the fiber, which in the use in composites is supposed to extend up to a temperature higher than the one necessary for thermoplastic polymer matrix processing. This appears particularly critical in the case of polypropylene (PP), where the processing interval is not very large, in the worst cases not exceeding a few degrees Celsius [34]. The way fibers are extracted may also has an effect on thermal stability; in particular this appears to be higher in the case of flax, when the fibers are obtained from the bast using a steam explosion method [35].

To evaluate the thermomechanical properties, hence studying the evolution of the viscoelastic behavior over temperature, also dynamical mechanical analysis (DMA) has also been often performed [36]. Also, in the case of DMA, the effect of fiber treatment proved significant in improving the dynamical mechanical properties, e.g. in flax/epoxy composites [37].

Another thermal issue that is important to investigate in NFCs is their fire behavior. In general terms, the introduction of lignocellulosic fibers into an oil-based polymer matrix does result in fire retardancy. The level of this effect can be ascertained by using cone calorimeter [38]. However, indications about the char yield from the NFCs can also be provided by thermogravimetric measurements performed up to 800 °C, as demonstrated by Dayo et al. [39].

6.3 General Properties and Composition of Flax Fibers

Flax fiber is constituted primarily by cellulose and then by hemicellulose and lignin with some amounts of wax and pectins. The amount of cellulose, which represents the most ordered structural material for the fibers, can be variable in a considerable range: a study [40] reported, for example, that cellulose content can be 47.2% in flax stalks. However, while extracting the short fibers, it can rise as high as 72.1%, in particular for the elimination of lignin; the amount of cellulose for flax fibers ready for application in composites may be even higher. The composition of course depends on the particular flax cultivar that is used; a comparative study of several varieties evidenced significant differences (in the particular case, $84 \pm 3.5\%$ for Hermes and $77 \pm 2.5\%$ for Oliver variety) [41]. A reasonable assumption for flax fibers used in composites is that they contain around 80% of cellulose.

In general terms, the mechanical properties of the flax fibers are influenced by the degree of crystallinity, the spiral angle of the fibrils, the degree of polymerization of the cellulose chains, the porosity content, the size and geometry of the lumen, and the presence of defects (kink bands) that leads to some deviation from the straightness in fibers [42]. Flax fibers have a quite high density, as related to other lignocellulosic fibers, around 1.5 $g\,cm^{-3}$, which are reported in various studies ([43, 44]). This is not far from the true density of fully crystalline cellulose, which has been measured as being between 1.582 and 1.599 $g\,cm^{-3}$ [45]. In practice, the crystallinity of cellulose in flax fibers has been recently measured in the order of 45% [46].

A number of studies reported on the mechanical properties of flax fibers, as compared with those of other lignocellulosic fibers. Some indications are given also in Figure 6.2 [47] on different fiber cultivars of French origin, from which an

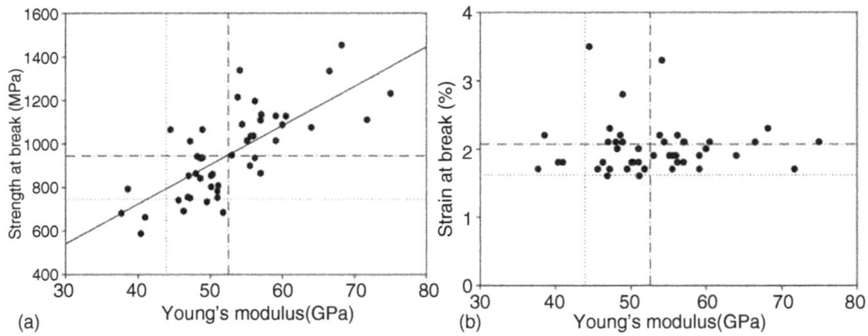

Figure 6.2 Stiffness vs. strength (a) and ultimate strain (b) of different batches of flax fibers. Source: Baley and Bourmaud [47].

average value of 950 MPa and 2.1% for ultimate strength and strain can be indicated, respectively, as well as a value of 52.5 GPa for stiffness with a considerable dispersion of values.

The extraction of flax fibers leads to them having normally a diameter in the order of 20 μm; therefore, it is not very different from that of glass fibers, though, differently from them, once again a large scattering of values has been measured [48].

6.4 Thermal Analysis of Flax Fibers

There is considerable literature first of all on thermal properties of pure flax fibers, which indicates the different peaks for degradation of their various components, the most important being that of cellulose. Early studies of thermal decomposition of flax fibers using TGA enabled elucidating the kinetic parameters of cellulose and lignin degradation. Also, the deconvolution of the peaks obtained in the TGA curve would allow measuring the respective quantities of cellulose, hemicellulose, and lignin in the specific flax variety used [49]. In particular, the formation of free radicals in a range of temperatures between 100 and 160 °C bears an essential role in the creation of hydroxyperoxides, which therefore lead to the process of depolymerization of cellulose, when TGA is performed in air. However, in nitrogen atmosphere, this hardly takes place at all [50]. Alimuzzaman et al. [51] suggested that the degradation of flax fibers progresses in three consecutive stages, in particular up to 150 °C releasing absorbed moisture, after which, cellulosic and hemicellulosic substances start degrading, with a more pronounced rate above 250 °C. The third and final stage was indicated to commence at 375 °C, which was indicated as the onset temperature for non-cellulosic materials, with the material being totally reduced into ashes at around 600 °C, with no more than a few percent remaining, when TGA is performed in nitrogen atmosphere. In general terms, the degradation of hemicellulose has been considered as a particularly critical aspect to be detrimental for the processing of thermoplastic matrix composites, especially at low heating rate [52].

The effect of treatments on the characteristics of flax fabrics has also been studied: from early studies, it has been clarified that mercerization, hence treatment with sodium hydroxide, results in a noteworthy decrease of the crystalline fraction of the fibers, especially in an enhanced degradation of the less structural parts of the yarns [53].

6.5 Thermal Analysis of Flax Fiber Composites

In composites, a large number of studies including flax fibers are available, most of which involve the use of traditional oil-based matrices. In practice, most frequently the first evaluation of the possibilities of natural fibers in a composite is carried out in a thermosetting matrix, such as epoxy or vinyl ester, owing to the relative simplicity of its fabrication, normally not involving severe heating and cooling cycles. As mentioned above, thermal studies are also able to reveal the effect of treatments on the progression of degradation with temperature. This is particularly relevant in the case of further complexity, for example, if the fibers are treated with combined methods, such as silanization with clay grafting, since in this case the glass transition temperature of the resin can be significantly modified [54].

There are some polymer matrices where the importance of assessing how processing affects the properties of flax fibers is particularly high, since it is unlikely that the production times and temperatures may be changed, but rather the natural fiber composite has to adapt to them. This is the case of polypropylene (PP), which is especially of interest in the automotive field. To be considered as the reinforcement of PP, a number of natural fibers have been experimented over time [55]. These include flax, which is one of those fibers that are more frequently used for the production of textiles and useful and more reliable as the reinforcement of composites and for rapid processing [56]. In the particular case of PP, a number of aspects were investigated, which are crucial for the intended application and whose modification/improvement has obviously an effect also on thermal properties of the obtained composite. These include the resistance to flammability [57] and wetting from the resin system [58]. It is worth noting that most often flax fibers to be applied in the composite need chemical treatment, such as maleinization and silanization, which equally modifies the thermal behavior of the fiber [59]. A study from Arbelaiz et al. [48], which performed both TGA and DSC thermal analysis on flax/PP composites, focused on the increase of PP crystallinity that was possibly obtained with the introduction of flax fibers, and on the effect of heating rate and of the different treatments on this process.

Another relevant objective toward sustainability is also to use flax fibers as the reinforcement for biodegradable matrices, such as poly(lactic acid) (PLA) [60]. Fillers, such as flax fibers, are deemed to improve the thermal stability of PLA matrices, in particular increasing the temperatures at which thermal degradation takes place [61]. The addition of flax fibers can also influence the glass transition temperature, increasing it in the case of PLA to up to 30 wt% of fiber content, which brings the possible range for use to above 100 °C [62]. A gradually increasing addition of flax fibers leads to the progressive matrix reinforcement. This has also an effect on the enhancement of crystallinity of PLA, which may be normally very

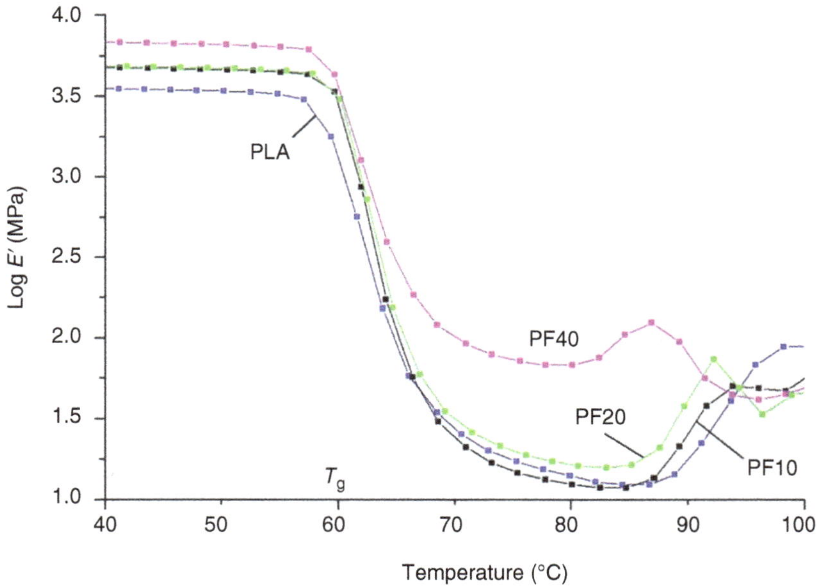

Figure 6.3 Dynamic modulus of PLA and PLA/flax composites with 10, 20, and 40 wt% fibers.

low, and improves the dynamic modulus of the matrix, as proven recently in a study carried out by [63], a result exposed in more detail in Figure 6.3.

Also, the introduction of flax into particular biomatrices that had been hardly ever used with this fiber required in-depth thermal studies prior to the fabrication. This is the case for poly(hydroxybutyrate) (PHB), where different treatments, including plasma and acetylation, were compared to establish their effectiveness for the creation of a sound fiber/matrix interface [64–67]. Other biomatrices were also used, such as poly(butylsuccinate) (PBS) [65]. This offered results that can be deemed applicable to the general case of biocomposites. In particular, the effect of small variations in the amount of flax fibers introduced on the residual mass at 700 °C appears very significant, leading as a whole to a lower thermal stability .

6.6 Conclusions

Thermal properties of flax fiber composites concentrate on a few aspects, which appear crucial for the use of the material in industrial applications, e.g. in the automotive industry. In particular, flax fibers, once extracted either by retting or steam explosion and treated to improve their performance and increase the removal of non-structural matter, contain around 80% of cellulose. This implies that it is important to know in which range of temperature the cellulose degradation takes place, which is normally between 250 and 400 °C, although with considerable differences depending on the flax variety, extraction method, and treatment applied. Another important consideration needs to be done on the use of matrices for the composite.

Of course, the study of degradation is mainly of interest for thermoplastic matrices, either oil based, such as polypropylene (PP), or bio-based, such as poly(lactic acid) (PLA), since these are processed in the softened state to prepare the composite. In this case, thermal properties assessment can conclude that the fibers are substantially degraded in the temperature range at which the matrix are processed and more in general, especially for PLA, which is in the glassy state at ambient temperature, establish a maximum temperature for use with the introduction of different amounts of fibers. More recently, DMA in temperature has been revealed to be important to clarify the viscoelastic behavior of flax fiber composites, on which there is quite limited information so far.

References

1 Bongarde, U.S. and Shinde, V.D. (2014). Review on natural fiber reinforcement polymer composites. *International Journal of Engineering Science and Innovative Technology* 3 (2): 431–436.

2 Saheb, D.N. and Jog, J.P. (1999). Natural fiber polymer composites: a review. *Advances in Polymer Technology* 18 (4): 351–363.

3 Westman, M.P., Fifield, L.S., Simmons, K.L., et al. (2010). Natural Fiber Composites: A Review. *U.S. Department of Energy Technical Report*. https://doi.org/10.2172/989448.

4 Ticoalu, A., Aravinthan, T., and Cardona, F. (2010). A review of current development in natural fiber composites for structural and infrastructure applications. *Southern Region Engineering Conference (SREC 2010)*, Toowoomba, Australia (10–12 November 2010). https://www.researchgate.net/profile/Francisco-Cardona-2/publication/277798012_A_Review_of_Current_Development_in_Natural_Fiber_Composites_for_Structural_and_Infrastructure_Applications/links/55da731e08aeb38e8a8a148a/A-Review-of-Current-Development-in-Natural-Fiber-Composites-for-Structural-and-Infrastructure-Applications.pdf.

5 Li, X., Tabil, L.G., and Panigrahi, S. (2007). Chemical treatments of natural fiber for use in natural fiber-reinforced composites: a review. *Journal of Polymers and the Environment* 15: 25–33.

6 Elanchezhian, C., Vijaya Ramnath, B., Ramakrishnan, G. et al. (2018). Review on mechanical properties of natural fiber composites. *Materials Today: Proceedings* 5 (1): 1785–1790.

7 Ku, H., Wang, H., Pattarachaiyakoop, N., and Trada, M. (2011). A review on the tensile properties of natural fiber reinforced polymer composites. *Composites Part B: Engineering* 42 (4): 856–873.

8 Ahmad, F., Choi, H.S., and Park, M.K. (2015). A review: natural fiber composites selection in view of mechanical, light weight, and economic properties. *Macromolecular Materials and Engineering* 300 (1): 10–24.

9 Joshi, S.V., Drzal, L.T., Mohanty, A.K., and Arora, S. (2004). Are natural fiber composites environmentally superior to glass fiber reinforced composites? *Composites Part A: Applied Science and Manufacturing* 35 (3): 371–376.

10 Sapuan, S.M., Kho, J.Y., Zainudin, E.S. et al. (2011). Materials selection for natural fiber reinforced polymer composites using analytical hierarchy process. *Indian Journal of Engineering and Materials Science* 18 (4): 255–267.

11 Goutianos, S., Peijs, T., Nystrom, B., and Skrifvars, M. (2006). Development of flax fibre based textile reinforcements for composite applications. *Applied Composite Materials* 13: 199–215.

12 Charlet, K., Jernot, J.P., Evea, S. et al. (2010). Multi-scale morphological characterisation of flax: from the stem to the fibrils. *Carbohydrate Polymers* 82: 54–61.

13 Pillin, I., Kervoelen, A., Bourmaud, A. et al. (2011). Could oleaginous flax fibers be used as reinforcement for polymers? *Industrial Crops and Products* 34 (3): 1556–1563.

14 Yao, F., Wu, Q., Lei, Y. et al. (2008). Thermal decomposition kinetics of natural fibers: activation energy with dynamic thermogravimetric analysis. *Polymer Degradation and Stability* 93 (1): 90–98.

15 Gu, J., Zhang, Q., Dang, J., and Xie, C. (2011). Thermal conductivity epoxy resin composites filled with boron nitride. *Polymers for Advanced Technologies*, Wiley https://doi.org/10.1002/pat.2063.

16 Mallick, P.K. (2007). *Fiber-Reinforced Composites: Materials, Manufacturing, and Design*, 3e. CRC Press.

17 Liu, K., Zhang, X., Takagi, H. et al. (2014). Effect of chemical treatments on transverse thermal conductivity of unidirectional abaca fiber/epoxy composite. *Composites Part A: Applied Science and Manufacturing* 66: 227–236.

18 Mohapatra, R.C., Mishra, A., and Choudhury, B.B. (2014). Measurement on thermal conductivity of pine wood dust filled epoxy composites. *American Journal of Mechanical Engineering* 2 (4): 114–119.

19 Harris, I.P., Yates, B., Batchelor, J., and Garrington, P.J. (1982). The thermal conductivity of Kevlar fibre-reinforced composites. *Journal of Materials Science* 17: 2925–2931.

20 Liu, K., Takagi, H., Osugi, R., and Yang, Z. (2012). Effect of lumen size on the effective transverse thermal conductivity of unidirectional natural fiber composites. *Composites Science and Technology* 72 (5): 633–639.

21 Idicula, M., Boudenne, A., Umadevi, L. et al. (2006). Thermophysical properties of natural fibre reinforced polyester composites. *Composites Science and Technology* 66: 2719–2725.

22 Ramiah, M.V. and Goring, D.A.I. (1965). The thermal expansion of cellulose, hemicellulose, and lignin. *Journal of Polymer Science Part C: Polymer Symposia* https://doi.org/10.1002/polc.5070110105.

23 Gassan, J. and Bledzki, A.K. (2001). Thermal degradation of flax and jute fibers. *Journal of Applied Polymer Science* 82: 1417–1422.

24 Azwa, Z.N., Yousif, B.F., Manalo, A.C., and Karunasena, W. (2013). A review on the degradability of polymeric composites based on natural fibres. *Materials and Design* 47: 424–442.

25 Sharma, H.S.S., Faughey, G., and McCall, D. (1996). Effect of sample preparation and heating rate on the differential thermogravimetric analysis of flax fibres. *The Journal of The Textile Institute* 87 (2): 249–257.

26 Faughey, G.J., Sharma, S.S., and McCall, R.D. (2000). Determining fiber fineness in flax using derivative thermogravimetric analysis, scanning electron microscopy, and airflow methods. *Journal of Applied Polymer Science* 75: 508–514.

27 Bourmaud, A., Morvan, C., and Baley, C. (2010). Importance of fiber preparation to optimize the surface and mechanical properties of unitary flax fiber. *Industrial Crops and Products* 32 (3): 662–667.

28 Zini, E., Scandola, M., and Gatenholm, P. (2003). Heterogeneous acylation of flax fibers. Reaction kinetics and surface properties. *Biomacromolecules* 4 (3): 821–827.

29 Vila, C., Barneto, A.G., Fillat, A. et al. (2011). Use of thermogravimetric analysis to monitor the effects of natural laccase mediators on flax pulp. *Bioresource Technology* 102 (11): 6554–6561.

30 Barneto, A.G., Vila, C., Ariza, J., and Vidal, T. (2011). Thermogravimetric measurement of amorphous cellulose content in flax fibre and flax pulp. *Cellulose* 18: 17–31.

31 Khalfallah, M., Abbès, B., Abbès, F. et al. (2014). Innovative flax tapes reinforced Acrodur biocomposites: a new alternative for automotive applications. *Materials & Design* 64: 116–126.

32 Srinivasan, V.S., Rajendra Boopathy, S., Sangeetha, D., and Vijaya Ramnath, B. (2014). Evaluation of mechanical and thermal properties of banana–flax based natural fibre composite. *Materials & Design* 60: 620–627.

33 Dong, Z., Ruo-Yao, D., Zheng, L. et al. (2015). Thermal properties of flax fiber scoured by different methods. *Thermal Science* 19 (3): 939–945.

34 Hine, P.J., Ward, I.M., and Teckoe, J. (1998). The hot compaction of woven polypropylene tapes. *Journal of Materials Science* 33: 2725–2733.

35 Wielage, B., Lampke Th, Marx, G. et al. (1999). Thermogravimetric and differential scanning calorimetric analysis of natural fibres and polypropylene. *Thermochimica Acta* 337 (1–2): 169–177.

36 Samal, S., Stuchlík, M., and Petrikova, I. (2018). Thermal behavior of flax and jute reinforced in matrix acrylic composite. *Journal of Thermal Analysis and Calorimetry* 131: 1035–1040.

37 Vinu Kumar, S.M., Senthil Kumar, K.L., Siddhi Jailani, H., and Rajamurugan, G. (2020). Mechanical, DMA and sound acoustic behaviour of flax woven fabric reinforced epoxy composites. *Materials Research Express* 7: 085302. https://doi.org/10.1088/2053-1591/abaea5.

38 Chapple, S. and Anandjiwala, R. (2010). Flammability of natural fiber-reinforced composites and strategies for fire retardancy: a review. *Journal of Thermoplastic Composite Materials* 23 (6): 871–893.

39 Dayo, A.Q., Gao, B., Wang, J. et al. (2017). Natural hemp fiber reinforced polybenzoxazine composites: curing behavior, mechanical and thermal properties. *Composites Science and Technology* 144: 114–124.

40 Bochek, A.M., Shevchuk, I.L., and Lavrent'ev, V.N. (2003). Fabrication of microcrystalline and powdered cellulose from short flax fiber and flax straw. *Russian Journal of Applied Chemistry* 76 (10): 16791682.

41 Bourmaud, A., Morvan, C., Bouali, A. et al. (2013). Relationships between micro-fibrillar angle, mechanical properties and biochemical composition of flax fibers. *Industrial Crops and Products* 44: 343–351.

42 Baley, C. (2004). Influence of kink bands on the tensile strength of flax fibers. *Journal of Materials Science* 39: 331–334.

43 Joseph, K., Toledo, R.D., James, B. et al. (1999). A review on sisal fiber reinforced polymer composites. *Revista Brasileira de Engenharia Agricola e Ambiental* 3: 367–379.

44 Hernandez Michelena, A., Graham-Jones, J., Summerscales, J., and Hall, W. (2017). Silane modification of the flax/epoxy system interface. *Procedia Engineering* 200: 448–456.

45 Sugiyama, J., Vuong, R., and Chanzy, H. (1991). Electron diffraction study on the two crystalline phases occurring in native cellulose from an algal cell wall. *Macromolecules* 24: 4168–4175.

46 Chen, F., Sawada, D., Hummel, M. et al. (2020). Unidirectional all-cellulose composites from flax via controlled impregnation with ionic liquid. *Polymers* 12: 1010.

47 Baley, C. and Bourmaud, A. (2014). Average tensile properties of French elementary flax fibers. *Materials Letters* 122: 159–161.

48 Arbelaiz, A., Cantero, G., Fernandez, B. et al. (2005). Flax fiber surface modifications: effects on fiber physico mechanical and flax/polypropylene interface properties. *Polymer Composites* 26: 324–332.

49 Titok, L.V., Shostak, L., and Khotyleva, L. (2006). Thermogravimetric analysis of the flax bast fibre bundle. *Journal of Natural Fibers* 3 (1): 35–41.

50 Van De Velde, K. and Kiekens, P. (2002). Thermal degradation of flax: the determination of kinetic parameters with thermogravimetric analysis. *Journal of Applied Polymer Science* 83: 2634–2643.

51 Alimuzzaman, S., Gong, R.H., and Akonda, M. (2013). Nonwoven polylactic acid and flax biocomposites. *Polymer Composites* 34 (10): 1611–1619.

52 Chaishome, J., Brown, K.A., Brooks, R., and Clifford, M.J. (2014). Thermal degradation of flax fibres as potential reinforcement in thermoplastic composites. *Advanced Materials Research* 894: 32–36.

53 Sharma, H.S.S. and Kernaghan, K. (1988). Thermogravimetric analysis of flax fibres. *Thermochimica Acta* 132: 101–109.

54 Wang, A., Xia, D., Xian, G., and Li, H. (2019). Effect of nanoclay grafting onto flax fibers on the interfacial shear strength and mechanical properties of flax/epoxy composites. *Polymer Composites* 40: 3482–3492.

55 Holbery, J. and Houston, D. (2006). Natural-fiber-reinforced polymer composites in automotive applications. *JOM* 58: 80–86.

56 Derbali, I., Terekhina, S., Guillaumat, L., and Ouagne, P. (2019). Rapid manufacturing of woven comingled flax/polypropylene composite structures. *International Journal of Material Forming* 12: 927–942.

57 Helwig, M. and Paukszta, D. (2000). Flammability of composites based on polypropylene and flax fibers. *Molecular Crystals and Liquid Crystals Science and Technology Section A* 354 (1): 373–380.
58 Aranberri-Askargorta, I., Lampke, T., and Bismarck, A. (2003). Wetting behavior of flax fibers as reinforcement for polypropylene. *Journal of Colloid and Interface Science* 263 (2): 580–589.
59 Cantero, G., Arbelaiz, A., Llano-Ponte, R., and Mondragon, I. (2003). Effects of fibre treatment on wettability and mechanical behaviour of flax/polypropylene composites. *Composites Science and Technology* 63 (9): 1247–1254.
60 Oksman, K., Skrifvars, M., and Selin, J.-F. (2003). Natural fibres as reinforcement in polylactic acid (PLA) composites. *Composites Science and Technology* 63 (9): 1317–1324.
61 Siengchin, S. (2014). Reinforced flax mat/modified polylactide (PLA) composites: impact, thermal, and mechanical properties. *Mechanics of Composite Materials* 50 (2): 257–266.
62 Rudnik, E. (2007). Thermal properties of biocomposites. *Journal of Thermal Analysis and Calorimetry* 88: 495–498.
63 Aliotta, L., Gigante, V., Coltelli, M.B. et al. (2019). Thermo-mechanical properties of PLA/short flax fiber biocomposites. *Applied Sciences* 9: 3797.
64 Lee, S.G., Choi, S.-S., Park, W.H., and Cho, D. (2003). Characterization of surface modified flax fibers and their biocomposites with PHB. *Macromolecular Symposia* 197: 89–99.
65 Dorez, G., Taguet, A., Ferry, L., and Lopez-Cuesta, J.M. (2013). Thermal and fire behavior of natural fibers/PBS biocomposites. *Polymer Degradation and Stability* 98 (1): 87–95.
66 Kannan, T.G., Wu, C.M., Cheng, K.B., and Wang, C.Y. (2012). Effect of reinforcement on the mechanical and thermal properties of flax/polypropylene interwoven fabric composites. *Journal of Industrial Textiles* 42 (4): 417–433.
67 Wong, S., Shanks, R., and Hodzic, A. (2004). Interfacial improvements in poly(3-hydroxybutyrate)-flax fibre composites with hydrogen bonding additives. *Composites Science and Technology* 64 (9): 1321–1330.

7

Thermal Properties of the Pineapple Leaf Fiber-Based Hybrid Composites

Bheemappa Suresha[1], Rajashekaraiah Hemanth[2], and Gurumurthy Hemanth[3]

[1] *Visveswaraya Technological University, The National Institute of Engineering, Department of Mechanical Engineering, Mysore, 570008, Karnataka, India*
[2] *Visveswaraya Technological University, NIE Institute of Technology, Department of Mechanical Engineering, Mysore, 570018, Karnataka, India*
[3] *MS/EAI3-PS, Robert Bosch Engineering and Business Solutions Pvt, Ltd, Bangalore, India*

7.1 Introduction

Composite materials have unlimited applications and are frequently an imaginative and innovative tools to improve and make more competitive products in many industrial sectors. Thermal properties include thermal conductivity, thermal diffusivity, heat capacity, thermal effusivity, and material density. Polymeric composite materials are entirely insulating materials that can easily be differentiated with metallic alloys and composites in which swift heat repression issues occur. To anticipate the thermal condition of planar structures (the related thermo-mechanical behavior) for design purposes, thermal properties and the related susceptibilities of the encompassed orthotropic composite are thus required.

Thermal conductivity testing can be performed utilizing hot plates with thermal conductivity instruments to address protection issues. Polymeric composites are commonly lighter than their unique partners, yet superior strength in the appraisal. A natural composite, for example, wood, is made of lignin and cellulose, two moderately frail elements considering separately, yet when joined, create a reliable and durable material. Fiber-reinforced polymeric composites (FRPCs) are the most well-known and most generally utilized human-made composites. To build the quality and solidness of polymers, fibers and inorganic fillers are included within the polymers. Given their lightweight attributes, fiber-strengthened polymeric composites are modified by micron-sized fillers and nanofillers, and these multiphase composites are being executed in regions of aviation and vehicles. All things considered, a lighter airplane is a more effective airplane, and the superior exhibitions of composites are a verifiable resource.

Composites are built materials framed out of at least two major constituents mostly used are nonmetallic materials. A large portion of the composites can be custom fitted to get properties superior to singular constituents. A polymer composite

Natural Fiber-Reinforced Composites: Thermal Properties and Applications, First Edition.
Edited by Senthilkumar Krishnasamy, Senthil Muthu Kumar Thiagamani, Chandrasekar Muthukumar, Rajini Nagarajan and Suchart Siengchin.

fortified with fiber is called fiber-strengthened polymer matrix composite (FRPC). With regard to a composite, which includes at least two large-scale constituent stages, the matrix/resin is alluded to as a continuous phase, and the fibers as the strengthening (discontinuous) phase [1]. The fibers increment the strength and stiffness, enhance the fatigue properties and thermal properties, and provide better dimensional stability and electrical resistivity. The essential capacity of the matrix is to transfer load and hold the fiber bundles, mechanical abrasion and shield fibers from the environment effects like moisture, oil, dust, and so on. Matrix materials also carry transverse loads and interlaminar shear stresses [2]. The upsides of composites incorporate high strength to weight proportion, non-destructiveness, lower maintenance cost, high electrical resistance, wear resistance, electromagnetic transparency, assuaging appearance, and so on. Composites can be made in various manners relying on the type of matrix, type of reinforcement, and application. Diverse manufacturing techniques include hand layup, compression molding, resin transfer molding (RTM), pultrusion, autoclave and filament winding, and so forth.

Regarding the mechanical and thermal properties, composites are anisotropic (for example, the properties differ with fiber direction), while their vast majority can be considered as orthotropic (for example, nine distinct constants are needed to depict an orthotropic material). This additionally allows the huge estimations of thermal stresses because of the temperature gradient over the area. This issue stresses the need for information on the thermal behavior of natural FRPC.

7.2 Thermal Properties of Polymers

7.2.1 The Hypothesis of Thermal Conductivity

The hypothesis of thermal conductivity was first introduced by Fourier in 1822. As indicated by Fourier, the essential heat conduction condition can be expressed as "For a homogeneous solid, the local heat flux is proportional to the negative local temperature gradient." For unidirectional steady heat transfer, the equation form of the above mentioned can be expressed as in Eq. (7.1):

$$Q = -k.A.\frac{\mathrm{d}T}{\mathrm{d}x} \tag{7.1}$$

where Q is the heat transfer rate, A is the area orthogonal to the direction of the heat transfer, k is the thermal conductivity of the material ($\mathrm{W\,m^{-1}\,K^{-1}}$) that is a positive second-order tensor expression, $\frac{\mathrm{d}T}{\mathrm{d}x}$ refers to delta temperature over the thickness of the conducting media, and the negative sign shows the thermal gradient from a surface of higher temperature to a surface of the cooler temperature. Upon rearrangement of Eq. (7.1), the thermal conductivity can be expressed as in Eq. (7.2) (assuming that heat flows in a direction normal to the face of the body and the other planes are neglected):

$$k = \frac{\left(\frac{Q}{A}\right)}{\left(\frac{\Delta T}{\Delta L}\right)} \tag{7.2}$$

where ΔT is the temperature gradient in Kelvin and ΔL is the distance in m for which heat transfer happens. Accordingly, k can be characterized as a rate at which heat energy is conducted through a given area of a material when the difference in temperature exists normal to the cross-sectional area of the body. The thermal conductivity of a composite material relies upon many factors, namely, the fiber, resin material, fiber volume fraction, the direction of the fiber, the direction of heat transfer, and temperature of operation.

7.2.2 Factors Influencing Thermal Conductivity of Composite Materials

The thermal conductivity of composites is non-uniform throughout the material system which is discussed in Section 7.2.2.5. The information on thermal conductivity is essential for effective design. Information on the thermal conductivity of resin facilitates to decrease stresses that arise due to dimensional shrinkage of composites that happens in the thermal curing stage and mismatch in the thermal expansion coefficient. Before directing trials to determine the thermal conductivity of different composites, information on the impact of various parameters affecting thermal conductivity is needed.

7.2.2.1 Fibers

Fiber is the fortifying phase of a composite material. The thermal conductivity of a composite relies on the thermal conductive nature of the fiber and matrix. Generally, utilized synthetic fibers for composites include glass, carbon, and aramid, and so forth. Glass fibers are normally used for designing FRPCs. Their utilizations incorporate the assembling of car parts, pipes, auxiliary components, and so on.

Glass fiber is accessible monetarily in bounty with great mechanical properties and thus generally utilized in composite structures [2]. Based on various applications, glass fiber (silica–oxygen arrange) is segregated into C glass, E glass, and S glass fibers [2]. E glass is utilized as a protector and generally utilized in the electrical firms, subsequently got the name "E" glass. E-glass additionally has substantial mechanical properties along with reduced cost and usability ease. The letter "S" in S-glass represents for structural applications. S-glass got diverse compound formulation, and it has a high strength-to-weight ratio and higher strain rate. S-glass is a bit costly and generally utilized in the structural usage. C-glass fibers are worthwhile in resisting chemical corrosion. Glass fibers are accessible in various structures like short, long, and continuous, chopped, and woven textures. The microstructure of any fiber assumes an indispensable job in carrying heat. The glass fiber has an amorphous structure. It comprises of SiO_2 particles and forms a three-dimensional silica polyhedral shape along the length of the fiber. It behaves nearly isotropic, bringing about almost the same conductivity properties toward any path of the fiber.

Carbon fibers are produced utilizing precursor materials like rayon, petroleum, or coal tar pitches and polyacrylonitrile (PAN) [3]. The transformation of pitch or PAN precursor to carbon fibers includes fabricating steps like fiber arrangement by spinning, stabilization to thermoset the fiber, carbonization, graphitization, surface treatment, and sizing [1]. During the graphitization stage at elevated temperatures, the crystallites are appropriately ordered and situated along the fiber direction. In PAN-based carbon fibers, during the process of graphitization, the direct structure morphology of carbon particles is modified into a basal planes (planar structure) that are situated or stacked along the length of the fiber. These basal planes are packed in a close vicinity and are liable for the increased modulus and larger electrical conductivities and thermal conductivities along the length direction. Natural fibers are becoming more popular and alternative to glass fiber strengthened polymeric composites in numerous applications. Use of natural fibers like hemp, jute, flax, cotton, and so on instead of synthetic fibers prompts increment in explicit properties like effect impact strength, crash behavior, sound assimilation, thermal insulation, and a considerable decrease in part's weight, contaminations, and ozone-depleting substance emanations making the composite more environmentally friendly. Natural fiber-reinforced polymer composites (NFRPCs) have numerous applications, including car parts like entryway trim boards, headliners/back panels. Natural fiber-loaded up with cellulose material, which goes about as a cover, in this way, a natural fiber composite shows much lower thermal conductivity when contrasted with a glass fiber reinforced polymer composite (GFRPC).

Pineapple fiber (PALF) are the fibers extracted from the leaves of the *Ananas comosus* that is usually known as the pineapple. The plant has a long and thorny leaves that looks like a shrub that grows about 1–1.5 m tall. The leaves are usually about 2–3 ft in length and the sharp edged and the typical pineapple plant is shown in Figure 7.1.

The fibers are separated from the leaves and then dried. The PALF thus extracted cannot be used directly in preparation of the composites. The fibers are usually porous and absorb the liquids it interacts with. However, it is usual that the natural fibers are water absorbant and are not ready to be used with the polymer matrix as they might induce the improper adhesion by inducing the shortage of the resin. Hence, the PALF should be treated with alkali or silane before using them with the polymer matrix. There are many types of the PALF that are available to the researchers to introduce them with the polymer matrix. To name a few, *Bromelia* sp., *Ananas comosus* var. comosus, Bilbergis sp., *Ananas comosus* var. bracteatus, *Ananas comosus* var. erectifolius, *Ananas macrodontes* × Primavera. The cross-sectional view of the natural fibers of PALF can be seen in Figure 7.2.

Figure 7.1 Photograph of a pineapple plant.

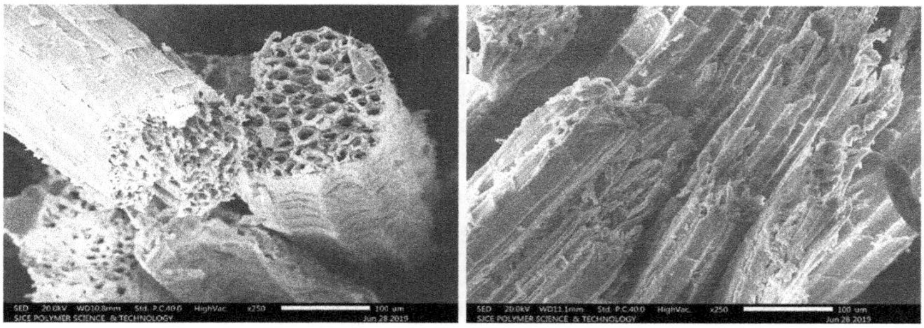

(a) (b)

Figure 7.2 (a) Cross-sectional view of the PALF and (b) PALF fiber bundle.

Table 7.1 Properties of PALF properties.

Properties	Pineapple fiber
Tensile strength (MPa)	387–1486
Youngs modulus (MPa)	29.8–81
Elongation at fracture (%)	2–3.5
Cellulose content (%)	64–67
Lignin content (%)	4
Diameter (mm)	0.6–1.1
Density ($g\,cm^{-3}$)	1.3

The tubular structure of the PALF and other natural fibers are responsible for the reduction of the tensile strength and other mechanical properties. However, the biodegradability aspects makes the natural fibers in general to gain the advantage for the automotive manufacturers to introduce them with the polymer matrix. Table 7.1 shows the PALF properties.

7.2.2.2 Resins

Resin or matrix materials are of various sorts like metal matrix, ceramic matrix, and polymer matrix. Polymer matrices are most usually utilized as a result of cost-effectiveness, simplicity of creating parts with higher structural complexity with the aid of less tooling cost, and likewise have superior room temperature properties when contrasted with ceramic and metallic counterparts.

Thermoplastic materials are shaped by expansion polymerization. Thermoplastics mollify or intertwine when heated, solidify, and become hard after cooling. Dissimilar to thermosets, thermoplastics can be adjusted or reused upon the need. They have a more extended timeframe of realistic usability and higher crack durability than thermoset matrices. Thermoplastic matrices have high viscosity and less creep resistance when contrasted with thermosets [2].

Thermoset resins are generated because of formation of amorphous cross-connected polymer matrix due to irreversible chemical transformation of resins. Because of substantial molecular structures, thermoset resins give outstanding electrical insulation and thermal insulation. Thermoset resins have lower viscosity that permits appropriate fibers to wet, notable thermal stability, and creep resistance. The most generally utilized thermoset resins are epoxy, unsaturated polyester, vinyl ester, and phenolic. Typically, the thermoset matrix can be detailed to give a broad scope of properties depending on the prerequisite.

Epoxy resin has great bond property contrasted with different resins. Moreover, it has low shrinkage after curing, excellent chemical resistance, astounding mechanical properties. Epoxies have been utilized for cutting-edge composites because of their attachment to a wide variety of fibers, with predominant mechanical and electrical properties, and excellent execution at elevated temperatures. Epoxies are costly contrasted with unsaturated polyester and less impervious to dampness. Unsaturated polyester has the benefits of minimal cost, ease of handling, and good chemical resistance with sensible mechanical properties. Unsaturated polyester and epoxy make around 85% of the fiber-strengthened polymer composites. Vinyl ester has the chemical backbone of epoxy and the curing mechanism of unsaturated polyester, thus its improved protection from compound assault and simplicity of manufacture, and it is stronger than unsaturated polyester and more affordable than epoxy. Vinyl ester has preferred protection from dampness assimilation over unsaturated polyester, and the holding capacity is not acceptable when contrasted with epoxy to a wide range of fibers. Vinyl ester offers great mechanical properties and magnificent corrosion resistance. The bonding ability of vinyl ester is acceptable to glass fiber yet less productive with carbon or Kevlar fibers. The phenolic resin cures through a condensation reaction, which produces water during the reaction. It has fantastic properties like high temperature and creep resistance, excellent thermal protection, and sound damping properties, in addition to top-notch fire properties.

7.2.2.3 Fillers

The essential preferred position of utilizing filler material in composites is to diminish the general expense of the composite. Besides, to reduce the cost, filler materials additionally fill in as significant ingredients, which improves the performance of the composite by increasing the crack resistance, diminishing shrinkage, affecting mechanical strength, improving imperviousness to fire, and so forth. Significant filler materials utilized in composite assembling are calcium carbonate, kaolin (earth), aluminum-trihydrate, and so on.

7.2.2.4 Additives

Additive materials are predominantly used to modify and tailor the material properties of the composite. Incorporating additives into the matrix/resin system enhances the processability and/or durability of the composite. Additives improves the performance or enhances a specific property. It decreases the overall cost of the product. Various additives widely used to boost the thermal and electrical conductivity of resins include graphite powder, chopped carbon fiber, or carbon nanotubes.

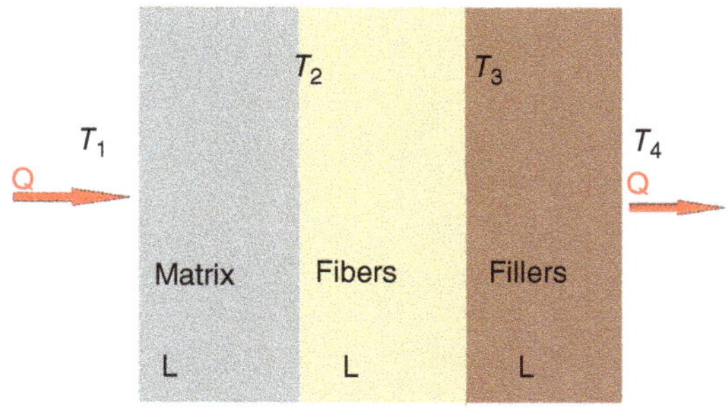

Figure 7.3 Heat transfer through rectangular slabs of three different materials.

7.2.2.5 Thermal Conductivity of the Composites

The thermal conductivity of the composite materials is very complicated to define in a theoretical equation. However, the concepts can be included to obtain the thermal conductivity. The composites contain matrix, fibers, and fillers in combination. Since there is a material change, this leads to the generation of thermal resistance. Hence, the polymer composites have wide applications in thermal insulations. It is a generalized phenomenon that the heat transfer always happens from the region of higher temperature to lower temperature. From Figure 7.2, it is evident that the heat transfer in a composite material involves a complicated structure. In order to define the thermal conductivities completely, the thermal conductivities of the individual components of the materials are required. Let us consider the composite to be a rectangular bar that has three layers, as shown in Figure 7.3.

Let us assume that the matrix is exposed to the higher temperature region, and the fillers are exposed to the lower temperature region. The heat transfer through the mentioned combination is the same through all the layers, and the overall conductivity can easily be defined as follows:

$$Q = \frac{K_{\text{Overall}} A\,(T_1 - T_4)}{3L} \tag{7.3}$$

Here, Q is the heat transfer, k is thermal conductivity W m^{-1} K^{-1}, A is the area normal to heat transfer, T is the temperatures in Kelvin, and L is the length of each rectangular bar.

Now, heat transfer through each bar is

$$Q_1 = \frac{K_{\text{matrix}} A(T_1 - T_2)}{L} \tag{7.4}$$

$$Q_2 = \frac{K_{\text{fiber}} A(T_2 - T_{3)}}{L} \tag{7.5}$$

$$Q_3 = \frac{K_{\text{filler}} A(T_3 - T_{4)}}{L} \tag{7.6}$$

The energy flow here is constant through the slabs; hence, upon expressing all the equations in terms of T_1 and T_4. We get

$$K_{\text{Overall}} = \frac{3}{\left(\frac{1}{K_{\text{matrix}}} + \frac{1}{K_{\text{fiber}}} + \frac{1}{K_{\text{filler}}}\right)} \tag{7.7}$$

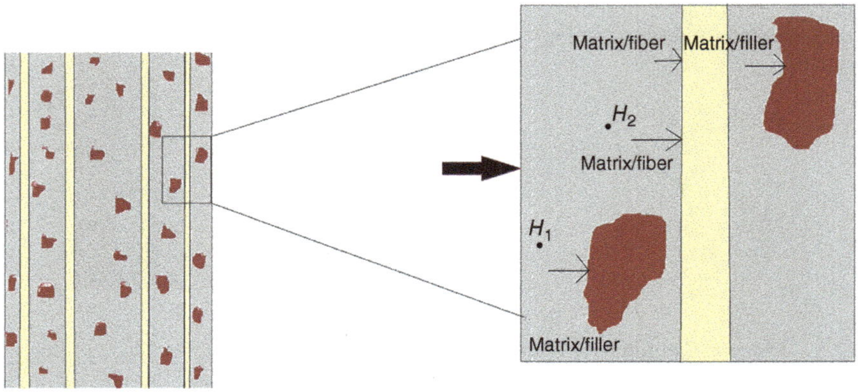

Figure 7.4 The composite material heat interaction.

Equation (7.7) is applicable only if the material system is like a composite bar in Figure 7.3, whereas the thermal conductivity still becomes complex if we have the system as we see in Figure 7.4. In Figure 7.4, the H_1, H_2 is the heat interactions between the matrix/filler and matrix/fiber heat interaction. This makes it a complex equation in determining the thermal conductivity theoretically. One more point to be observed here is that the nanoparticle is randomly distributed, and it is very difficult to estimate the thermal conductivity.

7.3 Improving the Thermal Properties of Epoxies

The thermal properties of epoxies can be modified suitably by alloying with polymers; or by reinforcing them by fibers (natural or synthetic); or by adding the fillers of micro-, nano-, or both micro- and nano-sized; or by the combination of the above three. However, for ease of understanding, this section is discussed under five sub-headings.

7.3.1 Modification of Epoxies with Other Polymers or Additives

Sun et al. [4] examined the thermal behavior of epoxy blended with thermoplastic polysulfone (PSF) and studied the differential scanning calorimetry (DSC), dynamic mechanical and thermal analysis (DMTA), and thermogravimetric analysis (TGA). With regard to DMTA, their results revealed that the increasing loading of PSF in the blends amplified the storage modulus. However, it is below that of the host epoxy resin. Further, they observed a single peak in tan δ curves, implying one T_g of the blends. T_g of PSF/epoxy blends is lower than that of host epoxy, and it augmented with the enhancement in PSF loading. Similar T_g results were observed under DSC characterization. TGA studies revealed two weight-loss stage patterns; initial weight reduction is importantly due to the degradation of the epoxy resin and the second reduction is due to the superior heat resistance of the PSF. Further, they observed that the thermal stability of the PSF/epoxy blends were enhanced by a small quantity in accordance with that of the pure epoxy resin.

Cavasin et al. [5] explored the thermal behavior of epoxy in the form of a foam that acts as an insulating thermal layer for a glass fiber-reinforced polymer under DMTA and TGA. They observed that the formation of the foam did not substantially dictate the thermal properties and mechanical properties of a crosslinked epoxy, and the T_g was not altered by the foam formation. The TGA plot did not present any substantial effect on the thermal stability of the foams when compared to the epoxy. This was a significant result as it is based on the fact that the foam should not influence the high thermal stability of the epoxy material.

Hong et al. [6] investigated the T_g of epoxidized natural rubber (ENR, 25 mol% epoxidation) modified epoxy systems by DSC. The presence of ENR in the epoxy matrix decreases the T_g of the epoxy phase, and the T_g decreases with the increasing ENR. The T_g of the blends (5, 10, and 15 parts of ENR) under study were 130, 121, and 117 °C, respectively, and are all lower than the T_g of neat epoxy, i.e. 134 °C.

7.3.2 Modification of Epoxies by Reinforcing Fibers

Sutrisno et al. [7], investigated the thermal properties of sago fiber-epoxy composites containing 0%, 9%, 13%, 17%, 20%, 23%, and 100% w/w epoxy resin. They reported that the thermal decomposition temperature of sago fiber is broad, as the sago fiber thermal damage is completely commanded by cellulosic deterioration that was later followed by hemicellulosic deterioration that occurs in the range of 275–370 °C. However, they reported that the epoxy resin decomposition starts above 350 °C.

Jain et al. [8] observed four steps of degradation in PALF reinforced epoxy composites with varying fiber loading of 5%, 10%, 15%, and 20%. They reported the first degradation temperature in the range of 50–100 °C that was primarily due to dehydration from the specimen studied. The second segment of weight reduction was noted at a temperature range of 255 –320 °C, in which this segment can be attributed to the decomposition of hemicellulosic or debonding of glycosidic linkages of natural fiber. A sudden reduction in weight of the fiber corresponding to the deterioration of cellulose, lignin, and other complex structures along with the strong linkages was noted within the temperature bandwidth of 320–395 °C. The final degradation was noted with of 395–800 °C, which includes the final deterioration of lignin and aromatic rings with various branches and may remain as ash and residue after thermal analysis. Further, they compared the investigations made by other researchers [9–12] and observed a nominal change in degradation temperature with an increase in PALF loading.

7.3.3 Modification of Epoxies by Micron-sized Particulate Fillers

Yasmin and Daniel [13] reported that the addition of 2.5 and 5 wt% of graphite platelets in epoxy demonstrated a gradual increase at the glass transition temperature from 143 to 145 and 146.8 °C, respectively, because of the excellent adhesion between the epoxy and graphite platelets, which restricted the segmental motion of crosslinking under loading.

7.3.4 Modification of Epoxies by Filling Nano-Sized Particulate Fillers

Many researchers have examined the influence of nanomaterials addition on the properties of polymer based nanocomposites [14–16]. However, many studies presented that the addition of nanomaterials augmented the polymer nanocomposites properties [17, 18]. For example, Wan et al. [19] reported that the incorporation of 0.5 wt% pristine graphene oxide enhanced the thermal degradation temperature of epoxy nanocomposites by 13 °C. Carbon nanotubes with 7 wt% in ENR has significantly increased the thermal degradation temperature of ENR composite by 18 °C [20]. Xu et al. [21] reported a 4 °C raise in the thermal degradation of epoxy filled with 15 wt% of nano-cellulose.

7.3.5 Modification of Epoxies by Hybridization

Jesuarockiam et al. [22] investigated the thermal behavior of Kevlar/Coconut sheath/epoxy (KCSE) composites laminates with graphene nanoplatelets (GNP). Different wt% such as 0.25–0.75 wt% in steps of 0.25 wt% of GNP were included with epoxy resin through ultra-sonication. KCSE composites with different weight ratios (100/0, 75/25, 50/50, 25/75, 0/100) were prepared alongside retaining a fiber to matrix ratio at 9 : 11. Thermal detorioration and visco-elastic properties were examined using TGA, DSC, and dynamic mechanical analysis (DMA). They revealed that 25/75 KCSE composites at 0.75 wt% of GNP and Kevlar/epoxy composite demonstrated similar thermal stability. It has been substantiated with the DSC experiment that GNP act as a thermal barrier. However, DMA results revealed that the 50/50 KCSE composites at 0.75 wt% of GNP unveiled similar visco-elastic behavior in contrast with Kevlar/epoxy (100/0) composites at 0 wt% GNP due to effective crosslinking, which improves the stress transfer rate. They concluded that the Kevlar could be effectively (50%) replaced with CS with an optimal GNP loading for structural applications.

Chinnasamy et al. [23] investigated the thermal behavior of glass, kevlar, and kevlar/glass fiber epoxy composite laminates with and without nanoclay. They concluded that the hybridization of base composites (epoxy with glass fiber and epoxy with Kevlar fiber) with modified Kevlar/glass fiber improved the thermal properties and increased the glass transition temperature without affecting the thermal stability. Detailed literature survey is reported in Table 7.2.

7.4 The Thermal Properties of PALF Composites

7.4.1 Thermogravimetric Analysis of PALF Composites

The TGA provides insight into the changes in the mass of the material under study with the temperature. In this chapter, we narrow our focus only upon PALF-based composites and consolidated the results in this section. Luo and Netravali studied the mechanical and thermal properties of the PALFs in poly(hydroxybutyrate-*co*-valerate) resin. From TGA curves, they observed that

Table 7.2 The literature survey of the thermal properties of the epoxy in the past 10 years.

Epoxy-based composites	Properties evaluated	Remarks	References
Aerospace-grade epoxy with single-walled carbon nanotubes	Mechanical and thermal properties	Functionalized composites noticeably beat the other two ratios studied in this experimental work by offering higher mechanical properties and T_g • Epoxy filled with 0.2 wt% SWCNT content offers the most improved mechanical properties by improving fracture toughness by 40%, maximum tensile strength by 34%, and tensile strain at ultimate load by 54%, respectively	[24]
DGEBA–DDS epoxy with single-walled carbon nanotubes	Mechanical and thermal expansion	• The functionalized SWNT-epoxy nanocomposite portrayed a 42% decrement in thermal expansion laterally in the nanotube direction than that observed in the nonfunctionalized SWNT/epoxy	[25]
Epoxy (ED-20) with boric acid	Thermal properties	The incorporation of fine powder of boric acid enriched the thermal stability of epoxy composites	[26]
Epoxy resin ED-20 with aluminum nanopowder and boric acid	Mechanical and thermal characteristics	Found enhanced thermal stability as well as mechanical characteristics of the epoxy modified with nano-aluminum and boric acid fillers • Boric acid modified epoxy composite showed the best thermal stability • Combined fillers, namely boric acid and nano-aluminum fillers in the epoxy, showed the highest bending properties	[27]
Modified epoxy with boron nitride modified by octadecyl trimethyl ammonium bromide and GNPs	Thermal properties	• For the mass ratio of 6 : 4 (BN/GNP) filled composite, the thermal conductivity attained 0.48 W m^{-1} K^{-1}, which is 108.7% more than the un-modified epoxy • The modified epoxy composite with boron nitride retains the exceptional electric-insulating property • TGA and DSC experimental data showed that the addition of fillers (BNOTAB/GNP) could improve the thermal stability of epoxy composites	[28]
Epoxy with 2D graphitic fillers	Thermal conductivity	• The outstanding thermal conductivity of 2D graphitic materials makes them promising candidates to obtain highly thermally conductive epoxy-based composites	[29]
Irradiated (doses of 30, 100, and 300 kGy) ED-20 cured by polyethylene polyamine	Thermal and mechanical properties	• The thermal properties of the epoxy marginally augmented after irradiation (heating in air) • Tensile properties such as tensile strength/modulus of the epoxy increased by the action of the electron beam up to a dose of 100 kGy and then diminished	[30]

(Continued)

Table 7.2 (Continued)

Epoxy-based composites	Properties evaluated	Remarks	References
Organoclay/epoxy	Thermal properties and corrosion resistance	• Organoclay loaded composites augmented the thermal conductivity and corrosion resistance • The thermal stability of the organoclay loading (1 phr) composite showed the best performance • The corrosion resistance of composite change with the inclusion of organoclay and 1 phr in the composite shows the best result	[31]
Montmorillonite/Epoxy	Electrical and thermal properties	• The MMT filler influence the thermal properties of the epoxy composites and the experimental results displayed improved thermal conductivity and thermal stability • Unmodified MMT and surface modified MMT, displays low dielectric permittivity, low dielectric loss, and enhanced dielectric strength	[32]
Nano-molybdenum disulfide/epoxy	Thermal expansion and mechanical properties	• Epoxy composite filled with 0.5 wt% MoS_2 shows a 35.8% lessening in CTE than that of neat epoxy • The inclusion of a small weight fraction of MoS_2 (i.e. 0.1%) in the composites augmented the tensile and flexural strength by 39.2% and 9.0% correspondingly • Epoxy filled with 0.1 wt% MoS_2 shows a 7.39% increase in T_g	[33]
Short hair fibers reinforced epoxy	Dielectric and thermal properties	• With increase in fiber weight % into the matrix, the thermal conductivity of the composites curtailed with increase in the fiber loading • The dielectric constant value of the composites was also influenced substantially by the fiber loading	[34]
Polydopamine-decorated hexagonal boron nitride (h-BN)	Thermal, mechanical, and wear behavior	• Thermal analysis presented for all h-BN-based composites possessed stability thermally • h-BN filler improved the mechanical and properties with the superior wear resistance of epoxy composites	[35]
Hexagonal boron nitride/epoxy	Thermal conductivity	• The incorporation of h-BN particles boosted the thermal conductivity of epoxy composite successfully • For the epoxy composite with h-BN of 15% by weight, the thermal conductivity was up to 0.6264 $W m^{-1} K^{-1}$, augmented by 223.72% than that of unfilled epoxy resin	[36]
WEDM microparticles/epoxy	Thermal and mechanical properties	• Epoxy based composite loaded with 6 wt% WEDM microparticles presented improved thermal properties by 87.3% • Furthermore, WEDM particles modified epoxy composites showed superior tensile as well as flexural strengths	[37]

(Continued)

Table 7.2 (Continued)

Epoxy-based composites	Properties evaluated	Remarks	References
C-class fly-ash/epoxy	Mechanical, thermal, and dielectric behavior	• Incorporation of fly-ash particle-modified epoxy enhanced the mechanical and thermal properties and declined the insulation property • The dielectric test results reported that fly-ash in epoxy significantly improved the dielectric constant	[38]
Hexagonal boron nitride/epoxy	Thermal conductivity	• The experimental results showed that the boron nitride processed by 8 h milling time had the superior thermal conductivity and it is around three times greater than that of neat epoxy • Furthermore, the modified BN was reported to have a better augmentation of the mechanical properties of epoxy composites than that of pure BN	[39]
Grafted montmorillonite-multiwalled carbon nanotube (MMT-MWCNT)/epoxy	Flame resistance, dynamic mechanical and the thermal degradation properties	• The flame resistance analysis discovered that, when compared to the pristine epoxy, smoke production rate, the intensity of heat release rate, and carbon monoxide release rate of EP/MMT and EP-MWCNT truncated by 7.4%, 26.3%, and 13.9% • Furthermore, the EP/MMT and EP-MWCNT showed higher mechanical properties against the pristine epoxy. The dynamic mechanical properties of EP-MMT and EP-MWCNTs were enhanced by 21.5% and declined by 34.2%, correspondingly	[40]
Graphene/epoxy	Thermal conductivity	• A novel multi stage predictive method was established for thermal conductivity of graphene/epoxy nanocomposites • The predicted and experimental results of as-received and surface unmodified graphene blended epoxy composites with various filling of graphene in the epoxy matrix were compared and analyzed	[41]

the moisture content of the fibers is higher, and the PALF has no effect on the decomposition of the poly(hydroxybutyrate-co-valerate) (PHBV) resin due to their differences in the temperature of decomposition. The thermal decomposition of PHBV occurs at 270–273 °C, whereas that of fibers at 334 °C. Also in the hybrid composites, the PALF has no effects on the decomposition of the matrix, which means it was easier to differentiate the deterioration of resin and fibers [42]. From the TGA thermo-plot of the host polycarbonate (PC), PALF and PC/PALF composites indicated that have PC has a one-stage thermal deterioration i.e. about 440–520 °C. TGA was done on the PALF alone and it degraded in three stages.

At first, it began from 40 to 130 °C as long as the consumed dampness was taken out from the filaments. The subsequent thermal change was in the temperature band width of 195 to 360 °C, which was ascribed to the debasement of cellulosic and hemi-cellulosic substances. The last stage was in the temperature scope of 360 to 470 °C, in which the degradation was a direct result of the debasement of non-cellulosic materials. The PALF composites delineated the beginning of thermal deterioration almost at 270 °C. This could be deciphered as the composites have low thermal dependability with host PC matrix on account of the low thermal stability of PALF. Almost every composite followed a similar pattern of thermal degradation contrasting with the one of the unfilled. PC with the high alkali-treated PALF content (i.e. 20 wt%) were thermally analyzed and discovered that the degradation initiation temperature occurred at lower temperature ranges when contrasted with different composites under study. However, 20 wt% silane treated PALF composites outperformed various combinations in better thermal stability [43].

The study was conducted with high impact polystyrene (HIPS)/PALF with two different compatibilizing agents (polystyrene-*block*-poly(ethylene-ranbutylene)-*block*-poly(styrene-*graft*-maleic anhydride) and poly-(styrene-*co*-maleic anhydride). The TGA was performed from 25 to 800 °C on HIPS/PALF composites. The HIPS has only a single stage of weight loss initiated at 341 °C and terminated at 483 °C with a peak at 418 °C. Ninety percentage of weight loss was found. The thermal decomposition of the fibers happens in the two-stage. The first stage was at 25–103 °C, resulting in weight loss due to the moisture content. The second stage occurred from 126 to 542 °C with peak transition at 339 °C – the decomposition of the cellulose. The composite decay happens with three steps. Later the effects of compatibilizer were also accounted. The reduction in weight in the first stage is 1.1%; the second stage is 31–36%, and in the third stage, 48–53%. The residual weight at 800 °C increased from 9% to 12–15%. However, the benefits of the PALF in improving the thermal properties can be observed in thermoplastics [44].

TGA of the polypropylene/PALF/nanoclay composites presented a furtherance in the composite's properties. The polypropylene had the first decomposition at 246 °C. The use of natural fibers usually has initial weight loss in temperature lesser than 100 °C because of moisture evaporation; however, in this study, the authors treated the fibers with the detergent that probably prolonged the temperature of decomposition. The addition of PALF had a small increment in the first decomposition temperature (249 °C). The addition of nanoclay improved the first decomposition temperature to 252.6 °C. However, significant improvement in the maximum decomposition temperature from 371.7 to 400.1 °C was observed [45]. Similarly, the efforts were made to improve the properties with epoxy resin. The literature findings are observed in Table 7.3.

7.4.2 Dynamic Mechanical Analysis of PALF Composites

The dynamic mechanical analysis of the composites helps to study the viscoelastic behavior of the composites, especially polymer matrix. From DSC analysis, the glass transition temperature of HIPS was found to be 105 °C, whereas that for PALF

Table 7.3 The details of the thermogravimetric analysis of epoxy/PALF composites.

Materials	Fabrication	TGA remarks	References
Six variety PALF	None	This article deals with the thermal properties of the six different varieties of PALFs. They incorporated a 10 °C min^{-1} heating rate. 7% reduction was observed as the volatile loss where the curve converts from the linear behavior to the decreasing curve. However, the usual literature reported 3–5%. The reason for the higher reduction was not drying the fibers before preparing the composites. The authors define the shoulder area where the onset of hemicellulose, lignin, pectin, and initiation of cellulose degradation. The second degradation was observed at a temperature of 375–500 °C. The *Ananas comosus* var. erectifolius has better thermal stability when compared to the other variants. Though the better performance was portrayed by *Ananas comosus* var. erectifolius, other species also were significantly closer to the actual value (within 10%)	[46]
Epoxy + PALF	Composites were prepared through hand layup using a mold	The PALF composites were subjected to thermal degradation. The primary reduction occurred at 35–100 °C. This reduction in mass can be attributed to the evaporation of absorbed moisture. 200–300 °C glycosidic linkage was decomposed, and post 400 °C; the lignin content was decomposed. The epoxy had the decomposition at the temperature range of 350 °C. The multiple branches were observed in the TGA curve in the temperature range of 400–500 °C because of the aromatic rings of the lignin	[47]
Epoxy+ PALF	Alkali treated PALF (3, 5, 7 and 10 wt% alkali treatment)	Alkali treated PALF was analyzed and also compared with untreated PALF. There were four steps of degradation. The first degradation initiated at a temperature range of 0–100 °C, the second degradation was in range of 230 °C–350 °C, third degradation was in the range of 350 °C–420 °C, and lastly after 450 °C. It can be observed that when fibers are treated with alkali, the thermal degradation of fiber varies. The plot for 7 wt% alkali treated PALF showed the maximum stretch up to around 720 °C. This affirms the best thermal resistance among all the composites studied. The intensity of the peak represents the extent of change in mass within a short period. The second stage gives information on the extent to which glycosidic bonds are broken, and the third stage follows with hemicellulose and cellulose degradation. At the final stage of degradation, lignin degrades until the specimens convert into carbon char	[48]

(Continued)

Table 7.3 (Continued)

Materials	Fabrication	TGA remarks	References
Epoxy+ PALF powder	The prepreg was done with two roll mill maintained at different temperatures. Later it was compression molded.	In this article, the PALF fibers, palf powder, and nonfibrous material were fabricated into a composite. Later, they were tested for the thermogravimetric analysis. First, reduction happened at 80 °C; this was attributed to the evaporation of the water content leading to a decrease of 6–7% reduction in weight. The fiber decomposition was initiated at 280 °C, and it was documented that the further increase in the temperature will lead to the faster reduction of the weight of the PALF fibers than that of the palf fiber powder and nonfibrous units. The larger percentage of residue in WGL and nonfibrous after the second step loss indicates a more significant remaining amount of non-cellulosic cell contents	[49]
Epoxy + PALF/ Flax +Cellulose micro filler	The compression molding technique was used for fabrication. The conditions are 14 MPa and 120 °C.	Hybrid composites of PF with epoxy that was incorporated with Cellulose micro filler (CMF) presented notable thermal stability in the different steps. Initially, unfilled fiber/resin composites of 30 wt% PALF showed lesser 50%, 70%, and 75% degradation temperature with 387.73, 434.8, and 468.81 °C in comparison to CMF added composites that had 387.73–391.08, 434.81–454.81, 468.81–553.36 °C and the residual percentage from 17.69 to 24.35%. The hybrid combinations with 30 wt% PALF/Flax observed noteworthy thermal stability enhancement at 2% CMF	[50]
Epoxy +PALF +banana +Glass fiber	Alkali treated fibers (NaOH soaking for 2 h), the layup technique was used to with the help of the hot press.	For banana based composite samples, the significant deterioration occurred in the temperature range of 273–427 °C with a peak at 349 °C; for the PALF composite specimens, degradation induced between the temperature ranges of 253–426 °C at a peak at 341 °C. It was noticed that there were four stages of thermal deterioration. The initial mass loss was recorded in the range of 30–100 °C, this could be due to water content evaporation. The succeeding stage is in range of 130 and 380 °C for composites, due to the degradation of the components like hemicellulose, cellulose, and lignin. The cellulose decomposition usually happens after the hemicellulosic decay. At the final stage, only the inorganic materials like NaOH will be remaining in the carbonaceous outcome of deterioration	[51]

(Continued)

Table 7.3 (Continued)

Materials	Fabrication	TGA remarks	References
Epoxy+ PALF+ $CaCO_3$	*n*-$CaCO_3$ was dispersed in epoxy by ultrasonication, and the compressive load was applied for 24 h, and post-curing was done	Thermogravimetric analysis indicated that *N*-$CaCO_3$ filled PF/Ep composite showing better thermal-stability compared to unfilled PF/Ep composite. Initial weight loss of 5% was observed for unfilled PF/Ep, and *N*-$CaCO_3$ filled PF/Ep below 200 °C. The initial weight loss of 10% happened below 300 °C; in this stage, water molecules are evaporated from the fibers. Beyond this temperature, thermal stability is slowly declining, and decomposition of fiber occurs. The first stage of decomposition occurs at a temperature ranging from 300 to 400 °C; at this stage, the thermal depolymerization of pectin, hemicellulose, and the cleavage of glycosides links of cellulose. In the first stage of decomposition, a mass reduction of 50% was observed. The second stage of decomposition occurs at a temperature ranging from 400 to 500 °C; at this stage, the decay of a-cellulose and corresponding mass reduction of 87.5% was observed. In the DTG curve of unfilled PF/Ep composite, it can be appreciated that maximum cellulose decay happens at temperature 351.95 °C and in N-$CaCO_3$ filled PF/Ep composite decomposition occurs at a temperature of 359.23 °C. Hence, nano-$CaCO_3$ filled PF/Ep composite degrades later than the Unfilled PF/Ep composite. As a consequence, the thermal-stability of *N*-$CaCO_3$ filled PF/Ep composite found to be greater than unfilled PF/Ep composite	[52]
Epoxy+ PALF+ Nano-silica+ Mahua oil	The Mahua oil was blended with epoxy by stirring at 80 °C. The nano-silica was dispersed in the matrix, and hand layup was used to fabricate the three fabric layered composites	The epoxy had three-stage of thermal decomposition, whereas the addition of mahua oil led to a decrease in mass stability because of the evaporation of water and the ether content of the matrix. The addition of surface-treated fibers enhanced the decay temperature mainly because of the higher heat capacity of the PALF. Upon the addition of the nano-silica, absorption of the heat resulted in enhancing the decay temperature	[53]
Epoxy+ PALF+ Boron carbide+ rubber	A hand layup technique was incorporated to prepare composites	The Thermal stabilities (initial, mid, and final) were found improved with the addition of PALF, B_4C, and rubber microparticles. The improvement in the thermal stability was attributed to the specific heat absorption of the B_4C particles. Carboxyl-terminated acrylonitrile butadiene addition further improved the absorptivity that led to an increase in temperature stability	[54]

was 75 °C. The composite has the highest of 123 °C. The compatibilizing agent has a detrimental effect on the T_g of the composites. The melting point also follows a similar trend, but the fibers treated with NaOH, the melting point increased by only 2 °C [42]. Polypropylene reinforced with PALF and nanoclay has shown notable results in the DMA. The addition of the PALF had improved the storage modulus. However, the addition of nanoclay improved this property further, and the attributed reason is that the exfoliation of the nanoclay. These reinforcements have helped to reduce the rate of decomposition of the polymer matrix because of increased resistance toward molecular mobility. In the reinforced composites, the peak widens because the fibers hold a lot of stress and reduce the strain on the composites [45]. Asim et al. discussed that the storage modulus of the pure resin was found to drop suddenly with the increase in the temperature. However, it is observed that the fibers improved the storage modulus – the 60 wt% of fiber loading lacked the proper wettability and had a poor performance, whereas 50 wt% had adequate bonding with the phenolic resin and projected the highest storage modulus. The treated fibers, on the other hand, improved the fiber surface roughness and the improved interfacial interaction between the fiber and the matrix material. This hindered the molecular motion and increased storage modulus. The silane treatment improved the dynamic glass transition temperature and high-temperature rubbery state was stabilized. The loss modulus showed higher peaks that eventually presented the polymer molecular chain diversity in the three-dimension [55]. Uma et al. have discussed different combinations of a hybrid with additional loading of PALF, and GF composites were studied for DMA. The storage modulus of glass fiber is greater than that of the PALF. The 0.2 vol% of fibers enhanced stiffness but reduced the damping effect. GF–PALF–GF combination even retained noteworthy stiffness at the higher temperature zone. Mainly concerning damping properties, significant differences were not observed among the different varieties of the GF/PALF hybrid. The composites with higher reinforcement efficiency result in higher activation energy – similarly, the composites with good interfacial bonding [56]. Table 7.4 shows the literature findings of the DMA of epoxy/PALF composites.

7.4.3 Differential Scanning Calorimetric Analysis of PALF Composites

The differential scanning calorimetric analysis is a thermal characterization technique used to measure the heat transfer in a material, usually with respect to temperature or time. In this chapter, the DSC analysis was scrutinized with care to the PALF reinforced polymer composites. Luo and Netravali have discussed that in the incorporation of PALF into PHBV resin, there was no improvement in the melting temperature and crystallization temperature when compared with neat PHBV resin [42]. Nadirah et al. have studied the glass and crystalline transition and melting temperature of the PALF. The DSC plot showed the endothermic reaction at 80 °C. The melting point of the composite was observed at around 290 °C [60]. In the studies performed by Siregar et al. the glass transition temperature of HIPS as measured through DSC was 105 °C, and that of PALF was 75 °C, whereas the combination of HIPS and PALF collectively yielded the temperature 18 °C more than

Table 7.4 The details of the dynamic mechanical analysis of epoxy/PALF composites.

Materials	Fabrication	DMA remarks	References
Epoxy+ PALF	PALF fiber composites were prepared using a hand layup technique	The fibers improved the storage modulus and dynamic glass transition temperature. The three dimensional crosslinking and the interfacial adhesion are responsible for the improvement in the property	[47]
Epoxy+ PALF+ Banana Fibers (BaF)+ GF	Alkali treated fibers (NaOH soaking for 2 h), the layup technique was used to with the help of the hot press	For PALF and BaF, the storage modulus decreased with temperature. However, the strength and rigidity were retained because of the molecular motion hindrance. There was a shift in the dynamic glass transition temperature. The lowest fiber content projected the higher glass transition temperature. This is attributed to the fact that there is an increase in the rigidity of the composite materials. The PALF fiber had absorbed the resin; hence, as the fiber content increased, there was a reduction in the damping coefficient.	[51]
Epoxy+ PALF	PALF powder, epoxy, and hardener were mixed and poured into the mold	The three different loadings of the PALF were analyzed. The storage modulus was enhanced with an increase in the fiber content. The reason that was attributed to the rise in the storage modulus is the interfacial bonding, molecular mobility hindrance that slowed down the stress relaxation for the composite. Storage modulus of 17 and 23 wt% of fiber had a similar storage modulus, whereas 34 wt% of the fiber loading had a considerable improvement and showed the best storage modulus. The shift was observed in the peak of the tan δ of the composite to that of epoxy. This was attributed to the increased interaction between polymer molecular chains and the PALF. 34 wt% of the PALF loading had a broader and shorter peak that was because of fair fiber distribution	[57]
Epoxy+ PALF+ Basalt fiber (BF)	The hand layup technique was incorporated to fabricate the composites	The PALF, BF, PALF/BF/PALF, and BF/PALF/BF combination of the composites were analyzed under different frequencies, 0.1, 1, and 10 Hz. As the temperature increases, the inter-fiber and intermolecular interaction varies, and the extent of interaction determines the improved thermal stability and rate of decrease in the storage modulus. The variation of the frequency affects the dynamic properties significantly. As the frequency increased, and enhancement of the storage modulus can be found, that is because of the rearrangement of the composite structure. This rearrangement helps to minimize localized stress	[58]
Epoxy+ PALF+ Coir fiber (CF)	The composites were prepared using the compression molding	The CF and PALF were oriented and properly aligned before fabrication. The improvement in the composite's dynamic thermomechanical stability increased with the addition of the reinforcements, mostly the temperatures above 100 °C, as indicated by the characteristic plateau region. The aligned PALF had a better stiffness than the CF, as demonstrated in the loss modulus curve of the composites that had a sharp peak. Additionally, the damping ability was hindered upon reinforcement as fibers restrict the epoxy mobility	[59]

the neat HIPS, which was 123 °C. The authors also introduced the melting point in the discussion. The melting point of the neat HIPS was 150 °C. Upon reinforcement, the melting point was observed to decrease by 1 °C. The effect of the compatibilizing agent was found to reduce the melting point of the PALF composites lower than that of the neat HIPS. However, when fibers are surface treated, the melting point was found to increase by a very minimal amount (2 °C) [44]. According to the studies conducted by Biswal and coauthors, on polypropylene and its composites filled with nanoclay, they observed that the addition of PALF and nanoclay with polypropylene reduced the melting point. The addition of the PALF reduced the rate of crystallization, whereas the addition of nanoclay increased the speed of crystallization [45]. Chollakup et al. studied the low-density polyethylene (LDPE) and poly polyethylene (PP) and observed that the length of the fibers had a significant impact on the enthalpy with only LDPE. In contrast, it has no substantial influence on the PP [61]. Kim et al. have discussed the PALF flour in the polylactic acid polymer and noted that the PALF flour/PLA (polylactic acid) composites showed two peaks in the DSC thermogram because of molecular weight decrease and the breakage of the chain [62]. Panyasart et al. have studied the DSC on the neat PA6 and the PALF reinforced PA6. Twenty percentage of PALF was found to be the optimum loading of the PALF as the other loading decreased the crystallization temperature. It was observed that the PALF might induce the PA6 to crystallize faster than expected. The melting temperature was slightly enhanced with the addition of PALF to PA6. However, the critical observation is that there was not a noteworthy improvement in the melting point with different fiber loading, even with or without fiber treatment [63]. Table 7.5 shows the literature findings of DSC of the epoxy/PALF composites.

7.4.4 Thermal Conductivity of PALF Composites

The thermal conductivity of a material is a material property that determines the rate of conductive heat transfer through the material held at a temperature differential. It is essential to determine the thermal conductivity of the material to choose it for the thermal applications. Mangal et al. have studied the variation of the thermal conductivity of the composites with varying fiber content and noted that fiber content reduced the thermal conductivity [65]. The PALF/GF/polyester hybrid composites were studied for the thermal conductivity and thermal resistance. The key

Table 7.5 The details of the differential scanning calorimetric of epoxy/PALF composites.

Materials	Fabrication	DSC remarks	Reference
Epoxy+ PALF	PALF short fibers were stirred with epoxy and hardener and later poured into the mold	The DSC analysis put forth the observation that the untreated PALF was having better dynamic glass transition temperature when compared to the NaOH-treated fibers because the treated fibers reduce the formation of hydrogen that was in correlation with the literature findings.	[64]

Table 7.6 The details of the Thermal conductivity of epoxy/PALF composites.

Materials	Fabrication	Thermal conductivity remarks	References
Epoxy+ PALF+ BaF	The hand layup technique was incorporated to fabricate the composites	The thermal conductivity of the material decreased with the addition of the PALF and BaF. This is probably attributed to the lower thermal conductivity of the fibers themselves. The authors also measured the thermal conductivities at the different temperatures and found that as the temperature increases, the thermal conductivity of the composites also increases	[68]
Epoxy+ PALF+ Jute+ BaF	—	The article deals with the measurement of the thermal conductivity of the various combinations of the natural fiber composites. The combinations are mentioned in Table 7.7. The thermal conductivity of the composites were observed to increase with increase in the temperature. The analysis was performed by mounting the thermocouples at various locations. Different thicknesses were also incorporated. However, in general the thermal conductivity of the material got enhanced with increasing PALF content with decreasing jute fibers and constant BaF	[69]

observation to be noticed here is that the thermal resistance decreased with the addition of the fiber. The hybridization of the composites improved the heat transfer of the composites [66]. The PALF reinforced with the natural rubber latex is very close to the commercially available thermal insulators with 1 : 3 of the fiber to matrix ratio [67]. Table 7.6 shows the literature findings of epoxy/PALF composites.

7.5 Concluding Remarks

To obtain appropriate ecofriendly materials, as of late, natural fiber-reinforced composites are considered as a viable option in contrast to traditional engineered fiber-based composites. A variety of natural fibers should be used in different polymer matrices, for example, kenaf, jute, oil palm, coir, pineapple, and other plant fibers, animal fibers, or mineral fibers. Obviously, utilizing them in structural applications or generally in the industry, it is important to test them and to determine their mechanical and thermal properties. Specifically, the thermal properties of NFRCs has gained attention in a novel exploration field given the limited available literature. Thus, this chapter intends to provide an overview of the logical outcomes on this aspect, in light of important and on-going research work, in order to provide for the perusers a structure of this logical issues. In particular, the authors focused on the influence of selected fillers and PALFs reinforced epoxy composites on thermal behavior.

Table 7.7 The details of the thermal conductivity of epoxy/natural fiber composites.

Materials	Weight/volume fraction	Thermal conductivity ($W\,m^{-1}\,K^{-1}$)	References
Epoxy+ PALF+BaF		0.2	[68]
Epoxy+ PALF+ Jute+ BaF	0/40/60 8/32/60 16/24/60 24/16/60 32/8/60 40/0/60 Shared across the volume fraction for 15 mm thickness.	0.015 0.01 0.015 0.017 0.013 0.021	[69]
Epoxy+ Kenaf	0.66	0.406	[70]
Epoxy+ Abaca	0.1 0.5	~0.29 ‘0.245	[71]
Epoxy+ Bamboo	0.1 0.5	~0.303 ~0.313	[71]
Epoxy+ Jute	0.3 0.4 0.5	0.251 0.239 0.231	[72]
Epoxy+ BaF	0.3 0.4 0.5	0.243 0.230 0.228	[72]
Epoxy+ Jute nanofibers	0 0.30	0.29 0.38	[73]
Epoxy+ Grass broom fiber	0.163 0.261 0.358	0.231 0.219 0.196	[74]
Epoxy+ Pseudo stem banana woven fibers	2 layers 4 layers 6 layers	0.025 0.028 0.033	[75]

The literatures indicated that the TGA and the dynamic mechanical analysis were predominantly studied to determine the thermal properties. This paves an avenue for the research community to identify the research gap in differential scanning calorimetric analysis and thermal conductivity of the PALF-reinforced polymer composites. The nanoparticle reinforcements can also be identified to fulfill the research gap. Clearly, the influence of nanoparticles on thermal properties have been investigated by thermogravimetric analysis.

References

1 Mirmira, S.R.R. (1999). Thermal conductivity of fibrous composites: experimental and analytical study. PhD Dissertation, Texas A&M, USA.

2 Barbero, E.J. (1998). *Introduction to Composite Materials Design*. Taylor & Francis.

3 (2001). *ASM Handbook: Composites*, vol. 21. Novelty, OH, USA: ASM International.

4 Sun, Z., Lei, X., Chen, Z. et al. (2019). Enhancing the mechanical and thermal properties of epoxy resin via blending with thermoplastic polysulfone. *Polymers* 11 (461): 1–17.

5 Cavasin, M., Giannis, S., Salvo, M. et al. (2018). Mechanical and thermal characterization of an epoxy foam as thermal layer insulation for a glass fiber reinforced polymer. *Journal of Applied Polymer Science* https://doi.org/10.1002/APP.46864.

6 Hong, S.G., Chan, C.K., Chuang, C.C. et al. (2005). The curing behavior and adhesion strength of the epoxidized natural rubber modified epoxy/dicyandiamide system. *Journal of Polymer Research* 12: 295–303.

7 Sutrisno, W., Rahayu, M., and Adhika, D.R. (2020). Thermal properties of sago fiber-epoxy composite. *Fibers* 08 (04): 1–13.

8 Jain, J., Jain, S., and Sinha, S. (2018). Characterization and thermal kinetic analysis of pineapple leaf fibers and their reinforcement in epoxy. *Journal of Elastomers and Plastics*: 1–20. https://doi.org/10.1177/0095244318783024journals.sagepub.com/home/jep.

9 Mittal, V. and Sinha, S. (2014). Effect of chemical treatment on the mechanical and water absorption properties of bagasse fiber-reinforced epoxy composites. *Journal of Polymer Engineering* 35 (6): 545–550.

10 Bilba, K., Arsene, M.A., and Ouensanga, A. (2007). Study of banana and coconut Fibers Botanical composition, thermal degradation and textural observations. *Bioresource Technology* 98: 58–68.

11 Ouajai, S. and Shanks, R.A. (2005). Composition, structure and thermal degradation of hemp cellulose after chemical treatments. *Polymer Degradation and Stability* 89: 327–335.

12 Prasad, N., Agarwal, V.K., and Shishir, S. (2016). Banana fiber reinforced low-density polyethylene composites: effect of chemical treatment and compatibilizer addition. *Iranian Polymer Journal* 25 (3): 229–241.

13 Yasmin, A. and Daniel, I.M. (2004). Mechanical and thermal properties of graphite platelet/epoxy composites. *Polymer* 45: 8211–8219. https://doi.org/10.1016/j.polymer.2004.09.054.

14 Baheri, B., Shahverdi, M., Rezakazemi, M. et al. (2015). Performance of PVA/NaA mixed matrix membrane for removal of water from ethylene glycol solutions by pervaporation. *Chemical Engineering Communications* 202: 316–321. https://doi.org/10.1080/00986445.2013.841149.

15 Rezakazemi, M., Dashti, A., Asghari, M., and Shirazian, S. (2017). H2-selective mixed matrix membranes modeling using ANFIS, PSO-ANFIS,

GA-ANFIS. *International Journal of Hydrogen Energy* 42: 15211–15225. https://doi.org/10.1016/j.ijhydene.2017.04.044.
16 Rezakazemi, M., Ebadi Amooghin, A., Montazer-Rahmati, M.M. et al. (2014). State-of-the-art membrane based CO_2 separation using mixed matrix membranes (MMMs): an overview on current status and future directions. *Progress in Polymer Science* 39: 817–861. https://doi.org/10.1016/j.progpolymsci.2014.01.003.
17 Rezakazemi, M., Sadrzadeh, M., and Matsuura, T. (2018). Thermally stable polymers for advanced high-performance gas separation membranes. *Progress in Energy and Combustion Science* 66: 1–41. https://doi.org/10.1016/j.pecs.2017.11.002.
18 Shahverdi, M., Baheri, B., Rezakazemi, M. et al. (2013). Pervaporation study of ethylene glycol dehydration through synthesized (PVA-4A)/polypropylene mixed matrix composite membranes. *Polymer Engineering and Science* 53: 1487–1493. https://doi.org/10.1002/pen.23406.
19 Wan, Y., Gong, L., Tang, L. et al. (2014). Mechanical properties of epoxy composites filled with silane-functionalized graphene oxide. *Composites Part A: Applied Science and Manufacturing* 64: 79–89. https://doi.org/10.1016/j.compositesa.2014.04.023.
20 Krainoi, A., Kummerlöwe, C., Nakaramontri, Y. et al. (2018). Influence of critical carbon nanotube loading on mechanical and electrical properties of epoxidized natural rubber nanocomposites. *Polymer Testing* 66: 122–136. https://doi.org/10.1016/j.polymertesting.2018.01.003.
21 Xu, S., Girouard, N., Schueneman, G. et al. (2013). Mechanical and thermal properties of waterborne epoxy composites containing cellulose nanocrystals. *Polymer* 54: 6589–6598. https://doi.org/10.1016/j.polymer.2013.10.011.
22 Jesuarockiam, N., Jawaid, M., Zainudin, E.S. et al. (2019). Enhanced thermal and dynamic mechanical properties of synthetic/natural hybrid composites with graphene nanoplateletes. *Polymers* 11: 1085. https://doi.org/10.3390/polym11071085.
23 Chinnasamy, V., Pavayee Subramani, S., Palaniappan, S.K. et al. (2020). Characterization on thermal properties of glass fiber and kevlar fiber with modified epoxy hybrid composites. *Journal of Materials Research and Technology* 9 (3): 3158–3167.
24 Ashrafi, B., Martinez-Rubi, Y., Khoun, L. et al. (2013). Influence of the reaction stoichiometry on the mechanical and thermal properties of SWCNT-modified epoxy composites. *Nanotechnology* 24 (26): 265701.
25 Fasanella, N. and Sundararaghavan, V. (2015). Atomistic modeling of thermomechanical properties of SWNT/Epoxy nanocomposites. *Modelling and Simulation in Materials Science and Engineering* 23 (6): 065003.
26 Visakh, P.M., Nazarenko, O.B., Amelkovich, Y.A., and Melnikova, T.V. (2015). Thermal properties of epoxy composites filled with boric acid. *IOP Conference Series: Materials Science and Engineering* 81: 1–6.
27 Nazarenko, O.B., Melnikova, T.V., and Visakh, P.M. (2016). Thermal and mechanical characteristics of polymer composites based on epoxy resin,

aluminium nanopowders and boric acid. *Journal of Physics: Conference Series* 671 (1): 012040. IOP Publishing Ltd.

28 Chen, H., Zhao, C., Xu, S., and Yang, X. (2017). Preparation and properties study of thermally conductive epoxy/modified boron nitride/graphene nanosheets composites. *IOP Conference Series: Materials Science and Engineering* 274 (1): 1–7.

29 Mu, M., Wan, C., and McNally, T. (2017). Thermal conductivity of 2D nano-structured graphitic materials and their composites with epoxy resins. *2D Materials* 4 (4): 042001.

30 Visakh, P.M., Nazarenko, O.B., Chandran, C.S. et al. (2017). Effect of electron beam irradiation on thermal and mechanical properties of aluminum based epoxy composites. *Radiation Physics and Chemistry* 136: 17–22.

31 Baiquni, M. and Soegijono, B. (2018). Thermal properties and corrosion resistance of organoclay/epoxy resin film. *Journal of Physics: Conference Series* 985 (1): 012037. IOP Publishing Ltd.

32 Jia, Z.R., Gao, Z.G., Lan, D. et al. (2018). Effects of filler loading and surface modification on electrical and thermal properties of epoxy/montmorillonite composite. *Chinese Physics B* 27 (11): 117806.

33 Madeshwaran, S.R., Jayaganthan, R., Velmurugan, R. et al. (2018). Mechanical and thermal properties of MoS2 reinforced epoxy nanocomposites. *Journal of Physics: Conference Series* 991: 1742–1749.

34 Nanda, B.P. and Satapathy, A. (2018). Effect of fiber content on the thermal conductivity and dielectric constant of hair fiber reinforced epoxy composite. *IOP Conference Series: Materials Science and Engineering* 338 (1): 012040.

35 Song, J., Dai, Z., Li, J. et al. (2018). Polydopamine-decorated boron nitride as nano-reinforcing fillers for epoxy resin with enhanced thermomechanical and tribological properties. *Materials Research Express* 5 (7): 075029.

36 Wu, X., Wang, Z., Wang, X., and Qu, X. (2018). Research on preparation and performance of epoxy resin-boron nitride thermal-conductive composite. *IOP Conference Series: Materials Science and Engineering* 394 (2): 022052.

37 Babu N.B, K. and Ramesh, T. (2019). Enhancement of thermal and mechanical properties of novel micro-wear debris reinforced epoxy composites. *Materials Research Express* 6 (10): 105358.

38 Gnanavel, M., Maridurai, T., and Kumar, K.M. (2019). Mechanical, thermal and dielectric behaviour of C-class fly-ash coarse and fine particles reinforced epoxy resin composite. *Materials Research Express* 6 (9): 095507.

39 Ma, J., Luo, N., Xie, Z. et al. (2019). Preparation of modified hexagonal boron nitride by ball-milling and enhanced thermal conductivity of epoxy resin. *Materials Research Express* 6 (10): 1050d8.

40 Xue, Y., Shen, M., Zeng, S. et al. (2019). A novel strategy for enhancing the flame resistance, dynamic mechanical and the thermal degradation properties of epoxy nanocomposites. *Materials Research Express* 6 (12): 125003.

41 Wang, D., Zhang, H., Wei, H., and Li, S. (2020). A thermal conductivity predictive method of graphene epoxy nanocomposites. *Journal of Physics: Conference Series* 1605: 012129.

42 Luo, S. and Netravali, A.N. (1999). Mechanical and thermal properties of environment-friendly "green" composites made from pineapple leaf fibers and poly(hydroxybutyrate-*co*-valerate) resin. *Polymer Composites* 20 (3): 367–378.

43 Threepopnatkul, P., Kaerkitcha, N., and Athipongarporn, N. (2009). Effect of surface treatment on performance of pineapple leaf fiber–polycarbonate composites. *Composites Part B: Engineering.* 40 (7): 628–632.

44 Siregar, J.P., Salit, M.S., Rahman, M.Z., and Dahlan, K.Z. (2011). Thermogravimetric analysis (TGA) and differential scanning calometric (DSC) analysis of pineapple leaf fibre (PALF) reinforced high impact polystyrene (HIPS) composites. *Pertanika Journal of Science and Technology* 19 (1): 161–170.

45 Biswal, M., Mohanty, S., and Nayak, S.K. (2009). Influence of organically modified nanoclay on the performance of pineapple leaf fiber-reinforced polypropylene nanocomposites. *Journal of Applied Polymer Science* 114 (6): 4091–4103.

46 Neto, A.R., Araujo, M.A., Souza, F.V. et al. (2013). Characterization and comparative evaluation of thermal, structural, chemical, mechanical and morphological properties of six pineapple leaf fiber varieties for use in composites. *Industrial Crops and Products* 43: 529–537.

47 Nagarajan, T.T., Babu, A.S., Palanivelu, K., and Nayak, S.K. (2016). Mechanical and thermal properties of PALF reinforced epoxy composites. *Macromolecular Symposia* 361 (1): 57–63.

48 Jain, J., Sinha, S., and Jain, S. (2019). Compendious characterization of chemically treated natural fiber from pineapple leaves for reinforcement in polymer composites. *Journal of Natural Fibers* 18 (2): 1–2.

49 Kengkhetkit, N. and Amornsakchai, T. (2014). A new approach to "Greening" plastic composites using pineapple leaf waste for performance and cost effectiveness. *Materials & Design* 55: 292–299.

50 Sumesh, K.R., Kanthavel, K., and Kavimani, V. (2020). Peanut oil cake-derived cellulose fiber: extraction, application of mechanical and thermal properties in pineapple/flax natural fiber composites. *International Journal of Biological Macromolecules* 150: 775–785.

51 Zin, M.H., Abdan, K., and Norizan, M.N. (2019). The effect of different fiber loading on flexural and thermal properties of banana/pineapple leaf (PALF)/glass hybrid composite. In: *Structural Health Monitoring of Biocomposites, Fibre-Reinforced Composites and Hybrid Composites* (eds. M. Jawaid, M. Thariq and N. Saba), 1–17. Woodhead Publishing.

52 Mahadevaswamy, H.S. and Suresha, B. (2020). Role of nano-$CaCO_3$ on mechanical and thermal characteristics of pineapple fibre reinforced epoxy composites. *Materials Today: Proceedings* 22: 572–579.

53 Dinesh, T., Kadirvel, A., and Thermo-Mechanical, H.P. (2020). Wear behaviour of surface-treated pineapple woven fibre and nano-silica dispersed mahua oil toughened epoxy composite. *Silicon* 12 (6): 1–10.

54 Vinothkumar, M., Jaiganesh, V., and Malarvannan, R.R. (2019). Effects of surface-modified pineapple fibre-reinforced micro B4C and CTBN rubber particle toughened epoxy hybrid composites in mechanical, impact damage, thermal and water absorption behavior. *Materials Research Express* 6 (11): 115343.

55 Asim, M., Jawaid, M., Nasir, M., and Saba, N. (2018). Effect of fiber loadings and treatment on dynamic mechanical, thermal and flammability properties of pineapple leaf fiber and kenaf phenolic composites. *Journal of Renewable Materials.* 6 (4): 383–393.

56 Devi, L.U., Bhagawan, S.S., and Thomas, S. (2010). Dynamic mechanical analysis of pineapple leaf/glass hybrid fiber reinforced polyester composites. *Polymer Composites* 31 (6): 956–965.

57 Mittal, M. and Chaudhary, R. (2018). Effect of fiber content on thermal behavior and viscoelastic properties of PALF/Epoxy and COIR/Epoxy composites. *Materials Research Express* 5 (12): 125305.

58 Doddi, P.R., Chanamala, R., and Dora, S.P. (2020). Effect of fiber orientation on dynamic mechanical properties of PALF hybridized with basalt reinforced epoxy composites. *Materials Research Express* 7 (1): 015329.

59 Luz, F.S., Monteiro, S.N., and Tommasini, F.J. (2018). Evaluation of dynamic mechanical properties of PALF and coir fiber reinforcing epoxy composites. *Materials Research* 21: 1–5.

60 Nadirah, W.W., Jawaid, M., Al Masri, A.A. et al. (2012). Cell wall morphology, chemical and thermal analysis of cultivated pineapple leaf fibres for industrial applications. *Journal of Polymers and the Environment* 20 (2): 404–411.

61 Chollakup, R., Tantatherdtam, R., Ujjin, S., and Sriroth, K. (2011). Pineapple leaf fiber reinforced thermoplastic composites: effects of fiber length and fiber content on their characteristics. *Journal of Applied Polymer Science* 119 (4): 1952–1960.

62 Kim, K.W., Lee, B.H., Kim, H.J. et al. (2012). Thermal and mechanical properties of cassava and pineapple flours-filled PLA bio-composites. *Journal of Thermal Analysis and Calorimetry* 108 (3): 1131–1139.

63 Panyasart, K., Chaiyut, N., Amornsakchai, T., and Santawitee, O. (2014). Effect of surface treatment on the properties of pineapple leaf fibers reinforced polyamide 6 composites. *Energy Procedia* 56: 406–413.

64 Fauzi, F.A., Siregar, J.P., and Tezara, C. (2016). Investigation of thermal behavior for natural fibres reinforced epoxy using thermogravimetric and differential scanning calorimetric analysis. *MATEC Web of Conferences* 78: 01042.

65 Mangal, R., Saxena, N.S., Sreekala, M.S. et al. (2003). Thermal properties of pineapple leaf fiber reinforced composites. *Materials Science and Engineering: A* 339 (1–2): 281–285.

66 Idicula, M., Boudenne, A., Umadevi, L. et al. (2006). Thermophysical properties of natural fibre reinforced polyester composites. *Composites Science and Technology* 66 (15): 2719–2725.

67 Tangjuank, S. (2011). Thermal insulation and physical properties of particleboards from pineapple leaves. *International Journal of Physical Sciences* 6 (19): 4528–4532.

68 Santosha, P.C., Gowda, A.S., and Manikanth, V. (2018). Effect of fiber loading on thermal properties of banana and pineapple leaf fiber reinforced polyester composites. *Materials Today: Proceedings* 5 (2): 5631–5635.

69 Muthukumar, K., Sabariraj, R.V., Kumar, S.D., and Sathish, T. (2020). Investigation of thermal conductivity and thermal resistance analysis on different combination of natural fiber composites of banana, pineapple and jute. *Materials Today: Proceedings* 21: 976–980.

70 Xia, C., Garcia, A.C., Shi, S.Q. et al. (2016). Hybrid boron nitride-natural fiber composites for enhanced thermal conductivity. *Scientific Reports* 6: 34726.

71 Liu, K., Zhang, X., Takagi, H. et al. (2014). Effect of chemical treatments on transverse thermal conductivity of unidirectional abaca fiber/epoxy composite. *Composites Part A: Applied Science and Manufacturing.* 66: 227–236.

72 Pujari, S., Ramakrishna, A., and Padal, K.T. (2017). Investigations on thermal conductivities of jute and banana fiber reinforced epoxy composites. *Journal of The Institution of Engineers (India): Series D* 98 (1): 79–83.

73 Sharma, A. and Patnaik, A. (2018). Experimental investigation on mechanical and thermal properties of marble dust particulate-filled needle-punched nonwoven jute fiber/epoxy composite. *Journal of Metals* 70 (7): 1284–1288.

74 Ramanaiah, K., Prasad, A.R., and Reddy, K.H. (2012). Thermal and mechanical properties of waste grass broom fiber-reinforced polyester composites. *Materials & Design* 40: 103–108.

75 Maleque, M.A., Saiful, A., and Sapuan, S.M. (2005). Effect of volume fraction of pseudo-stem banana fiber on the properties of epoxy composite. In: *Proceedings of the International Conference on Mechanical Engineering*, 28–30. Dhaka, Bangladesh: Web of Science.

8

Thermal Properties of the Grass/Cane Fiber-Based Hybrid Composites

Manickam Ramesh[1], Jaganathan Maniraj[1], and Sengottuvelu Ramesh[2]

[1] *KIT-Kalaignarkarunanidhi Institute of Technology, Department of Mechanical Engineering, Coimbatore, 641402, Tamil Nadu, India*
[2] *Jerusalem College of Engineering, Department of Mechanical Engineering, Chennai, 600100, Tamil Nadu, India*

8.1 Introduction

The biocomposites are an excellent material owing to their distinctive characteristics. It opened up an avenue for academic researchers and manufacturing engineers to manufacture a sustainable composites for industrial applications. Manufacturing of materials are one of the most energy consumable things in the globe [1]. To fulfill the requirement of the conventional material supply, the composite materials have been introduced [2]. Material usage for specific application is another type of the demand to meet requirement with viable cost, properties, etc. [3]. Therefore, the researchers tend to develop new materials to fulfill such demands [4]. Composite materials have been proposed as an alternative to other traditional materials due to their improved properties [5–7]. Several research findings have documented the advantages of cellulose fibers, such as being richly abundant in nature, nonhazardous, reusable, and low cost and supplying the requisite bonding to the matrix to dramatically boots ductile properties, flexural strength, and material impact tolerance [8–10]. Several types of plant fibers exist, for example, cotton, jute, fleece, silk, sisal, bamboo, hemp, and flax removed from stem and leaf of the plants. A lot of raw materials are likewise a concern to the manufacturing assembly, and the obligation to utilize always greener innovations has made this territory of examination fascinating [11]. Regular fibers that are classified (Figure 8.1) by their source of origin: plants, animals, or minerals.

Natural fibers are to uncover some finest material characteristics like biodegradability, high tensile strength, and stiffness. Natural fiber-based composite offers specified unique properties against man-made fiber composites such as lightweight, optimum quality, durability, environmental degradation, and recyclability. Such outcomes and the economic advantages of these composites make them major reinforcements over man-made composites for contemporary applications [12].

Natural Fiber-Reinforced Composites: Thermal Properties and Applications, First Edition.
Edited by Senthilkumar Krishnasamy, Senthil Muthu Kumar Thiagamani,
Chandrasekar Muthukumar, Rajini Nagarajan and Suchart Siengchin.

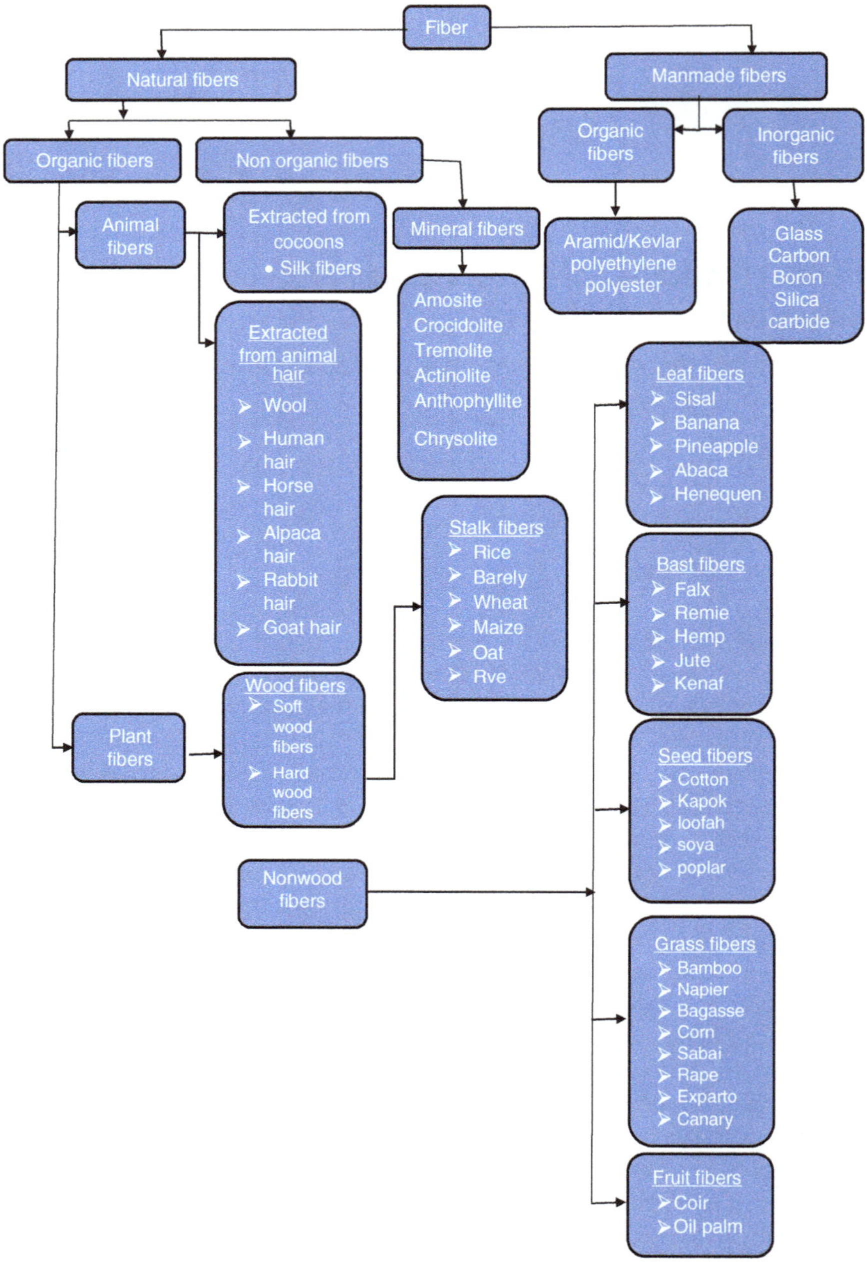

Figure 8.1 Classification of fibers.

Natural fibers have similar internal structures, with various chemical compositions, depending on the type. Their inclusion in the thermoset/plastic polymers has led to their usage in various industrial applications [13, 14]. Plant fiber-reinforced composites may also be an alternative to man-made fibers. They can provide an excellent solution to the waste minimization issue of synthetic fibers due to their improved recyclability. The main requirements to achieve the goal is sustainable crop growth, adequate facilities for fiber extraction, processing conditions, interfacial properties, part manufacturing, product design, and healthy service life. Their distinct individual characteristics should be a tough foundation for exploring new biocomposite technologies and opportunities in the coming centuries [3].

In the past, many countries have forced guidelines on green material usage to protect the environment. Common fibers are the inexhaustible assets to keep the feasibility while nonrenewable energy sources, for example, coal, oil, and gaseous petrol, are fossil fuels and can be exhausted out soon [15]. Researchers have provided details regarding the enhancement of heat conductivity of engineered fiber-based polymeric composites [16–18]. Few hypothetical and experimental models were created to foresee the heat conducting ability of materials [19–23]. Thermometer was utilized to quantify the heat conductivity of polymeric materials [19, 24]. Explicit conduction limit of composite materials can be estimated by differential scanning calorimetry (DSC) analysis [25, 26]. Truth to be told, restricted data is accessible on thermomechanical characteristics of fiber-based composites at or above atmospheric conditions [27, 28]. Henceforth, studies were at that point of the investigation on processing and properties of these fibers and their composites [29]. In an examination, the idea is to utilize grass/stem fiber in the improvement of mostly biodegradability nature with lower thermal conductivity. This is a fundamental prerequisite so as to lessen energy utilization in building hardware structures.

8.2 Hybrid Composite Materials

Hybrid composites are those where acceptable properties of at least two types of fibers are utilized to strengthen in a structure [30]. There has been a substantial increment in the quantity of experimental works that are identified with hybrid composites since 2013. The development of hybrid composites was focused on enhancing various material properties or achieving desired properties and a superior exhibition of such composite. The main point was to defeat the restriction of single filler-reinforced laminate with the different filler(s) that have comparative or better properties when compared with the underlying fiber. This idea is known as "hybridization," in which at least two fillers can be joined into a typical network to beat the inadequacies of the other filler. Notwithstanding, this idea additionally applies to the mixing of at least two polymers and strengthening with one or more fillers and including a similar filler type that has at least two sizes or measurements, known as auto-hybridization.

Hybridization is common across many domains, for example, mechanics, science, designing, physical science, science and innovation, polymer science,

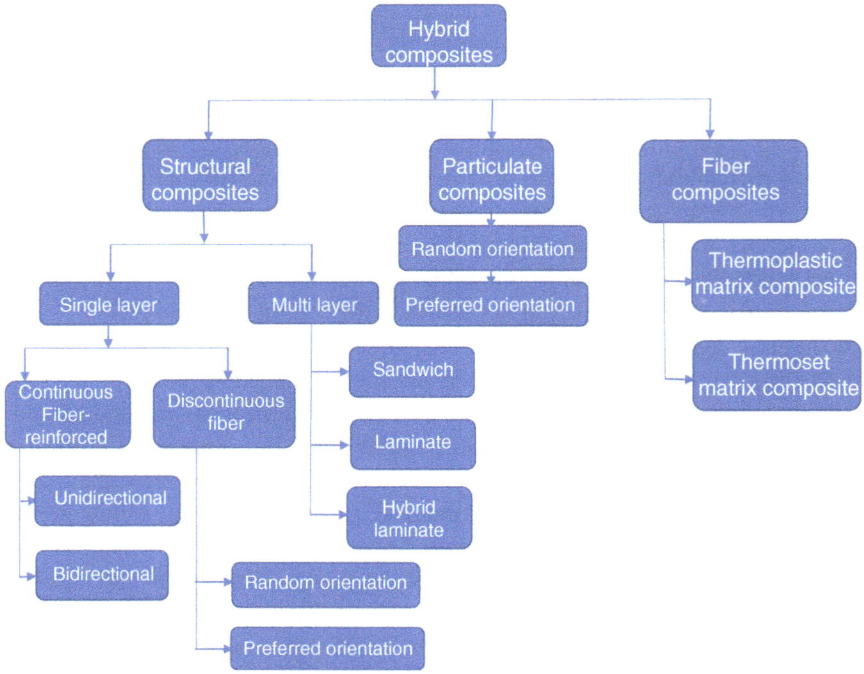

Figure 8.2 Classification of hybrid composite materials.

electrochemistry, metallurgy, energy, and building sciences [31]. Independent of the domain, the main objective was to mix at least three materials that have better execution/properties when compared with every material for a planned application. When two fillers are added into a polymer matrix as well as mixing is to give guarantee that they can gain from one another and in coordination, so they have produce superior properties at lower cost [32]. Figure 8.2 shows a structural composite, reflecting a hybrid laminate structure. The reinforcement is arranged layer by layer in various orientations in a multilayer composite. In a laminate, the constituent material stays the same in all layers. In the composite structure, hybrid laminates, on the other hand, consist of more than one constituent.

8.3 Cane/Grass Fiber Hybrid Composites

The processing of fiber-reinforced composites includes handling of fiber preforms and afterward blending with the matrix material by different strategies. Fiber preforms include weaving, sewing, and twisting in the form of sheets or laminates [33–35]. These were utilized to accomplish a significant level of computerization by using mechanical methods, which offers authority over fiber direction and volume fraction on each layer of the component to be shaped [36]. The hand layup technique is mainstream and the most established strategy for manufacturing composites. At first, dry fibers such as woven, weaved, sewed, or bond textures are set in the shape where a layer of gel coating is applied for simple removal. Resin is applied over the reinforcing fibers. To accomplish the necessary thickness, hand brush is utilized to smoothen the textures of an upgraded bonding between the progressive layers of the fibers and the resin [37–39].

Spray-up strategy is the same as the hand layup process. Nevertheless, it utilizes a hand gun that showers resin and hacked fibers on a shape. At the same time, a roller is utilized to mold these filaments into the resin. It is an open-type method, where fibers give great similarity and wetted very quickly compared with hand layup [40, 41]. During the process, the vacuum bagging method is utilized to make mechanical weight on a cover. Regularly, the vacuum bag shaping procedure is done with the guide of the hand layup method [42, 43]. With the guide adaptable film made of a material, the component is encased and fixed to prevent the external air from getting into the form. A vacuum syphon catches the air between the form and the vacuum bag while compacting the fiber layers for effective load transmission between fiber packages and preventing fiber movement during processing.

In the case of resin transfer molding (RTM), preform fiber reinforcement pads or quilted rampaging configured at the bottom of the mold cavity and preheated resin are supplied via an inflator. RTM [44, 45] can accomplish a wider range of pairings of fiber material including its inclination, along with 3D strengthening. It makes high-strength, superior quality, composite components that fit the mold's quality characteristics [46]. Vacuum infusion or RTM is now a commonly used method for the fabrication of plant fiber-based composites, as it has the advantage of producing large and complex composite components of good quality [47]. Initially dry preform fibers are mounted on a mold, and a tube is mounted in between vacuum bag and resin container. Use the difference between the ambient pressure and the vacuum pressure to satisfy the particular requirements.

Compression molding method includes step-by-step methodology for composite fabrication, utilizing preheated molds fastened on a press. A prepared pre-impregnated reinforcement kit is put between the mold plates and pressed against each other to obtain the required shape of the mold. It offers shorter turnaround times, a high degree of reliability, and a focus on quality mechanization; it therefore finds numerous applications in the automotive industry [48–50]. Continuous fibers are passed through a resin bath that is further reorganized in a hot press in this process. This is a process that is worthwhile for the manufacture of composites with a slightly wider uniform cross section, enabling manufacturing with a higher performance level and lower production costs [51–53].

Injection molding is one of the most crucial composite processing methods of producing near-net shape products that have unique mechanical and physical properties. It has the potential to construct composites with greater accuracy and low processing times [54]. The equipment is used to transport, melt, and pressurize the fiber-filled biocomposite materials that are fed into the machine in the form of granules supplied by a hopper and then shipped by a heated barrel screw. Once it is melted, the screw infuses the melt into the mold via a nozzle, where it cools down and takes on the intended shape [55]. Enhancement in fiber-matrix consistency and homogeneity in fiber dispersion in the matrix is gained during surface modifications. [56].

8.4 Properties of Cane/Grass Fiber Hybrid Composites

In the growing environmental insight and understanding in the contemporary time, numerous researchers have described the employment of natural fibers as components of reinforcement for polymer matrices [57]. The disadvantages of natural fibers, such as process deterioration, undesirable dimensions, and high water intake, make them unacceptable for real-time applications [58]. Poor bonding between natural hydrophilic fibers and hydrophobic polymer is another problem [59]. These issues have been resolved by fiber surface treatment or reconfiguration of polymer matrices. Boosting the fiber properties includes enhancing the fiber-matrix bonding and limiting the fiber content. In this section, authors coined out the properties of cane/grass fiber-reinforced hybrid composites.

The chemically treated fibers (HCl and NaOH) are greatly enhanced by the matrix and fiber bonding of the composites. Mechanical bonding is enabled by increasing the surface roughness of the fibers by breaking the hydrogen bonding in the structure. The strength of sacred grass composite materials that are NaOH treated is substantially higher than that of the plain test specimen [11]. This is because the stress applied to the fibers of broom grass is transmitted and distributed by the polyester resin, yielding greater strength. Residue grass broom composite is stronger than that of vakka and polyester composites reinforced very similar to the banana fiber [60].

Kommula et al. [57] studied the moisture content, chemical properties, and flexural characteristics of Napier grass fiber composites and observed that both raw and alkali-treated fiber composites had maximum strengths with 20 wt% loading. Accordingly, the long fibers exhibit higher mechanical properties than those of short randomly oriented fibers, and composite had good adhesive bonding. However, the flexural strength of composites of NaOH-modified sacred grass is significantly less than that of untreated one. The higher weight fraction of treated fiber composites showed high flexural strength owing to the strong interfacial bonding between the fibers and the matrix. Compared with untreated and alkali-modified sacred grass fiber composites, the flexural strength of hybrid chemically treated sacred grass fiber composites is significantly improved at all volume fractions. The energy absorption of composites of untreated sacred grass fibers is significantly higher than that of hybrid sacred grass composites treated with sodium hydroxide [61].

8.5 Thermal Properties

By hand layup technique, elephant grass fibers and glass fiber-enhanced hybrid composites have been prepared. The thermal characteristics of composites were tested using a cone calorimeter. The addition of elephant grass fiber has effectively decreased the matrix's average heat release rate and peak heat release rate by 28% and 36%, respectively. Compared with that of the matrix, the maximum average heat release rate and carbon monoxide yield of the biocomposites decrease significantly. However with the addition of elephant grass fiber, the average carbon dioxide yield and the overall smoke release values of the matrix are marginally increased [62].

8.5.1 Thermal Conductivity

The guarded heat flow meter test method was applied to measure the thermal conductivity of *Typha angustifolia* grass fiber-reinforced composites using a Unitherm Model 2022 instrument in accordance with ASTM-E1530. The sample of size 50 mm in diameter was placed between two polished surfaces, and the pressure of 10 psi was applied on the top portion of the stack [63]. The effect of nanoclay on the thermal stability of wild cane grass fiber-reinforced composites was studied by Gudapati et al. [64]. The fundamental heat conducting ability of these composites without nanoclay was found to be significantly higher than that of composites with nanoclay. This is because of the moisture content in the fibers of wild cane grass. Cane fiber-reinforced composites are manufactured with unsaturated polyester resin and further processed with alkaline solution [65]. Holding down flame-retardant and smoke production is the most difficult feature of natural fiber composite materials. The lightweight of the wild cane grass fiber is porous, and air is trapped due to the wild cane grass fiber's good thermal insulation properties [64]. Figure 8.3 shows that with increasing volume fraction, thermal conductivity decreases in composites with and without nanoclay. This phenomenon is observed from the DTC-300 direct thermal conductivity meter.

Ramanaiah et al. [66] experimented thermomechanical characteristics of Sansevieria grass fiber composites. The property analysis revealed that the basic heat ability of fibers is increased with temperature while the thermal conductivity and diffusiveness of fibers is decreased. This is due to the reduction of the fiber's moisture content. They concluded that in developing grass fiber composites, Sansevieria fibers serve as a good thermal insulating material. The same research group [67]

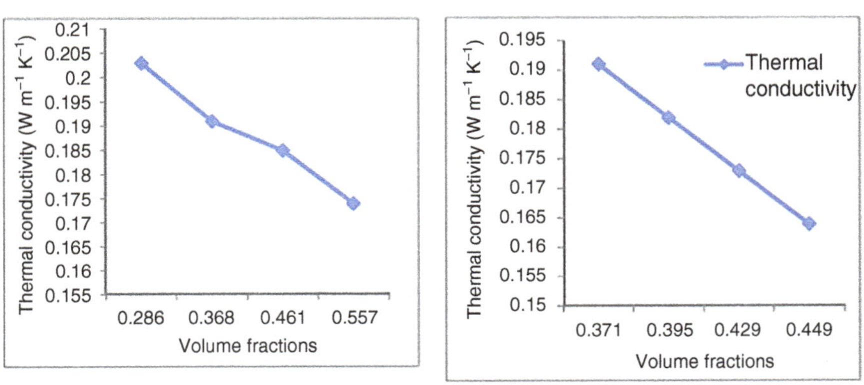

Figure 8.3 Thermal conductivity of hybrid composites. Source: Gudapati et al. [64].

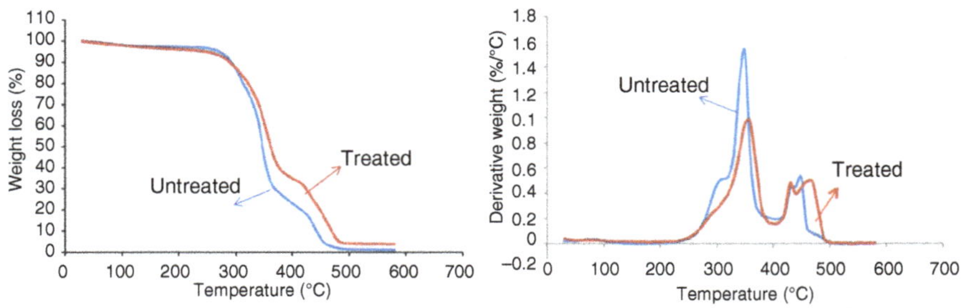

Figure 8.4 TGA thermograms of Sansevieria grass fibers. Source: Ramanaiah et al. [66].

found that the decrease in heat conductive capacity with increase in fiber wt% and the reverse pattern were noted while testing the thermophysical and fire-retardant characteristics of Sansevieria fiber composites. During testing of fire properties, they found that peak heat release rate of Sansevieria fiber is less than the polyester.

8.5.2 Thermogravimetric Analysis

Thermogravimetric analysis (TGA) is the determination of molecular degradation due to overheating, which is weight loss [68]. TGA test is also performed to analyze the transition in mass as the temperature increases. In terms of percentage weight loss, a change in thermal stability was investigated as a function of temperature, as estimated in this method [49]. The alkali-treated and alkali-untreated primary and derivative thermograms of the grass fibers of Sansevieria are shown in Figure 8.4. Parameters of thermal degradation were calculated from TGA, and the peak, where the highest decomposition rate was observed, determined the peak temperature. The final tailing area suggested an end to the decomposition of cellulose [66].

8.5.3 Differential Scanning Calorimetry Analysis

The thermal transitions of surface-modified grass fibers are analyzed by DSC. The samples were subjected to a temperature scan varying from 30 to 500 °C under nitrogen flow at a heating rate of 10 °C min^{-1}. In Figure 8.5, the DSC curve is presented. The endothermic peak was observed in the 250–480 °C temperature range and can be due to the moisture loss from the fibers [49].

8.5.4 Flammability Test

In accordance with ASTM D-635, flammability test specimens have been prepared to assess how readily the specimens can fire. The 127 mm-long, 3 mm-deep, and 12.7 mm-wide specimens were prepared to conduct the test. A collection of 3 bars were measured with a maximum fraction of fiber volume, and their average burning rate per minute has been calculated [68].

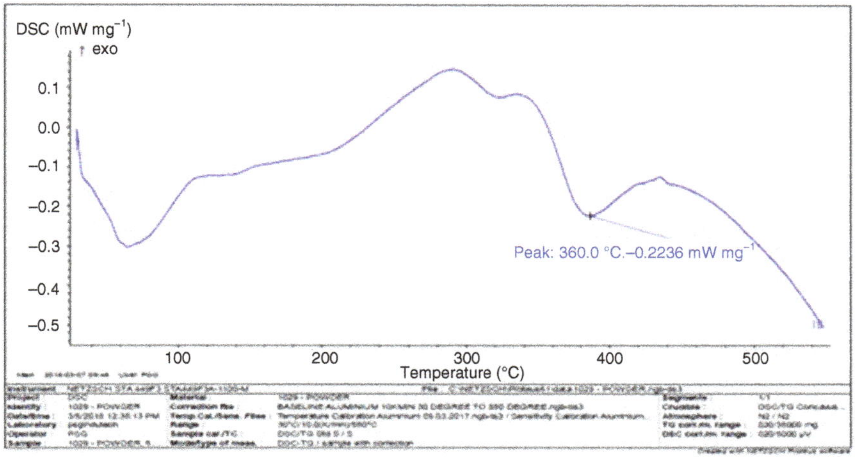

Figure 8.5 DSC analysis of the grass fiber composites. Source: Ramesh et al. [49].

8.5.5 Heat Deflection Temperature Test

In accordance with ASTM D-646, heat deflection temperature (HDT) test specimens have been prepared to determine the temperature at which the material deflects by a given amount under the action of a specified load. The 127 mm-long, 3 mm-deep, and 12.7 mm-wide specimens were prepared to conduct the test. A set of three values with a maximum fraction of fiber volume have been checked, and their average temperature of heat deflection has been calculated [68].

8.5.6 Specific Heat Capacity Measurement

A differential scanning calorimeter at a heating rate of 10 °C min^{-1} was used to measure the specific heat capacity of the composite samples [66]. The basic heat capacity values are estimated to be in the range of 30–120 °C of fiber-reinforced composite at maximum volume fraction (0.362), and pure resin is shown in Figure 8.6. It was observed that with the temperature, the specific heat capacity of all samples increased linearly. The composite heat capacity is influenced by the content of the fiber and is lower than that of the matrix. The difference is smaller at high temperatures and is due to the decrease in the mean free path of phonons [69].

8.5.7 Thermal Diffusivity

The physical significance of thermal diffusivity during temperature changes over time is correlated with the propagation of heat through the medium. The lower the thermal diffusivity, the more time it takes to penetrate the solid for heat. Thermal diffusivity (α) is a function of thermal conductivity (k), heat power (C_p), and density

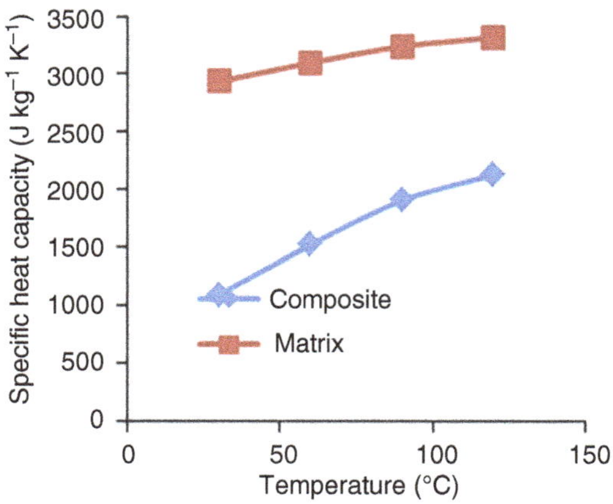

Figure 8.6 Specific heat capacity of polymer matrix and elephant grass fiber composite. Source: Ramanaiah et al. [69].

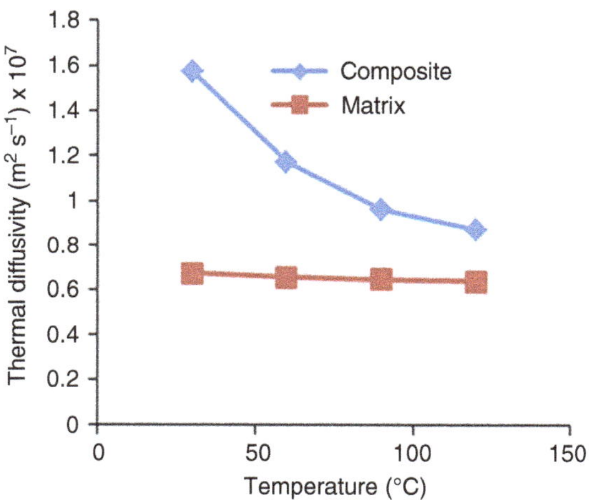

Figure 8.7 Thermal diffusivity of polymer matrix and elephant grass fiber composite. Source: Ramanaiah et al. [69].

(ρ) and can be measured using the equation below [66]:

$$\alpha = k/\rho C_p \tag{8.1}$$

Over a temperature range of 30–120 °C, the thermal diffusivity of the composite and resin was calculated, and the values are presented in Figure 8.7. The thermal diffusivity of the composite is higher than that of the matrix since; with the addition of fiber, the composite density decreases. The results showed that the thermal diffusivity decreased as the temperature increased [69].

8.6 Applications of Grass/Cane Hybrid Composites

Growing environmental awareness demands a change in the range of materials from synthetic materials to non-polluted materials in the automobile, construction, and packaging industries to take advantage of end-of-life biodegradability of products. Owing to their high mechanical strength and CO_2 neutrality, low weight, and minimized carbon fiber-reinforced materials, natural fibers replace manmade fiber-reinforced polymer composites [70]. These are used in door frames, seat covers, dashboard liners, auto doors, headliners, pallets, packaging, decking, parcel shelves, spare wheel panels, car inside covers, vessels, inner panels, and architectural moldings. They are currently being used by most car manufacturers, and this is projected to be uptrend in the coming years [71]. Napier grass fiber-reinforced epoxy composites are chemically treated with strong mechanical properties, interfacial bonding, and composite chemical resistance. It can be used in the manufacture of agricultural goods, car interiors, and household items [57]. Surface-treated wild cane grass fiber have significantly high tensile modulus over the sisal/banana fibers and greater than coir fiber. Also, density of wild cane fiber is low compared with bamboo and banana fibers, and it possessed high strength, making an impressive factor in fabricating high toughness and lightweight materials [11]. Nanoclay blended wild cane grass fiber composites with low thermal diffusivity; it can be used to manufacture components in industries like packaging, construction, and aerospace applications [64].

8.7 Conclusion

The bio-based hybrid composites are compatible in terms of processing, usage, and recycling or disposal without damaging the earth, which has been and still is the key motive for developing biocomposites. Thermal properties of the grass/cane fiber-based hybrid composites are discussed in this chapter. From the literature, it is found that the thermal conductivity of grass fiber composites decreases with fiber content. These composites display remarkable thermal behavior and have found that there is no significant change in the thermal stability of the composites up to 280 °C and begin to lose weight at 300 °C from TGA studies. It is also found that the effect of temperature on the composite heat absorbing capability was close to that of thermal conductivity. The flammability values suggested that the composite burning rate is very high and the difference between temperature and thermal diffusivity was only contrary to that of thermal conductivity. This chapter further revealed that the composites reinforced with grass fiber are light in weight and economical and have strong thermal insulating properties. These composite materials can therefore be used for applications such as construction facilities, interior parts of vehicles, electronic goods packages, etc.

References

1 Murali, B., Ramnath, B.V., and Chandramohan, D. (2019). Mechanical properties of *Boehmeria nivea* reinforced polymer composite. *Materials Today: Proceedings* 16: 883–888.

2 Asrofi, M., Sapuan, S.M., Ilyas, R.A., and Ramesh, M. Characteristic of composite bioplastics from tapioca starch and sugarcane bagasse fiber: effect of time

duration of ultrasonication (Bath-Type). *Materials Today: Proceedings* https://doi.org/10.1016/j.matpr.2020.07.

3 Ramesh, M. and Deepa, C. Properties of cellulose based bio-fibres reinforced polymer composites. *Biofibers and Biopolymers for Biocomposites: Synthesis, Characterization and Properties.* https://doi.org/10.1007/978-3-030-40301-0_3.

4 Radhakrishnan, B., Vijayan, V., Parameshwaran Pillai, T., and Sathish, T. (2019). Influence of surface roughness in turning process-an analysis using artificial neural network. *Transactions of the Canadian Society for Mechanical Engineering* 43 (4): 509–514.

5 Chukov, D., Nematulloev, S., Torokhov, V. et al. (2019). Effect of carbon fiber surface modification on their interfacial interaction with polysulfone. *Results in Physics* 15: 102634.

6 Vimal, R., Subramanian, K.H.H., Aswin, C. et al. (2015). Comparisonal study of succinylation and phthalicylation of jute fibres: study of mechanical properties of modified fibre reinforced epoxy composites. *Materials Today: Proceedings* 2: 2918–2927.

7 Ramesh, M. and Gopinath, A. (2017). Measurement and analysis of thrust force in drilling sisal-glass fiber reinforced polymer composites. *IOP Conference Series: Materials Science and Engineering* 197: 1–7.

8 Ramesh, M., Sudharsan, P., and Palanikumar, K. (2015). Processing and mechanical property evaluation of flax-glass fiber reinforced polymer composites. *Applied Mechanics and Materials* 766: 144–149.

9 Ardanuy, M., Claramunt, J., García-Hortal, J.A., and Barra, M. (2011). Fiber-matrix interactions in cement mortar composites reinforced with cellulosic fibers. *Cellulose* 18: 281–289.

10 Balea, A., Fuente, E., Blanco, A., and Negro, C. (2019). Nanocelluloses: natural-based materials for fiber-reinforced cement composites. A critical review. *Polymers* 11: 518.

11 Prasad, A.V.R., Rao, K.M., Gupta, A.V.S.S.K.S., and Reddy, B.V. (2011). A Study on flexural properties of wildcane grass fiber-reinforced polyester composites. *Journal of Materials Science* 46: 2627–2634.

12 Nair, A.B. and Joseph, R. (2014). Eco-friendly bio-composites using natural rubber (NR) matrices and natural fiber reinforcements. In: *Chemistry, Manufacture and Applications of Natural Rubber* (eds. S. Kohjiya and Y. Ikeda), 249–283. Sawston, UK/Cambridge, UK: Woodhead Publishing.

13 Omrani, E., Menezes, P.L., and Rohatgi, P.K. (2016). State of the art on tribological behavior of polymer matrix composites reinforced with natural fibers in the green materials world. *Engineering Science and Technology, an International Journal* 19: 717–736.

14 Ouarhim, W., Zari, N., Bouhfid, R., and Qaiss, A. (2019). Mechanical performance of natural fibers–based thermosetting composites. In: *Mechanical and Physical Testing of Biocomposites, Fibre-Reinforced Composites and Hybrid Composites* (eds. M. Jawaid, M. Thariq and N. Saba), 43–60. Sawston, UK/Cambridge, UK: Woodhead Publishing.

15 Osamu, N. (2004). The role of bamboo in the sustainable economic development: from a view point of social common capital. *JSME International Journal* 47 (4): 527–532.
16 Bhoopathi, R., Ramesh, M., Rajaprasanna, R. et al. (2017). Physical properties of glass-hemp-banana hybrid fiber reinforced polymer composites. *Indian Journal of Science and Technology* 10 (7): 1–6.
17 Pattanayak, S., RajuramanV, S., and Kanagaraj, S. (2000). Insulating property of polyurethane-kevlar composites at low temperatures. *International Plastic Engineering Technology* 4: 47–58.
18 Kanagaraj, S. and Pattanayak, S. (2003). Measurement of thermal expansion of metals and FRPs. *Cryogenics* 43 (7): 399–424.
19 Kalaprasad, G., Pradeep, P., Mathew, G. et al. (2000). Thermal conductivity and thermal diffusivity analyses of low-density polyethylene composites reinforced with sisal, glass and intimately mixed sisal/glass fibres. *Composites Science and Technology* 60 (16): 2967–2977.
20 Tavman, I.H. (1998). Effective thermal conductivity of isotropic polymer composites. *International Communications in Heat and Mass Transfer* 25 (5): 723–732.
21 Zou, M., Yu, B., Zhang, D., and Ma, Y. (2003). Study on optimization of transverse thermal conductivities of unidirectional composites. *Journal of Heat Transfer* 125 (6): 980.
22 Zhou, H., Zhang, S., and Yang, M. (2007). The effect of heat-transfer passages on the effective thermal conductivity of high filler loading composite materials. *Composites Science and Technology* 67 (6): 1035–1040.
23 Zhao, X., Wang, X., Wu, Z. et al. (2018). Temperature effect on fatigue behavior of basalt fiber-reinforced polymer composites. *Polymer Composites* 40: 2273–2283.
24 Kim, S.W., Lee, S.H., Kang, J.S., and Kang, K.H. (2006). Dependence of mechanical and thermal properties of thermoplastic composites on electron beam irradiation. *International Journal of Thermophysics* 27: 1873–1881.
25 Xue, L., Tabil Lope, G., Oguocha Ikechukwuka, N., and Satyanarayana, P. (2008). Thermal diffusivity, thermal conductivity, and specific heat of flax fiber–HDPE biocomposites at processing temperatures. *Composites Science and Technology* 68: 1753–1758.
26 Ghazanfari, A., Emami, S., Panigrahi, S., and Tabil, L.G. (2008). Thermal and mechanical properties of blends and composites from HDPE and date pits particles. *Journal of Composite Materials* 42 (1): 77–89.
27 Khedari, J., Watsanasathaporn, P., and Hirunlabh, J. (2005). Development of fibre-based soil–cement block with low thermal conductivity. *Cement and Concrete Composites* 27 (1): 111–116.
28 Binici, H., Aksogan, O., Bodur, M.N. et al. (2007). Thermal isolation and mechanical properties of fibre reinforced mud bricks as wall materials. *Construction and Building Materials* 21 (4): 901–906.
29 Rao, K.M.M., Prasad, A.V.R., Babu, M.N.V.R. et al. (2007). Tensile properties of elephant grass fiber reinforced polyester composites. *Journal of Materials Science* 42 (9): 3266–3272.

30 Bhoopathi, R., Deepa, C., Sasikala, G., and Ramesh, M. (2015). Experimental investigation on mechanical properties of hemp-banana-glass fiber reinforced composites. *Applied Mechanics and Materials* 766: 167–172.

31 Makisima, A. (2004). Possibility of hybrids materials. *Ceramic Japan* 39: 90–91.

32 Sanchez, C. and Gomez-Romero, P. (2004). *Functional Hybrid Materials.* Willey-VCH.

33 Chung, D.D.L. (2017). Polymer-matrix composites: structure and processing. In: *Carbon Composites: Composites with Carbon Fibers, Nanofibers, and Nanotubes* (ed. D.D.L. Chung), 161–217. Amsterdam, The Netherlands: Elsevier Science.

34 Boisse, P. (ed.) (2015). *Advances in Composites Manufacturing and Process Design.* Sawston, UK/Cambridge, UK: Woodhead Publishing.

35 Balasubramanian, K., Sultan, M.T.H., and Rajeswari, N. (2018). Manufacturing techniques of composites for aerospace applications. In: *Sustainable Composites for Aerospace Applications* (eds. M. Jawaid and M. Thariq), 55–67. Sawston, UK/Cambridge, UK: Woodhead Publishing.

36 Gascons, M., Blanco, N., and Matthys, K. (2012). Evolution of manufacturing processes for fiber-reinforced thermoset tanks, vessels, and silos: a review. *IIE Transactions* 44: 476–489.

37 Gunge, A., Koppad, P.G., Nagamadhu, M. et al. (2019). Study on mechanical properties of alkali treated plain woven banana fabric reinforced biodegradable composites. *Composites Communications* 13: 47–51.

38 Elkington, M., Bloom, D., Ward, C. et al. (2015). Hand layup: understanding the manual process. *Advanced Manufacturing: Polymer & Composites Science* 1: 138–151.

39 Ramesh, M. and Arpitha, G.R. (2019). Thermal and mechanical characterization of biocomposites. In: *Processing of Natural Fibre Reinforced Bio-composites* (ed. P.K. Bajpai), 39–70. Woodhead Publishing India.

40 Perna, A.S., Viscusi, A., Astarita, A. et al. (2019). Manufacturing of a metal matrix composite coating on a polymer matrix composite through cold gas dynamic spray technique. *Journal of Materials Engineering and Performance* 28: 3211–3219.

41 Marques, A.T. (2011). Fibrous materials reinforced composites production techniques. In: *Fibrous and Composite Materials for Civil Engineering Applications* (ed. R. Fangueiro), 191–215. Sawston, UK/Cambridge, UK: Woodhead Publishing.

42 Ervina, J., Ghaleb, Z.A., Hamdan, S., and Mariatti, M. (2018). Colloidal stability of water-based carbon nanotube suspensions in electrophoretic deposition process: Effect of applied voltage and deposition time. *Composites Part A: Applied Science and Manufacturing* 117: 1–10.

43 Carruthers, J. (2018). Vacuum bagging process overview, coventive composites. https://coventivecomposites.com/explainers/what-is-vacuum-bagging/ (accessed 5 August 2020).

44 Meola, C., Boccardi, S., and Carlomagno, G. (2017). Composite materials in the aeronautical industry. In: *Infrared Thermography in the Evaluation of Aerospace*

Composite Materials: Infrared Thermography to Composites, 1–24. Amsterdam, The Netherlands: Elsevier Science.

45 Ahmad, N., Bilal, I., Khattak, S. et al. (2018). Polyester usage in manufacturing of electrical and mechanical products and assemblies. In: *Polyester – Production, Characterization and Innovative Applications*. https://doi.org/10.5772/intechopen.74368.

46 Davis, D.C. and Mensah, T.O. (2017). Fabrication and fatigue of fiber-reinforced polymer nanocomposites-a tool for quality control. In: *Nanotechnology Commercialization: Manufacturing Processes and Products* (eds. T.O. Mensah, B. Wang, G. Bothun, et al.). Hoboken, NJ: Wiley https://doi.org/10.1002/9781119371762.ch10.

47 Hsiao, K.-T. and Heider, D. (2012). Vacuum assisted resin transfer moulding (VARTM) in polymer matrix composites. In: *Manufacturing Techniques for Polymer Matrix Composites (PMCs)* (eds. S.G. Advani and K.-T. Hsiao), 310–347. Woodhead Publishing Series in Composites Science and Engineering.

48 Mitschang, P. and Hildebrandt, K. (2012). Polymer and composite moulding technologies for automotive applications. In: *Advanced Materials in Automotive Engineering* (ed. J. Rowe), 210–229. Sawston, UK/Cambridge, UK: Woodhead Publishing.

49 Ramesh, M., Deepa, C., Tamil Selvan, M., and Reddy, K.H. Effect of alkalization on characterization of ripe bulrush (*Typha Domingensis*) grass fiber reinforced epoxy composites. *Journal of Natural Fibers* https://doi.org/10.1080/15440478.2020.1764443.

50 Matveenko, V.P., Kosheleva, N.A., Shardakov, I.N., and Voronkov, A.A. (2018). Temperature and strain registration by fibre-optic strain sensor in the polymer composite materials manufacturing. *International Journal of Smart and Nano Materials* 9: 99–110.

51 Correia, R. (2013). Pultrusion of advanced fibre-reinforced polymer (FRP) composites, Chapter 9. In: *Advanced Fibre-Reinforced Polymer (FRP) Composites for Structural Applications* (ed. J. Bai), 207–251. Sawston, UK/Cambridge, UK: Woodhead Publishing.

52 Verma, D., Joshi, G., Dabral, R., and Lakhera, A. (2019). Processing and evaluation of mechanical properties of epoxy-filled E-glass fiber–fly ash hybrid composites. In: *Mechanical and Physical Testing of Biocomposites, Fibre-Reinforced Composites and Hybrid Composites* (eds. M. Jawaid, M. Thariq and N. Saba), 293–306. Sawston, UK/Cambridge, UK: Woodhead Publishing.

53 Joshi, S.C. (2012). The pultrusion process for polymer matrix composites. In: *Manufacturing techniques for polymer matrix composites (PMCs)* (eds. S.G. Advani and K.-T. Hsiao), 381–413. Sawston, UK/Cambridge, UK: Woodhead Publishing.

54 Liu, S.-J. (2012). Injection moulding in polymer matrix composites. In: *Manufacturing Techniques for Polymer Matrix Composites (PMCs)* (eds. S.G. Advani and K.-T. Hsiao), 15–46. Woodhead Publishing Limited.

55 Leong, Y.W., Thitithanasarn, S., Yamada, K., and Hamada, H. (2014). Compression and injection molding techniques fornatural fiber composites. In: *Natural Fibre Composites*, 216–232. Sawston, UK/Cambridge, UK: Woodhead Publishing.

56 Gonzalez-Lopez, M.E., Perez-Fonseca, A.A., Manriquez-Gonzalez, R. et al. (2018). Effect of surface treatment on the physical and mechanical properties of injection molded poly(lactic acid)-coir fiber biocomposites. *Polymer Composites* 40: 2132–2141.

57 Kommula, V.P., Reddy, K.O., Shukla, M. et al. (2014). Mechanical properties, water absorption, and chemical resistance of Napier grass fiber strand- reinforced epoxy resin composites. *International Journal of Polymer Analysis and Characterization* 19: 693–708.

58 Athijayamani, A., Thiruchitrambalam, M., Natarajan, U., and Pazhanivel, B. (2009). Effect of moisture absorption on the mechanical properties of randomly oriented natural fibers/polyester hybrid composite. *Materials Science and Engineering A* 517: 344–353.

59 Torres, F.G. and Cubillas, M.L. (2005). Study of the interfacial properties of natural fiber reinforced polyethylene. *Polymer Testing* 24 (6): 694–698.

60 Rao, K.M.M., Prasad, A.V.R., and Rao, K.M. (2010). Fabrication and testing of natural fibre composites: vakka, sisal, bamboo, and banana. *Materials & Design* 31: 508–513.

61 Srinivasababu, N. (2019). Understanding the durability of long sacred grass/Imperata cylindrica natural/hybrid FRP composites. In: *Durability and Life Prediction in Biocomposites, Fibre-Reinforced Composites and Hybrid Composites* (eds. M. Jawaid, M. Thariq and N. Saba), 287–300. Woodhead Publishing Series in Composites Science and Engineering.

62 Ramanaiah, K., Prasad, A.V.R., and Reddy, K.H. (2014). Fire properties of elephant grass fiber and glass fiber reinforced polyester composites. *Applied Mechanics and Materials* 592-594: 380–384.

63 Ramanaiah, K., Ratna Prasad, A.V., and Chandra Reddy, K.H. (2011). Mechanical properties and thermal conductivity of *Typha angustifolia* natural fiber–reinforced polyester composites. *International Journal of Polymer Analysis and Characterization* 16 (7): 496–503.

64 Gudapati, S.P.K., Chidambaranathan, S., and Ratna Prasad, A.V. (2020). Influence of nanoclay on thermal properties of wild cane grass fiber reinforced polyester composites. *Materials Today: Proceedings* 23 (3): 632–636.

65 Begum, M.H.A. and Tanvir, N.I. (2019). Fabrication and characterization of cane fiber reinforced unsaturated polyester resin composites. *Bangladesh Journal of Scientific and Industrial Research* 54 (3): 247–256.

66 Ramanaiah, K., Prasad, A.V.R., and Reddy, K.H. (2011). Thermal and mechanical properties of sansevieria green fiber reinforcement. *International Journal of Polymer Analysis and Characterization* 16: 602–608.

67 Ramanaiah, K., Prasad, A.V.R., and Reddy, K.H. (2013). Mechanical, thermophysical and fire properties of sansevieria fiber-reinforced polyester composites. *Materials & Design* 49: 986–991.

68 Mounika, M., Anusha, P., Ramanaiah, K. et al. (2014). Thermal properties of bamboo fiber reinforced polyester composite. *International Journal of Engineering Sciences & Research Technology* 3 (4): 87–93.

69 Ramanaiah, K., Ratna Prasad, A.V., and Hema Chandra Reddy, K. (2012). Thermo physical properties of elephant grass fiber-reinforced polyester composites. *Materials Letters* 89: 156–158.

70 Faruk, O., Bledzki, A.K., Fink, H.P., and Sain, M. (2014). Progress report on natural fiber reinforced composites. *Macromolecular Materials and Engineering* 299: 9–26.

71 Ramesh, M., Palanikumar, K., and Reddy, K.H. (2017). Plant fibre-based bio-composites: Sustainable and renewable green materials. *Renewable and Sustainable Energy Reviews* 79: 558–584.

9

Thermal Properties of the Banana Fiber-Based Hybrid Composites

Nasmi Herlina Sari[1] *and Senthil Muthu Kumar Thiagamani*[2]

[1] *University of Mataram, Department of Mechanical Engineering, Faculty of Engineering, Jl. Majapahit No.62, Gomong, Kec. Selaparang, Kota Mataram, Nusa Tenggara, 83115, Indonesia*
[2] *Kalasalingam Academy of Research and Education, Department of Mechanical Engineering, Anand Nagar, Krishnankoil, Tamil Nadu, 626102, India*

9.1 Introduction

Composite material technology has been used in many industrial sectors to improve and create more competitive and innovative products. They are made to replace conventional materials. Currently, the usefulness of composites is very high. Several technologies related to automobiles, and aerospace and underwater require some typical combinations that are not possible using metals and alloys. In a cutting-edge domain such as aeronautics, the high performance of composite materials is an undeniable asset. Metals were eventually replaced by composite materials in aircraft structures, owing to some of their disadvantages related to heat transfer, although these have advantages in terms of mechanical properties. Composite materials have better insulation properties than metallic materials due to the problem of controlling heat quickly. Polymer composites find failures under variable temperature conditions for different applications such as military weapons, automobile sector, medical instruments, aerospace bodybuilding, etc. As a solution, reinforcing a combination of two types of fibers into one type of resin was tested in polymer composites, known as hybrid polymer composites (HPCs). Components and products designed and manufactured from hybrid composite materials as a replacement to metallic materials have been reported by many industries in the engineering market such as medical instruments and equipment, military weapons, aerospace, household, buildings, vehicles, and construction materials [1–5].

Banana fiber is a cellulose fiber that can be extracted from banana stems and has many advantages such as high physical and good mechanical strength. They are available in tropical countries such as Indonesia, Malaysia, and South India [6]. The physical and mechanical properties of banana fibers are shown in Table 9.1. This has been widely used in the textile, paper, and decoration industries. The banana fiber contains 63–64% cellulose, 5% lignin, 10% hemicellulose, and a little pectin and pigments. Besides, banana fibers are thermally stable at temperatures

Natural Fiber-Reinforced Composites: Thermal Properties and Applications, First Edition.
Edited by Senthilkumar Krishnasamy, Senthil Muthu Kumar Thiagamani, Chandrasekar Muthukumar, Rajini Nagarajan and Suchart Siengchin.

Table 9.1 Physical and mechanical properties of banana fiber and other fibers.

Fibers	Density ($g\,cm^{-3}$)	Moisture content (%)	Length (cm)	Diameter (cm)	Tensile strength (MPa)	Strain break (%)	Young's modulus (GPa)	References
Banana	1.35	10–11	0.94–0.4	12–30	529–914	5–6	27–32	[7]
Pineapple leaf	1.520	—	—	—	413	1.6	34	[8]
Glass	2.550	—	—	—	3400	3.4	71	[8]
Cornhusk	0.43	11.961	0.186 ± 0.02	0.186 ± 0.02	160.49 ± 17.12	21.08 ± 2.860	4.57 ± 0.54	[9]
Jute	1.23	1.2˙	0.8–6	5–25	187–773	1.5–3.1	20–55	[7]

of 25–250 °C [10]. The experimental thermal conductivity value of the fibers was found to be 0.044 15 $W\,m^{-1}\,K^{-1}$ at a density of 70.4 $kg\,m^{-1}$ [11].

Hybridization of banana fiber with other materials in composite materials can improve mechanical properties such as flexural and impact strength and low water absorption, for example, the hybridization of banana fibers and sisal fibers in epoxy composites [12]. The thermal stability of composites depends on the matrix such as the high-density polyethylene (HDPE), maleic anhydride-*grafted* polypropylene (MAgPP), epoxy, polyester, and resistant starch matrices at 525, 450, 400, 245–289, and 250 °C, respectively [13, 14]. Nowadays, banana fiber-based hybrid composites are used in building and industrial production.

The role of natural and manmade fiber hybrid composite materials is growing at a faster rate in engineering and technology because they have good calorific value and are harmless to health and safety during handling and inexpensive and have low density and excellent mechanical properties [15, 16]. This composite is made by mixing two or more different fiber types into the same matrix. Hybrid composites offer several advantages over using only one type of fiber in the polymer.

In this chapter, an understanding of the thermal properties of banana fiber-based hybrid composites are presented. The discussion covered the thermal properties of banana fiber and banana fiber-based hybrid composites in terms of thermogravimetric analysis (TGA), thermal conductivity, and limiting oxygen index (LOI) of banana fiber-based hybrid composites. In addition, this study also discusses the thermal properties of banana hybrid composites with various other fibers such as jute, pineapple leaf, flax, and fiber glass, as well as fabrication and their applications.

9.2 Fabrication of Banana Fiber-Based Hybrid Composite

9.2.1 Hand Layup

The hand layup method can be carried out in the manufacture of fiber-based hybrid composites by pouring a matrix (for example, polyester resin or epoxy) into a mold that has been filled with hybrid fibers (for example, banana fiber, pineapple leaves) and ending with a layer of the matrix resin. The production of a good-quality hybrid composite can be done by minimizing the deformation and movement of the fibers in the composite. When curing, a pressure of 0.05 MPa for 24 hours can be applied to the mold. Then, before being removed from the mold, the composites should be post-cured at 70 °C for two hours to ensure drying of the composites.

9.2.2 Vacuum Bag-Assisted Resin Infusion Technique

According to Vivek and Kanthavel [17], this technique can be used to make fiber hybrid composites. Fibers with the same weight percentage (length 10 mm) are randomly mixed homogeneously in a mold measuring $150 \times 150 \times 5\,mm^3$ and then wrapped in plastic sheets and vacuum-packed. One side of the port mold is connected through a resin trap with a vacuum gauge to the vacuum pump and to the resin inlet for the other port. For example, epoxy resin and filler are mixed in percentage by weight and then hardened by manual stirring. During the vacuum infusion process, vacuum pressure can be applied from 0.7 to 0.8 bar. Furthermore, the composites were allowed to dry for 12 hours at room temperature. This technique has been used in making composites by hybridizing the fibers of banana/kenaf (HBK), banana/flax (HBF), sisal/kenaf (HSK), and sisal/flax (HSF).

9.3 Thermal Properties of Banana Fiber-Based Composites

9.3.1 Thermal Stability

The thermal stability of banana fiber-reinforced composites and other natural fibers depends on the material matrix and chemical concentration. The specific chemical modification of natural fibers can increase the thermal stability of the composites. Moreover, the thermal stability of the natural fiber-based composites is listed in Table 9.2.

9.3.2 Limiting Oxygen Index (LOI)

Hybrid composites made of glass fiber-reinforced polymer (GFRP) and banana and hemp have good flame resistance. This is because this type of composite has a LOI value close to 26 [25]. The result also has been confirmed by Van Krevelen [26] who stated that LOI is greater than 26, indicating that the composite has good flame resistance.

The LOI value of hybrid composite can be determined using Krevelen's relation as shown in Eq (9.1) [26, 27]:

$$LOI = 17.5 + 0.4CR \tag{9.1}$$

where CR is the char residue in wt%.

Table 9.2 Thermal stability of natural fiber-based polymer composites.

Item	Thermal stability (°C)	References
Jute matrix	250–365	[18–20]
Treated jute matrix	250–400	[21]
Banana matrix	25–250	[10]
Coir matrix	25–250	[22]
Treated coir matrix	250–400	[23]
Treated coir matrix	250–400	[23]
Treated banana matrix	250–300	[24]

9.3.3 Thermogravimetric Analysis (TGA)

9.3.3.1 Jute and Banana Fiber Hybrid Composites

According to Boopalan et al. [28], TGA of banana and hemp fibers with epoxy hybrid composites showed a reduction in the weight of the constituents (hemicellulose, lignin, pectin, wax, moisture) at different temperature ranges (shown in Figure 9.1). The initial weight loss occurs at 200 °C as indicated by the loss of the solvent in the polymer matrix. The largest weight reduction occurred between 75.64 and 82.14% at 376.51–380 °C caused by the degradation of the epoxy matrix and fibers. With 50% banana and 50% jute, 25% banana and 75% jute fiber, and 75% banana and 25% jute fiber with epoxy hybrid composite, the major weight loss of 75.64%, 82.14%, and 79.01%, respectively, occurs at 380, 377.72, and 376.51 °C, respectively. The degradation shifted to higher temperature, and final residue occurs at 710 °C [28].

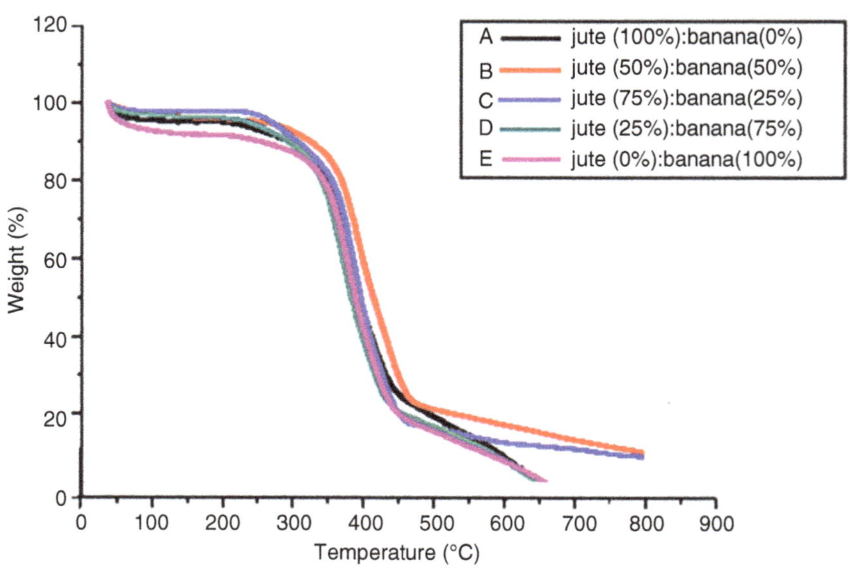

Figure 9.1 Banana hybrid composite material. Source: Boopalan et al. [28].

9.3.3.2 Glass and Banana Fiber Hybrid Composites

The TGA results of the banana glass fiber-based composite showed that the main degradation occurred at a temperature range of 273–427 °C with a peak at 349 °C (shown in Figure 9.2a,b). Zin et al. [8] stated that banana–glass fiber-based composites experience loss of initial mass at temperatures between 30 and 100 °C due to the moisture evaporation. A constituent such as a hemicellulose and lignin occurs between 130 and 380 °C, especially in bio-hybrid composites. This result implies that the different load of banana hybrid fibers in the composite material causes the thermal properties of the composite to be different.

After hemicellulose decomposed, cellulose followed, which has better thermal stability because they have a crystalline structure on their cellulose chains. Lignin has the highest thermal stability because it decomposes slowly around 160–900 °C since lignin is the toughest component in plant material. The residue from the fiber after being treated with sodium hydroxide (NaOH) is an inorganic material that can only decompose at very high temperatures, namely, 1700 °C and above.

Derivative thermogravimetric (DTG) curves for banana fiber and PALF hybrid composites are shown in Figure 9.2c,d. Minor peaks are obtained in the temperature range of 310–350 °C where the moisture content of the composite evaporates due to the heating process. For the hybrid banana composite, the peak temperature

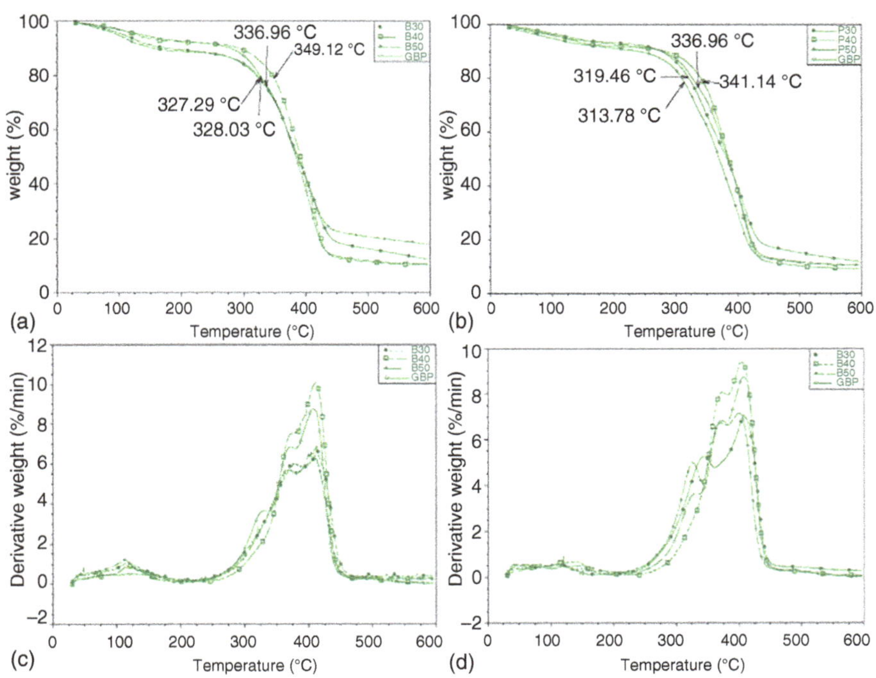

Figure 9.2 Effect of fiber loading on (a) thermal stability of the banana glass hybrid biocomposites, (b) thermal stability of pineapple leaf (PALF) glass hybrid biocomposites, (c) derivative thermogravimetric analysis of the banana glass hybrid biocomposites, and (d) derivative thermogravimetric analysis of the PALF glass hybrid biocomposites. Source: Zin et al. [8].

Table 9.3 The result TGA of banana, PALF, and glass hybrid composites.

Composites	Samples codes	% Mass loss from first to second inflection points	% Mass loss change from 27 °C to first inflection	% Mass loss from first to second inflection points	Total % of mass loss	Onset of degradation (°C)
Glass banana PALF	GBP	81.75	7.108	81.75	88.86	336.96
30% glass banana	GB30	72.89	10.44	72.89	83.33	327.79
40% glass banana	GB40	81.73	7.479	81.73	89.21	349.12
50% glass banana	GB50	69.38	10.60	69.38	79.98	328.03
30% glass PALF	GP30	81.41	5.559	81.41	86.97	319.46
40% glass PALF	GP40	82.20	7.656	82.20	89.86	341.14
50% glass PALF	GP50	80.34	7.479	80.34	87.82	313.78

Source: Zin et al. [8].

is obtained at 411.48 °C with a speed of 10.07 °C min^{-1} where the maximum weight loss is found to be 0.4 V_f. However, for PALF the maximum peak temperature was obtained at 403.24 °C with a rate of 9.36 °C min^{-1}, as shown in Table 9.3.

9.3.3.3 Banana/Flax-Based Glass Fiber Hybrid Composite

Figure 9.3 shows that the flame resistance and thermal stability of the hemp/banana-based natural fiber composites higher than that of hemp/GFRP and GFRP/banana. This can be seen by the two stages of decomposition experienced by this composite at 372 and 374 °C. In the second stage, the decomposition associated with hemp and banana composites occurred at 365 and 654 °C.

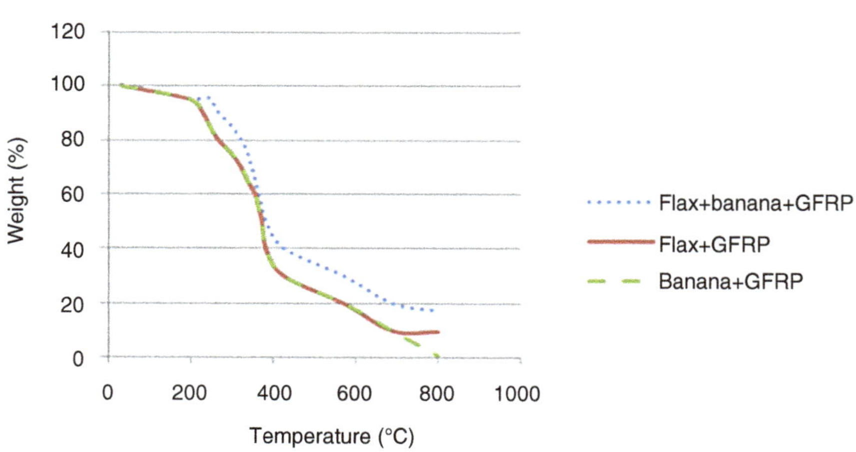

Figure 9.3 Weight (%) temperature for TGA test. Source: Srinivasan et al. [25].

9.4 Specific Heat of Banana Fiber Hybrid Composites

9.4.1 Banana and Pineapple Leaf Fiber Hybrid Composites

The amount of heat needed to change the temperature from the mass unit of a substance is called the specific heat capacity of the material. It can be measured using a differential scanning calorimeter at a heating rate of 10 °C min^{-1} according to ASTM standard E1269-11. The specific heat capacity values of banana fiber and pineapple leaf fiber hybrid-reinforced polyester composites in the range of 30–120 °C [29]. The specific heat capacity values of fiber-reinforced polyester composites are shown in Figure 9.4a,b.

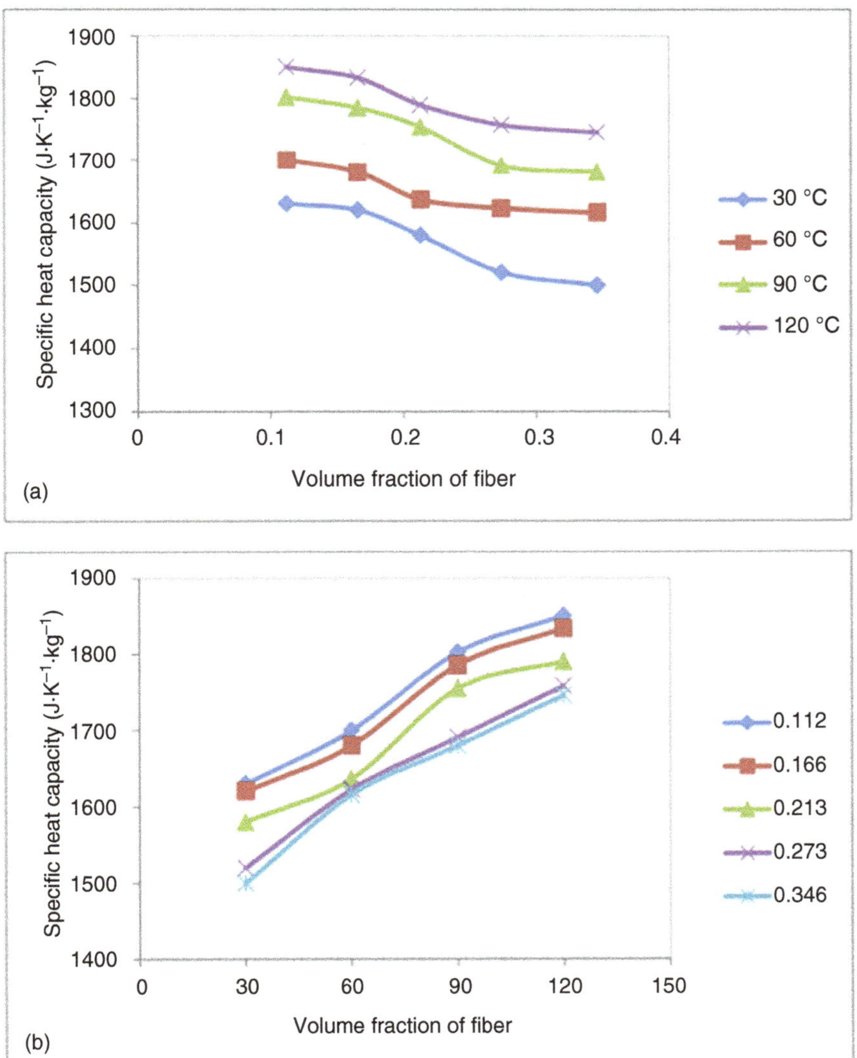

Figure 9.4 Specific heat capacity of composite with (a) volume fraction of fiber at different temperature and (b) temperature at different volume fractions of fiber. Source: Santosha et al. [29].

Moreover, Figure 9.4a,b shows that the specific heat of the hybrid composite decreases with increasing fiber content that is associated with a lower specific heat capacity.

9.5 Thermal Conductivity of Banana Fiber Hybrid Composites

The use of two or more materials in hybrid composites can have a different thermal conductivity that can be used in different engineering situations. We can know the thermal conductivity of the composite with a combination of reinforcement by looking at the highest strength obtained in the tensile and flexural tests using the heat flow meter method in accordance with ASTME1530 standards and carried out in steady conditions. In general, the thermal conductivity test is carried out at composite thermal resistance and thermal conductivity (k) of 4.16e – 0002 $m^2 K W^{-1}$ and 0.241 $W mK^{-1}$, respectively, and an average temperature of 60.96 °C [30].

9.5.1 The Pineapple Leaf and Banana Fiber Hybrid Composites

The thermal conductivity of banana fiber hybrid composites is influenced by the fiber content. The increase in fiber content causes the thermal conductivity of the hybrid composite to decrease. For example, the thermal conductivity of the pineapple leaf and banana fiber hybrid composites decreases with increasing fiber count. The thermal conductivity of pineapple leaf fiber and banana fiber hybrid composites was found to vary from 0.171 to 0.213 $Wm^{-1} K^{-1}$ at 30–120 °C [29].

Numerically, the thermal conductivity of the hybrid composite can be determined by simulating steady-state heat transfer and the effective thermal conductivity with an acceptable experimental value agreement of 0–6.5% and an error of 0–11%, respectively, we can predict with the finite element method. Furthermore, the percentage error is related to the value of the effective thermal conductivity of the different models, as shown in Table 9.4 where the marginal deviation occurs due to the assumptions when decreasing the correlation.

9.6 Thermal Diffusivity

An important property involving heat transfer in unstable conditions is thermal diffusivity that can be used to store heat energy from the relative heat energy delivered. In the heat transfer analysis, the thermal diffusivity value of the hybrid composite material can be calculated from the specific heat capacity and thermal conductivity and density values, as given in Eq. (9.2) [32]:

$$\alpha = \frac{k}{\rho C_p} \tag{9.2}$$

Table 9.4 Percentage errors related to effective thermal conductivity values from different models.

		Percentage errors					
Composite	Experimental	Proposed model	FEM	Halpin–Tsai	Rule of hybrid mixture	Lewis and Nielsen	Geometric mean
S1	0.363	0	0	0	0	0	0
S2	0.331	1.751	0.607	0.858	1.933	1.403	6.672
S3	0.303	3.153	1.689	3.000	3.274	3.483	14.082
S4	0.271	5.045	2.264	5.336	4.588	5.622	21.411
S5	0.245	6.13	3.813	9.502	4.207	9.685	30.636
S6	0.329	1.908	0.920	0.970	2.254	1.419	7.973
S7	0.301	1.986	3.793	4.773	2.665	4.877	18.414
S8	0.266	4.693	5.138	6.401	5.336	5.953	26.193
S9	0.239	5.007	10.648	11.401	4.985	10.460	37.829
S10	0.333	1.362	0.604	0.935	1.613	1.385	5.385
S11	0.304	2.564	1.672	2.729	2.916	3.135	11.017
S12	0.274	4.463	1858	4.211	4.547	4.521	15.955
S13	0.25	4.689	2.459	8.105	3.828	8.480	23.314

Source: Devireddy et al. [31].

Table 9.5 The specific heat and thermal diffusivity values of the hybrid composites.

Composites	Composition (wt%)	Specific heat	Thermal diffusivity
S1	Epoxy (100)	3020	1.048×10^{-7}
S2	Epoxy (90) and banana (5) + jute (5)	2780	1.036×10^{-7}
S3	Epoxy (80) and banana (10) + jute (10)	2610	1.007×10^{-7}
S4	Epoxy (70) and banana (15) + jute (15)	2430	9.655×10^{-7}
S5	Epoxy (60) and banana (20) + jute (20)	2250	9.411×10^{-7}
S6	Epoxy (90) and banana (2.5) + jute (7.5)	2850	9.994×10^{-7}
S7	Epoxy (80) and banana (5) + jute (15)	2720	9.548×10^{-7}
S8	Epoxy (70) and banana (7.5) + jute (22.5)	2500	9.148×10^{-7}
S9	Epoxy (60) and banana (10) + jute (30)	2290	8.958×10^{-7}
S10	Epoxy (90) and banana (7.5) + jute (2.5)	2730	1.063×10^{-7}
S11	Epoxy (80) and banana (51) + jute (5)	2560	1.034×10^{-7}
S12	Epoxy (70) and banana (22.5) + jute (7.5)	2420	9.854×10^{-7}
S13	Epoxy (60) and banana (30) + jute (10)	2230	9.748×10^{-7}

Source: Devireddy et al. [31].

Table 9.6 Various applications of banana-based hybrid composite and other composites.

Banana-based composite	Coir-based composite	Jute-based composite
Sanitary napkin, building panels diaper, packing material, projector cover, furniture, brushes and brooms, helmets and postboxes, storage tank, paper industry, textile	Flush door shutters, roofing sheets, building panels, paperweights, voltage stabilizer cover, filling material for the seat upholstery, mirror casing, ropes and yams (for nets, bags, and mats), padding for mattresses and seat cushions	Door frames, packing, device, door shutters, transportation, building panels, geotextiles, and chipboards, rooting, absorbent cotton, storage sheets, transport, door frames, door shutters, household accessories, shopping bag

Sources: Shahinur and Hasan [24], Barman et al. [33].

where α is the thermal diffusivity, k is thermal conductivity, and ρ and C_p are density and specific heat capacity, respectively. C_p and α values of the hybrid composites are shown in Table 9.5.

Table 9.5 shows that the specific heat capacity and thermal diffusivity values are influenced by the fiber contents. The increasing number of fibers in the hybrid composite causes specific heat capacity and thermal diffusivity to decrease. These results have been proven by Devireddy and Biswas [31] on banana fiber and hemp hybrid composites that when heated or cooled take longer than neat epoxy [31].

9.7 Applications

Due to the good thermal property of the banana fiber-based hybrid composites, the composite is used in building components, agriculture sector in the cold region, and automobiles [31]. Furthermore, for various applications of banana hybrid composite in construction and other industries, see in Table 9.6.

9.8 Conclusions

In this chapter, the thermal properties of banana-based hybrid composites have been reviewed. Hybridization of banana with other materials improves the functional properties of the ensuing composite materials, especially the enhancement of thermal properties. Banana-based hybrid composites possessed excellent flame resistance. Cellulosic components of the banana fiber could improve the thermal stability of the composites. Owing to the better thermal properties, the banana-based hybrid composites are widely used in applications such as the construction industry and automotive industry, as well as in household applications.

References

1 Mohammed, L., Ansari, M.N.M., Pua, G. et al. (2015). A review on natural fiber reinforced polymer composite and its applications. *International Journal of Polymer Science* 15: 1, 243947–8. https://doi.org/10.1155/2015/243947.

2 Chandrasekar, M., Shahroze, R.M., Ishak, M.R. et al. (2019). Flax and sugar palm reinforced epoxy composites: effect of hybridization on physical, mechanical, morphological and dynamic mechanical properties. *Materials Research Express* 6 (10): 105331.

3 Krishnasamy, S., Thiagamani, S.M.K., Muthu Kumar, C. et al. (2019). Recent advances in thermal properties of hybrid cellulosic fiber reinforced polymer composites. *International Journal of Biological Macromolecules* 141: 1–13. https://doi.org/10.1016/j.ijbiomac.2019.08.231.

4 Krishnasamy, S., Muthukumar, C., Nagarajan, R. et al. (2019). Effect of fibre loading and $Ca(OH)_2$ treatment on thermal, mechanical, and physical properties of pineapple leaf fibre/polyester reinforced composites. *Materials Research Express* 6 (8): 085545. https://doi.org/10.1088/2053-1591/ab2702.

5 Thiagamani, S.M.K., Krishnasamy, S., and Siengchin, S. (2019). Challenges of biodegradable polymers: an environmental perspective. *Applied Science and Engineering Progress* 12 (3): 1. https://doi.org/10.14416/j.asep.2019.03.002.

6 Sapuan, S.M., Leenie, A., Harimi, M., and Beng, Y.K. (2006). Mechanical properties of woven banana fibre reinforced epoxy composites. *Materials and Design* 27: 689–693.

7 Dubey, N. and Agnihotri, G. (2013). Evaluation of tensile and interfacial strength of coconut palm leaf midrib as a potential reinforcement for plastics. *International Journal of Scientific and Research Publications* 5 (2): 1–6.

8 Zin, H.M., Abdan, K., and Norizan, M.N. (2019). The effect of different fiber loading on flexural and thermal properties of banana/pineapple leaf (PALF)/glass hybrid composite, Chapter 1. In: *Structural Health Monitoring of Biocomposites, Fibre-Reinforced Composites and Hybrid Composites* (eds. M. Jawaid, M. Thariq and N. Saba), 1–17. Woodhead Publishing https://doi.org/10.1016/B978-0-08-102291-7.00001-0.

9 Sari, N.H., Wardana, I.N.G., Irawan, Y.S., and Siswanto, E. (2018). Characterization of the chemical, physical, and mechanical properties of NaOH-treated natural cellulosic fibers from corn husks. *Journal of Natural Fibers* 15 (4): 545–558.

10 Indira, K.N., Jyotishkumar, P., and Thomas, S. (2012). Thermal stability and degradation of banana fibre/PF composites fabricated by RTM. *Fibers and Polymers* 13: 1319–1325. https://doi.org/10.1007/s12221-012-1319-x.

11 Muthukumar, K., Sabariraj, R.V., Dinesh Kumar, S. et al. (2019). Investigation of thermal conductivity and thermal resistance analysis on different combination of natural fiber composites of banana, pineapple and jute. *Materials Today: Proceedings* https://doi.org/10.1016/j.matpr.2019.09.140.

12 Venkateshwaran, N., Elayaperumal, A., Alavudeen, A., and Thiruchitrambalam, M. (2011). Mechanical and water absorption behaviour of banana/sisal reinforced hybrid composites. *Materials & Design* 32: 4017–4021.

13 Bansal, R.K. and Agarwal, R. (1984). Studies on thermal stability of epoxy resins. *Angewandte Makromolekulare Chemie* 127 (1): 43–47.

14 Bora, C., Bharali, P., Baglari, S. et al. (2013). Strong and conductive reduced graphene oxide/polyester resin composite films with improved mechanical strength, thermal stability and its antibacterial activity. *Composites Science* 127 (1): 43–47.

15 Chandrasekar, M., Siva, I., Kumar, T.S.M. et al. (2020). Influence of fibre inter-ply orientation on the mechanical and free vibration properties of banana fibre reinforced polyester composite laminates. *Journal of Polymers and the Environment* 28: 2789–2800.

16 Senthilkumar, K., Ungtrakul, T., Chandrasekar, M. et al. (2020). Performance of sisal/hemp bio-based epoxy composites under accelerated weathering. *Journal of Polymers and the Environment* 29: 624–636.

17 Vivek, S. and Kanthavel, K. (2018). Effect of bagasse ash filled epoxy composites reinforced with hybrid plant fibres for mechanical and thermal properties. *Composites Part B: Engineering* https://doi.org/10.1016/j.compositesb.2018.10.038.

18 Kabir, M.M., Wang, H., Lau, K.T., and Cardona, F. (2012). Chemical treatments on plant-based natural fibre reinforced polymer composites: an overview. *Composites Part B: Engineering* 43 (7): 2883–2892.

19 Kabir, M.M., Wang, H., and Islam, M.M. (2013). Mechanical and thermal properties of jute fibre reinforced composites. *Journal of Multifunctional Composites* 1 (1): 71–77.

20 Rahman, M.R., Huque, M.M., Islam, M.N., and Hasan, M. (2009). Mechanical properties of polypropylene composites reinforced with chemically treated abaca. *Composites Part A: Applied Science and Manufacturing* 40 (4): 511–517.

21 Singh, A.A. and Palsule, S. (2014). Thermal properties of jute fiber reinforced chemically functionalized high-density polyethylene (JF/CFHDPE) composites. *Applied Polymer Composites* 2 (2): 97–108.

22 Haque, M.M., Hasan, M., Islam, M.S., and Ali, M.E. (2009). Physico-mechanical properties of chemically treated palm and coir fiber reinforced polypropylene composites. *Bioresource Technology* 100 (20): 4903–4906.

23 Haque, M., Islam, N., Huque, M. et al. (2010). Coir fiber reinforced polypropylene composites: physical and mechanical properties. *Advanced Composite Materials* 19 (1): 91–106.

24 Shahinur, S. and Hasan, M. (2020). Reference module in materials science and materials engineering. *Encyclopedia of Renewable and Sustainable Materials* 2: 751–756. https://doi.org/10.1016/B978-0-12-803581-8.10987-7.

25 Srinivasan, V.S., Rajendra Boopathy, S., Sangeetha, D., and Vijaya Ramnath, B. (2014). Evaluation of mechanical and thermal properties of banana-flax based natural fibre composite. *Materials & Design* 60: 620–627. http://dx.doi.org/10.1016/j.matdes.2014.03.014.

26 Van Krevelen, D.W. (1975). Some basic aspects of flame resistance of polymeric materials. *Polymer* 16: 615–620.

27 Vijaya Ramanth, B., Elanchezhian, C., Nirmal, P.V. et al. (2014). Experimental investigation of mechanical behavior of jute-flax based glass fiber reinforced composite. *Fibers and Polymers* 15: 1251–1262.

28 Boopalan, M., Niranjanaa, M., and Umapathy, M.J. (2013). Study on the mechanical properties and thermal properties of jute and banana fiber reinforced epoxy hybrid composites. *Composites Part B: Engineering* 51: 54–57. https://doi.org/10.1016/j.compositesb.2013.02.033.

29 Santosha, P.V. Ch, .R.K., Gowda, S.S.S.A., and Manikanth, V. (2018). Effect of fiber loading on thermal properties of banana and pineapple leaf fiber reinforced polyester composites. *Materials Today: Proceedings* 5: 5631–5635.

30 Vijaya Kumar, N., Sai Krishna, B., and Sai Chandrika, N. (2019). Evaluation of properties of glass-banana-fiber reinforced hybrid fiber polymer composite. *Materials Today: Proceedings* 18: 2137–2141.

31 Devireddy, S.B.R. and Biswas, S. (2016). Thermo-physical properties of short banana–jute fiber-reinforced epoxy-based hybrid composites. *Proceedings of the Institution of Mechanical Engineers, Part L: Journal of Materials: Design and Applications* 232 (11): 939–951. https://doi.org/10.1177/1464420716656883.

32 Richard Hetnarski, B. and Reza Eslami, M. (2009). Heat conduction problems. In: *Thermal Stresses – Advanced Theory and Applications*, edisi ke-Online-Ausg.e (ed. G.M.L. Gladwell), 132–218. Dordrecht: Springer Netherlands. ISBN: 978-1-4020-9247-3.

33 Barman, A., Katkar, P., and Asagekar, S. (2017). Natural and sustainable raw materials for sanitary napkin. *Journal of Textile Science and Engineering* 7 (3): 1000308.

10

Thermal Properties of Kenaf Fiber-Based Hybrid Composites

Chinnaiyan Deepa[1], Lakshminarasimhan Rajeshkumar[2], and Manickam Ramesh[3]

[1] *KIT-Kalaignarkarunanidhi Institute of Technology, Department of Computer Science and Engineering, Coimbatore, 641402, Tamil Nadu, India*
[2] *KPR Institute of Engineering and Technology, Department of Mechanical Engineering, Coimbatore, 641407, Tamil Nadu, India*
[3] *KIT-Kalaignarkarunanidhi Institute of Technology, Department of Mechanical Engineering, Coimbatore, 641402, Tamil Nadu, India*

10.1 Introduction

Plastic products seem to be the period of the present due to their widespread use up to now. In various advanced applications, thermoplastic polymers are commonly used. However, they have limited applications in many fields due to their lower thermal stability and lower strength. Despite of these polymer uncertainties, reinforcing materials are combined with polymer to improve its stability and strength. Increased cost and degradation of the fuel component restricts the use of synthetic fiber in polymer composites as a reinforcement material. Natural fibers are often used instead of synthetic fibers that are the preferred alternative for polymer matrix composites these days [1]. Natural fiber composites (NFCs) have become the obvious choice of many researchers and are broadly applied in many fields of engineering and technology due to the growing environmental concerns and sentience toward the use of biodegradable materials. NFC materials were continuously being employed as substitutes in many fields in place of carbon fiber and epoxy matrix composites by numerous researchers owing to their numerous advantages. NFCs were currently used in industrial and domestic applications that are progressing rapidly. They are mostly bio-based composites posing advantages including renewability, ecofriendliness, low weight and cost of manufacture, recycling ability, and biodegradability. Advantages of NFCs over the conventional synthetic fiber composites are that these are partially or completely biodegradable and is easily available from the natural resources. Their application focused on construction, aerospace, automotive, packing, and transportation [2].

Many fields of engineering utilizes advanced composite materials containing natural fibers specifically due to high values of strength and modulus compared with

Natural Fiber-Reinforced Composites: Thermal Properties and Applications, First Edition.
Edited by Senthilkumar Krishnasamy, Senthil Muthu Kumar Thiagamani,
Chandrasekar Muthukumar, Rajini Nagarajan and Suchart Siengchin.

their counterparts. Apart from the material-related characteristics like mechanical, thermal, chemical, and electrical characteristics, few other parameters such as ecofriendliness during the composite fabrication and disposal after the termination of fiber composite life cycle and their utilization are also to be accounted to have a hazard-free environment. Over a period of time, the emergence of NFCs play a significant role in defining the users' psychology to shift from a nonrenewable source to a renewable and natural fiber source for manufacturing composites. Normally biocomposites are the materials that have natural fibers such as kenaf, sisal, flax, etc. incorporated within bio-degradable polymers. Meanwhile those composites in which synthetic fibers like aramid and carbon are reinforced within biopolymers or natural fibers reinforced within nonbiodegradable matrix materials are termed as partial biocomposites [3–5].

Kenaf is one such natural fiber that has gained the limelight. Many researchers and industries utilize it to manufacture fiber-reinforced composites and many products to maintain the environmental stability and awareness and to satisfy the government norms laid by many countries. Kenaf is a major crop that takes the second place in cultivation next to cotton all around the globe but particularly is a native crop of Africa. This can be readily applied in the fields like construction and automobile since its length is very high as it is derived from the bast or the outer bark of the plant. It is dicotyledon type of plant contained in *Hibiscus* (Family: Malvaceae), a crop originating from ancient Africa before 400 decades that spread around the globe and reached farther most parts of the world. Kenaf fibers had a higher content of cellulose and were biodegradable. They possess a unique feature of lacking silica content in it and plays a vital role in minimizing the abrasion in processing machinery [6]. They also possess widespread applications as reinforcements in numerous areas of engineering and technology. Foremost among its utilization is that it is found to be an alternative material for paper and wood pulp composites that indirectly governs the conservation of forests. Alongside, kenaf fibers are also employed as reinforcements within broader classes of matrix materials to form composites that are tailored as per the requisite application. They are mostly employed in woven or nonwoven form as reinforcements in composite materials that finds its application in textile industry, fiber boards, and automobile industries [7, 8]. Thermal properties like thermogravimetric analysis (TGA), differential scanning calorimetry (DSC), derivative thermogravimetric analysis (DTG), dynamic mechanical analysis (DMA), flammability behavior, and heat deflection temperature (HDT) of kenaf fiber hybrid composites are elaborated in this chapter.

10.2 Hybrid Composites

Hybrid composites are the tailored materials for engineering applications that comprises more than two reinforcements of similar or dissimilar molecular structures within same matrix. Individual fiber-reinforced composites may cumulate only the characteristics of one fiber contained in the composite, whereas hybrid composites bring in the key characteristics of all the reinforcements that pave the way for many advanced solutions to technological problems. Researchers determined that these hybrid composites evolved from single fiber-reinforced composites and are potentially advantageous than individual fiber-reinforced composites [9].

Few researchers manufactured polymer composites with fiber combinations such as kenaf/glass, banana/carbon, and so on and found these combinations to be very much effective in terms of improving strength and modulus. Analysis were made by hybridizing kenaf with jute and hemp fibers in epoxy matrix, and it has been reported that the mechanical properties of hybrid composites were higher when compared with the individual fiber composites in the case of treated and untreated fibers. Hybrid composites with kenaf and seaweed fibers in thermoplastic matrix were assessed for their moisture absorption behavior by few researchers. Results revealed that soil-buried hybrid fibers with higher kenaf content rendered weight loss for the composites, owing to the lower hydrophilicity of the kenaf fibers than seaweed fibers [10–12].

10.3 Thermal Properties

Studies on thermal properties of kenaf-based composites have become essential in the current scenario so as to have a distinction in the behavior of composites when kenaf fiber is used as an alternative. Many experimental properties like specific heat, coefficient of thermal expansion (CTE), and thermal conductivity of both fiber and composites are lagging and are yet to be precisely determined. Normally, thermal analysis is carried out in the kenaf-based composites to determine the volatile constituent content and moisture absorption capability of the fibers. Such experiments are of considerable significance because the presence of moisture and reactive materials renders a degrading impact on composite characteristics [12]. Standard characterization methods including TGA, DTG analysis, DSC, DMA/dynamic mechanical thermal analysis (DMTA), differential thermal analysis (DTA), thermal imaging, HDT, thermal stability, thermal degradation, thermal conductivity, thermal diffusion, and fire reaction/flame-retardancy properties were used for this purpose. In Sections 10.3.1–10.3.7, we will discuss in detail about the above standards.

10.3.1 Thermogravimetric Analysis

Usually TGA is performed on polymer composites to evaluate the change in mass of the composites over a period of time when the temperature to which they are exposed changes continuously. This is broadly applied to evaluate the thermal reaction kinetics and related thermal characteristics of the composites. TGA has become one of the thermal analysis methods used to calculate composite material's mass transition, thermal stability, and thermal decomposition [13]. This can also be used to find the char content of the composite developed due to the change in temperature, and the composite samples were made to undergo TGA with the aid of TGA-Q500 instrument. TGA was conducted within a temperature range of 25–700 °C, at rate of heat supplied as 20 °C per minute, in N_2 environment [14]. Table 10.1 summarizes the

Table 10.1 Mass loss percentage of composites from kenaf fiber/PVC/TPU.

	Percentage loss of weight (%)						
Sample (Fiber %)	$T_{101.84}$ (°C)	$T_{253.51}$ (°C)	$T_{300.18}$ (°C)	$T_{352.67}$ (°C)	$T_{399.43}$ (°C)	$T_{451.91}$ (°C)	$T_{502.43}$ (°C)
Resin only	0.59	5.27	51.11	65.18	74.71	84.41	90.85
20%	0.13	4.55	47.81	62.12	69.45	78.36	84.81
30%	0.83	5.86	49.56	61.81	68.27	75.77	81.63
40%	0.91	6.68	51.78	63.15	69.12	76.48	82.14
Fiber only	8.59	10.95	20.92	50.68	76.55	80.05	82.83

Source: Based on El-Shekeil et al. [15].

loss in weight of kenaf fiber reinforced in composites as a percentage with respect to different temperatures replicated from [15].

TGA was measured for woven kenaf fiber mat-reinforced in epoxy matrix hybridized with carbon fibers. Experiments were performed with the help of TGA Q500 model of thermogravimetric analyzer for recording the variation of time and temperature of the hybrid composites in a controlled environment. Heating temperature for the samples was maintained within the range of 25–800 °C, while the rate of heating was 10 °C min^{-1} with nitrogen as carrier gas flowing at 50 ml min^{-1}. After heating a 10 mg hybrid composite sample in a heating pan, the weight of the heated sample was measured and plotted as a graph against temperature. TGA tests were carried out for the silane-treated and silane-untreated pineapple leaf fiber and kenaf fiber-reinforced hybrid epoxy composites [16]. Thermal properties evaluation of hybrid fiber-reinforced samples were carried out with the aid of TGA Q500 instrument, and the heating temperature range was maintained between 30 and 800 °C with the heating rate of 20 °C per minute in atmospheric conditions. Results indicated that the treated hybrid composite materials exhibited low thermal degradation but high when the untreated hybrid fiber composites had higher degradation with relatively lower loss in weight. Authors attributed this behavior with the incorporation of treated pineapple leaf fiber into the composite, and its combined action with kenaf fiber enhanced the thermal stability of composite material [17].

TGA was conducted on bamboo and kenaf fiber-reinforced hybrid epoxy composites with the aid of a Mettler Toledo TGA2 analyzer, and the thermal stability along with the decomposition rate of the hybrid composites was determined conforming ASTM E1131-03 standards. Weight of the samples was maintained at 40 mg and were kept in an alumina crucible for heating. Temperature range for heating was maintained at 30–800 °C in a nitrogen-filled environment. Results revealed that the samples containing equal proportions of bamboo and kenaf fibers in epoxy matrix showcased better thermal stability when compared with individual fiber-reinforced composites and other hybrid composites. TGA was carried out for the kenaf and recycled carbon-reinforced hybrid cardanol composites. Cardanol is the matrix material used in the work that was a derivative liquid material from cashew nut shell [18]. TGA was performed by few researchers in a Q500 model TGA equipment within a temperature range of 30–900 °C at a heat supply rate of 10 °C min^{-1}. Various stages of decomposition of kenaf/carbon cardanol composite specimens were determined from the analysis as a function of change in temperature. The influence of individual constituent over the thermal and decomposition behavior along with the influence fiber volume fraction of individual fibers over the various phases of decomposition of hybrid composites was also determined from TGA. The results observed that the composites containing recycled carbon and kenaf in the ratio of 70 : 30 had maximum thermal stability compared with any other composites, owing to the increased presence of recycled carbon content [19].

Few works were performed on TGA for the pineapple and kenaf fiber-reinforced hybrid vinyl ester composites manufactured by hand layup method. TGA was carried out using TGA Universal V3.9A instrument, INTROP, Malaysia. The heating rate of the sample was maintained at 10 °C min^{-1}. From the experimental results, the thermal degradation of pineapple and kenaf took place within a temperature range of 250–370 °C, and the degradation of pure vinyl ester, individual pineapple, and kenaf fiber composites took place between 300 and 450 °C that were possible owing to the physical characteristics of the matrix material. Results showcased the fact that pineapple and kenaf hybridization improved the composites' thermal stability in comparison with the individual pineapple and kenaf fiber-reinforced vinyl ester composites. Such enhanced thermal stability of the hybrid composites unveil the promising fact of utilizing the hybrid composites for many broader applications in structural, aerospace, and automotive fields [20]. Experimenters determined the thermal properties of short kenaf fibers reinforced with thermoplastic polyurethane in natural rubber hybrid composites using TGA. From the results it was noticed that composites containing one part of kenaf fiber, four parts of natural rubber, and the remaining with thermoplastic polyurethane behaved in a better way in terms of thermal degradation, and maximum degradation temperature was noted to be 362 °C. Authors stated the appreciable biodegradability of the kenaf-based hybrid composites rendered the highest degradation temperature for the composites [21].

Figure 10.1 depicts the thermograph of kenaf-based hybrid composites [22]. Decomposition graphs were observed to be similar for all the kenaf hybrid composites, but the curve of individual kenaf-reinforced composite stood alone. Three phases of decomposition could also be observed from the graph for both types of composites (individual and hybrid composites). As a first phase, composite degradation could be noticed within 100–200 °C that was due to the loss of moisture content from the fibers as a result of thermal degradation. At a temperature range of 300–450 °C, the aromatic group of epoxy resin decomposed and might have undergone pyrolysis that rendered a significant weight loss of the composites. Additionally, microconstituents like pectin, lignin, and glycoside links of cellulose and hemicellulose of kenaf fiber might have decomposed, supplementing the loss of weight in the same temperature range [23, 24].

The thermal stability of the magnesium hydroxide filler and kenaf-reinforced epoxy hybrid composites was determined. The results of the analysis exhibited

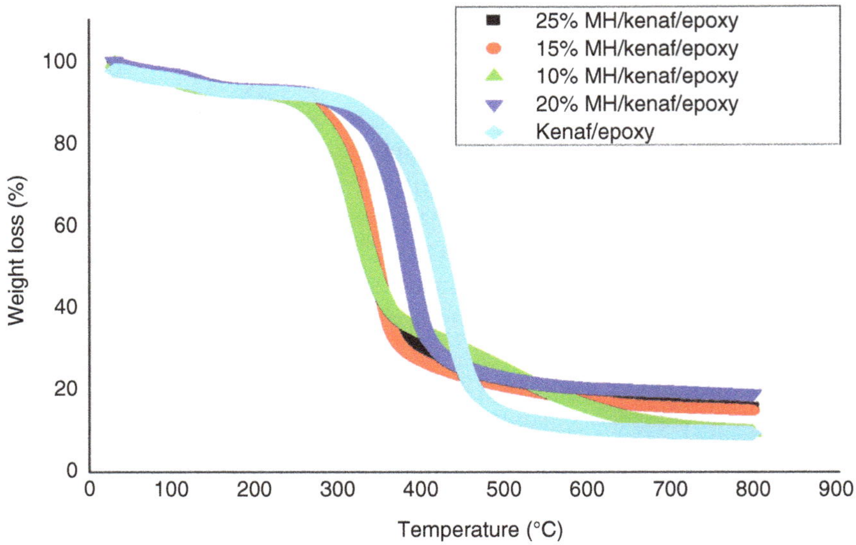

Figure 10.1 TGA curves of magnesium hydroxide/kenaf/epoxy hybrid composites. Source: Saba et al. [22].

the fact that the infusion of stiff magnesium hydroxide particulates into the epoxy matrix containing kenaf fibers augmented both storage and loss moduli and also reduced the damping behavior of the hybrid composites [25].

10.3.2 Dynamic Mechanical Analysis

DMA is performed on the polymer matrix composites to determine the viscoelastic properties of the samples by evaluating the complex modulus of the composites. Complex modulus is determined by applying a sinusoidal stress over the samples and measuring the strain induced due to the stress. DMA was carried out to evaluate the storage modulus (E'), loss modulus (E''), and $\tan\delta$ as a function of temperature [26]. Three-point bending method of analysis was performed to evaluate the dynamic mechanical characteristics of kenaf and magnesium hydroxide filler-reinforced hybrid epoxy composites by means of DMA Q800 instrument at a frequency level of 1 Hz and a temperature range of 30–200 °C with a heating rate of 5 °C min^{-1}. Specimen dimensions were maintained at $45 \times 10 \times 3\,\text{mm}^3$ [27]. DMA was carried out for pineapple and kenaf fibers treated with silane coupling agent hybrid composite materials as per ASTM D4065-01 standards and evaluated the viscoelastic properties as a function of temperature. Tests were carried out with TA DMA Q800 equipment in a three-point mode of bending with an oscillation frequency of 1 Hz. Temperature was increased according to the sinusoidal strain curve from 30 to 150 °C at an incremental rate of 50 °C min^{-1} under a constrained environment. Results revealed that the hybrid composite samples containing untreated pineapple and kenaf fibers in the ratio of 70 : 30 exhibited minimum storage modulus, while the samples with treated fibers in the ratio of 70 : 30 showcased maximum

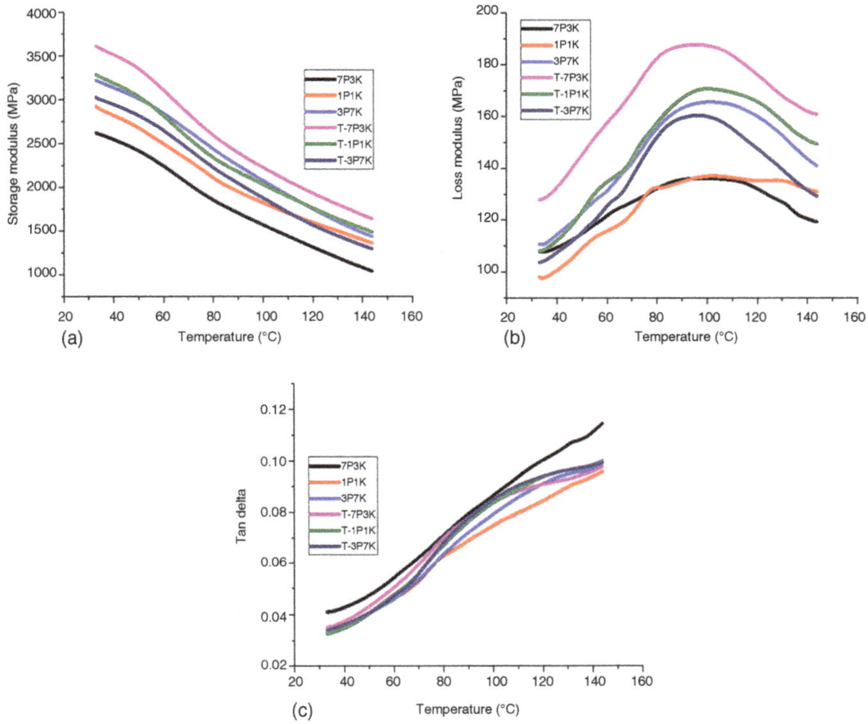

Figure 10.2 (a) Storage modulus, (b) loss modulus, and (c) damping factor of pineapple leaf/kenaf fiber hybrid composites. Source: Asim et al. [29].

value of loss and storage modulus. Samples with 50 : 50 composites experienced a low peak in tan delta curve, while the samples with 70 : 30 had maximum peak [28].

The investigations of storage modulus with respect to temperature variation for pineapple leaf/kenaf fiber hybrid composites were illustrated in Figure 10.2a, where the value was inversely proportional to the temperature and decreases with temperature rise. Storage modulus value, at lower temperatures, for the untreated hybrid fiber-reinforced composites was lower than the treated hybrid fiber-reinforced composites. A larger difference in corresponding values was also observed from the results. Figure 10.2b depicts the loss modulus graph for pineapple leaf/kenaf fiber hybrid composite material. Maximum value of glass transition temperature could be observed for treated fiber-reinforced composites, while untreated fiber hybrid composites exhibited lower values. From the loss modulus curve, broader peaks could be observed for treated fiber composites compared with that for the untreated fiber composites, which could be attributed to the relaxation induced due to the addition of treated pineapple leaf and kenaf fibers with phenolic resin. Figure 10.2c illustrates the temperature-dependent damping factor curve for pineapple leaf/kenaf fiber-reinforced phenolic matrix composites at a frequency value of 1 Hz. At the earlier stages, owing to the frozen chain segments in composite constituents, the damping factor curve was beneath the glass transition temperature for all composites while it gradually increased with temperature rise and attained its highest value at the vicinity of transition zone and decreased in the rubbery region. This could be attributed to the maximum molecular mobility within the composites. It was also noticed that the damping ratio increased with the increase in pineapple leaf fiber content, and when the proportion proportionally varied with kenaf fiber, the value of damping ratio varied accordingly. This could possibly be due to the noncompatibility of both fibers and poor adhesion at the interfacial zone. From all the test results, it is observed that the similarity in peaks for all hybrid samples and the deviation between highest and lowest values in peak is very minimum, owing to the unconstrained motion of the fiber polymer chain within the matrix. This rendered a saturated damping characteristics for all the hybrid composite samples [29].

DMA was performed on kenaf/basalt fiber-reinforced hybrid epoxy composites using the aforementioned parameters with the specimen dimensions of $60 \times 12 \times 3\,mm^3$. Operating temperature range was maintained at 30–200 °C. Curves like storage modulus, loss modulus, and tan δ were plotted using the results obtained. From the results, it was found that the storage modulus, loss modulus, and tan delta values were maximum for the kenaf fiber-reinforced epoxy composites when compared with the hybrid composites. This could be attributed to the ability of kenaf fibers to undergo better interfacial bonding with the matrix and its possession of high value of stiffness [30]. Dynamic mechanical properties were investigated for bamboo/kenaf hybrid epoxy composites prepared through molding technique. DMA was performed using a Mettler Toledo DMA1 instrument, and the complex modulus (E^*) was evaluated in a three-point bending approach as per ASTM D5023-01 standards. Dimensions of the sample were maintained at 45 mm unclamped length, 5 mm width, and 4 mm thickness. Parameters for the analysis were maintained at temperature of 25 °C, amplitude of 20 μm, frequency of 1 Hz, and current force of 3 N. At the time of recording data, the authors neglected the readings during the initial four minutes as they were highly fluctuating accounting the adaptability of composites to the conditions and the data was recorded every 10 minutes thereafter. Average of all the data of the E^* values was taken for the discussion. Results indicated that the E^* for the hybrid composites containing the bamboo and kenaf fibers in the ratio of 50 : 50 was high, due to the accelerated weathering of the samples and the cross-links induced by the ultra-voilet (UV) irradiation. Other DM properties like tan delta were also high for the composites with same composition [31].

Dynamic mechanical characteristics were evaluated with respect to the effect of hybridization of the cardanol hybrid composites reinforced with recycled carbon and kenaf fibers. ASTM D4065-01 standards was followed to perform the analysis with a specimen dimension of $35 \times 12 \times 3\,mm^3$ in an INTROP DMA Q 800 equipment. From the experimental results, it was noticed that the value of storage modulus of composites containing carbon and kenaf fibers in the ratio of 70 : 30 was maximum due to their presence in the rubbery region while its tan delta value was the second maximum with the highest value for pure cardanol composites indicating the dominating brittleness of the cardanol samples [32].

10.3.3 Derivative Thermogravimetric Analysis

DTG was also performed on the samples to evaluate the loss of weight in terms of temperature. Results indicated that the individual carbon fiber-reinforced

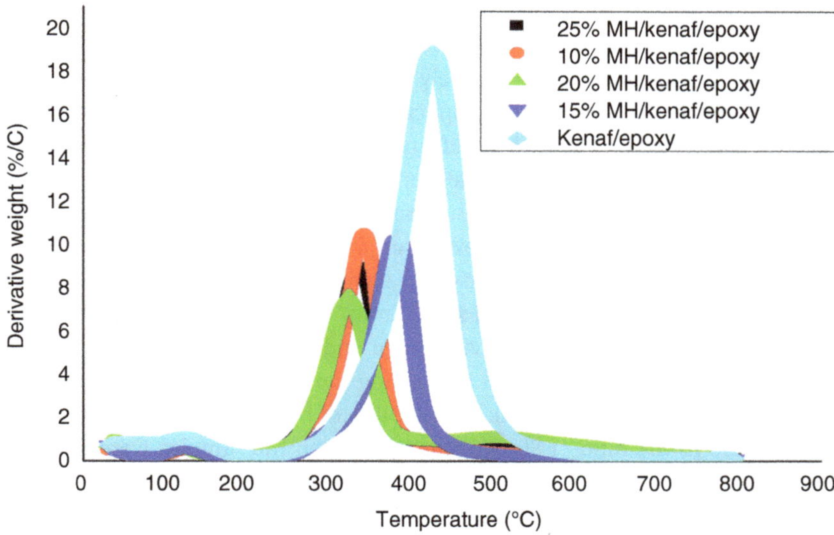

Figure 10.3 DTG plot of kenaf fiber hybrid composites. Source: Saba et al. [22].

composites exhibited better thermal stability in TGA and DTG analysis than the hybrid composite material, but among the hybrid composites, the sample containing more amount of kenaf fiber showcased appreciable thermal stability [25]. The DTG analysis was used to evaluate the weight loss and adjudge the thermal degradation properties of the pineapple leaf/kenaf fiber hybrid phenolic composites. The DTG curves of kenaf fiber-based hybrid composites are presented in Figure 10.3. Two characteristic peaks could be witnessed in the DTG graph within a temperature range of 130-–150 °C and 320-–500 °C. These two prominent peaks in the DTG graph depict the nonhomogeneous attributes of the kenaf fiber and the hybrid system in which it was present. At around 100 °C, the first peaks rise and denoted degradation that was primarily due to the hydroxyl molecule present within the kenaf fiber. Second peak, at a temperature of 400 °C, clearly depicts the weight loss due to the increase in temperature and was observed to be relatively broader than the first DTG peak. The rise of this peak was substantiated by decomposition of epoxy matrix in all the specimens [22].

10.3.4 Differential Scanning Calorimetry

DSC is performed on the polymer composites to determine the heat gradient necessary to increase the temperature of a composite represented as a function of temperature itself. DSC analysis was performed on the woven kenaf and carbon fiber mats reinforced within epoxy matrix to evaluate the heat flow changes that induce the material phase changes with the aid of DSC Q20 instrument. In nitrogen atmosphere, within a temperature range of 25–350 °C, three hybrid composite samples each weighing 4 mg were heated in aluminum crucible, having a pinhole in it. The heat was supplied at a rate of 10 °C min^{-1}, and the average of the individual values was considered finally. Results showed that when the woven mat was of simple weave arrangement, the fiber-matrix interfacial adhesion during DSC analysis was found to be substantial. When compared with all other samples, the decomposition of hybrid composites occurred at higher temperatures depicting its good thermal stability. From this the authors conclusively stated that the hybrid composites were most suitable for higher-temperature applications [33]. DSC tests were performed for nonwoven bamboo and woven kenaf fiber mat-reinforced hybrid epoxy composites using a Mettler Toledo DSC1 analyzer, and the oxidation onset temperature (OOT) hybrid composite samples were evaluated as per ASTM E2009-02 standards. In an aluminum crucible pan, a composite sample weighing 30 mg was taken, and they were heated at 25–250 °C in an oxygen environment while the heating rate and the gas flow rate were maintained at 2 °C min^{-1} and 50 ml min^{-1}, respectively. To promote oxidation between the environment and the heated composite sample at elevated levels, the aluminum pan containing the samples was left uncovered. Results of DSC showed that the oxidation stability of hybrid samples with bamboo and kenaf fibers in the ratio of 30 : 70 was high when the samples were made to undergo accelerated weathering while the sample with the ratio of 70 : 30 bamboo and kenaf fibers exhibited better oxidation stability when they were buried in the soil [34].

DSC analysis was carried out for the kenaf and recycled carbon-reinforced hybrid cardanol composites with the aid of DSC Q20 machine at a temperature range and heating rate of 30–280 °C and 10 °C min^{-1}, respectively [23]. Thermal properties of woven kenaf fiber mat and glass fiber mat-reinforced in unsaturated polyester resin were analyzed by DSC experiments. Tests were carried out for the samples before and after samples weathering. Composite showed slow and steady initial weight loss until major weight loss is detected at a temperature of 390 °C. While weathering increased, the weight loss started at a lowest temperature of 290 °C. It was also determined from the results that the interfacial bond between the molecules of fibers and matrix weakened by the moisture absorption capability of the both glass and kenaf fibers [26].

10.3.5 Thermal Mechanical Analysis

Thermomechanical analysis (TMA) is performed on the polymer composites to evaluate the change in dimensions of the samples as a function of temperature when a load is applied over them. This test is performed according to ASTM E831-30 standards, and the stress developed in the specimen is evaluated as a function of temperature. Few researchers carried out TMA, adopting ASTM D696 standard to analyze the CTE with the aid of TA Q400 equipment in N_2 atmosphere with a temperature range of 30–200 °C. Rate of heating was maintained at 5 °C min^{-1}. Variation in the specimen dimensions were assessed in the lateral direction to evaluate the CTE of hybrid composite materials [35]. TMA of the treated and untreated pine-apple leaf fiber (PALF)/kenaf fiber composites was evaluated by varying the fiber volume fraction and subsequently CTE, and changes in dimensions were evaluated in both rubbery and glass transition regions. Results showcased that the untreated fiber-reinforced composites exhibited a steep curve towards the downward direction at the rubbery region that could be possibly due to the presence of low cross-linked materials. Meanwhile, treated fiber-reinforced composites had exhibited an upward curve, illustrating a better compatibility between fiber and matrix, resulting in the enhanced mechanical characteristics of the composite materials [36, 37].

Thermomechanical properties of nonwoven bamboo mat fiber- and woven kenaf fiber-reinforced hybrid epoxy composites were evaluated. TMA was carried out utilizing Mettler Toledo TMA/SDTA1+ equipment of Schwerzenbach. Specimens were subjected to the evaluation of CTE under a small applied load in dilatometric mode. Test samples have a width of 3–6 mm and thickness of 4–6 mm and were manufactured with flat ends for easier load application and precise test results. A silica disc was kept between the probe (made of quartz with a ball of 3 mm diameter) and the composite sample to ensure the homogeneity of applied force (of magnitude 0.02 N) over the sample surface. For measuring the CTE, the specimens were subjected to a temperature ranging from −30 to 100 °C while the increment of temperature was 5 °C min^{-1}. CTE of the fiber-reinforced composites were evaluated both in transverse and longitudinal directions of fibers as the composite was an anisotropic material and the CTE would be influenced by the fiber orientation. Results revealed that the woven kenaf mat-reinforced composites exhibited better dimensional stability than the other composites. When the content of kenaf was increased in the hybrid composites, the dimensional stability was observed to improve, and especially at bamboo and kenaf fibers with the ratio of 50 : 50 in the hybrid composites, by having lowest rate of shrinkage, a better dimensional stability was noticed [38].

10.3.6 Flammability Tests

Flammability tests are usually carried out to assess the quickness of the polymer composite to burn when exposed to a flame. From this test, the reaction of the material while burning is physically evaluated. Flame retardance or flammability behavior of treated and untreated fibers of bamboo- and kenaf-reinforced epoxy composites was carried out by means of UL 94 tests in horizontal and vertical mode conforming ASTM D635 and ASTM D3801 standards, respectively. In the case of horizontal test, the specimen mounting orientation was also horizontal, while the flame was introduced at one end. Flammability of composites was measured when the flame traverses across a length of 25–100 mm from one end. In the case of horizontal tests, all the composite samples behaved as good fire-resisting materials, and specifically composite with a fiber ratio of 30 : 70 for treated pineapple leaf and kenaf fibers had maximum fire resistance. In vertical tests, the composite samples containing the treated pineapple leaf and kenaf fibers in the ratio of 30 : 70 and 50 : 50 had better flame retardance [35, 36].

Flammability tests were conducted on the recycled carbon and kenaf fiber-reinforced cardanol oil matrix hybrid composites as per ASTM 3801 standard in UL 90 HB. Before subjecting the specimens to testing, they were initially treated at a temperature of 23 °C for a time period of 48 hours at a relative humidity of 50%. Following this, the specimen was held in position by a boss clamp, and a flame of length 20 mm was applied by means of a Bunsen burner at an inclination of 45° at the thickness end of the specimen. Entire length of the specimen, as per standards, was maintained at 125 mm and was divided into three portions with the length of first, second, and third parts of 25, 75, and 25 mm, respectively. Specimen was exposed to the flame for 30 seconds, and then the time for the flame to extinguish itself from the composite specimen and the length of burnt area were noted. As a general case, the flame has to be stopped from application when it reaches the end of first portion, and the burning rate was calculated accordingly. Same procedure was adopted for five samples of same composition, and the average of the five was ultimately considered. Results indicated that the pure cardanol samples exhibited better fire retardance when compared with the fiber-reinforced samples, which could be due to the development of more amount of char and less heat transfer, while the hybrid composite samples possessed better flame resistance when compared with individual fiber-reinforced cardanol composites but lesser than the values of pure cardanol composites [37–39].

10.3.7 Heat Deflection Temperature

HDT of a polymer material denotes the temperature at which the deflection of the composite specimen reaches a saturated and optimum value under a particular rate of heating. It is a parameter measured to assess the heat proof of a polymer composite material [40]. HDT was performed on kenaf-based hybrid composites containing graphene nanoparticles (GNPs), ammonium poly phosphate (APP), magnesium hydroxide (MgOH), and magnesium oxide (MgO) suspended within the hybrid polypropylene composites with maleic anhydride polypropylene (MAPP) as compatabilizer. Tests were carried out following ASTM D648 standards using GOTECH GT-HV2000 W HDT/VICAT instrument, and the specimens were prepared with dimensions of $120 \times 4 \times 4\,mm^3$. Specimens were applied with an initial load of 0.46 MPa, and then they were enforced with a higher load of 1.80 MPa. Values of deflection and temperature were recorded under each magnitude of load. Results revealed that the HDT of kenaf-GNP with MgOH exhibited highest values when compared with all other quaternary hybrid composites. Such high value for this specific composition could be attributed to the synergy obtained between GNP and MgOH at this particular load, high flame retardance of GNP, high interfacial connectivity of all the constituents, and high modulus of kenaf and GNP [41, 42]. Figure 10.4 depicts the field emission scanning electron microscopy (FESEM) images of fractured composite sample containing kenaf fibers, GNP, and MgOH.

10.4 Conclusion

From all the above discussions. it is evidently witnessed that kenaf fibers have an enormous potential to be applied in many field like automotive, aerospace, and other related fields due to many advantageous characteristics such as less cost, better mechanical and thermal properties, and low weight. Hybridization of kenaf with synthetic fibers like carbon and glass has resulted in positive characteristics of

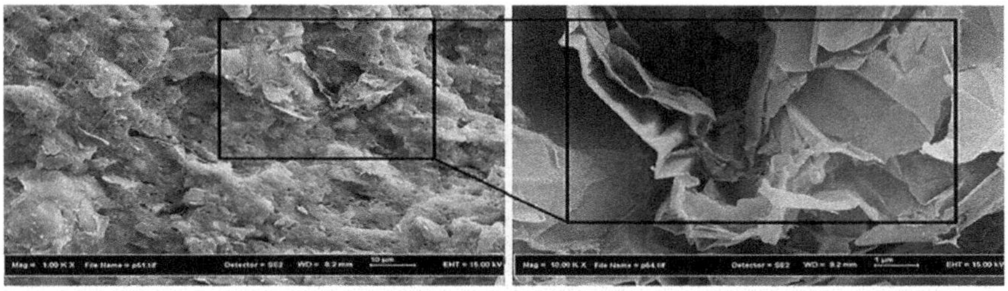

(a) (b)

Figure 10.4 FESEM of K-GNP-MgOH quaternary composites at (a) 100× and (b) 1000×. Source: Idumah et al. [42].

the composites due to the fact that glass fibers have excellent mechanical properties and kenaf has better thermal properties. Thermal stability of the hybrid composites improved while hybridizing kenaf fibers with various natural or artificial fibers in the proportion range of 30–50%. Decomposition temperature also increases due to the effect of incorporation of kenaf fibers. Owing to the good compatibility between kenaf fiber and other natural fibers, dimensional stability with minimum shrinkage of hybrid composites was also obtained. Due to their thermal conductivity, flame retardance of the hybrid composites containing kenaf as one of its constituents has moved to positive zone [23, 28, 34, 36, 38]. Yet, it was found from the above discussions that very limited literature or experiments were carried out to evaluate the thermomechanical behavior or HDT and other thermal properties of kenaf-based hybrid composites. If the experiments widely cover the aforementioned data, the scope for using kenaf fibers as one of the constituents in hybrid composites may expand broadly and may reach new heights.

References

1. Manral, A. and Bajpai, P.K. (2019). Static and dynamic mechanical analysis of geometrically different kenaf/PLA green composite laminates. *Polymer Composites* 41: 691–706.
2. Ramesh, M. (2018). Flax (*Linum usitatissimum* L.) fibre reinforced polymer composite materials: a review on preparation, properties and prospects. *Progress in Materials Science* 102: 109–166.
3. Aisyah, H.A., Paridah, M.T., Sapuan, S.M. et al. (2019). Thermal properties of woven kenaf/carbon fibre-reinforced epoxy hybrid composite panels. *International Journal of Polymer Science* 2019: 5258621. https://doi.org/10.1155/2019/5258621.
4. Ramesh, M., Palanikumar, K., and Reddy, K.H. (2017). Plant fibre based bio-composites: sustainable and renewable green materials. *Renewable and Sustainable Energy Reviews* 79: 558–584.
5. Ramesh, M. and Rajeshkumar, L. (2018). Wood flour filled thermoset composites. In: *Thermoset Composites: Preparation, Properties and Applications*, vol. 38

(eds. A. Khan, S.A. Bhawani, A.M. Asiri and I. Khan), 33–65. Materials Research Forum http://dx.doi.org/10.21741/9781945291876-2.
6 Ramesh, M. (2016). Kenaf (*Hibiscus cannabinus* L.) fibre based bio-materials: a review on processing and properties. *Progress in Materials Science* 78-79: 1–92.
7 Sanjay, M.R., Siengchin, S., Parameswaranpillai, J. et al. (2018). A comprehensive review of techniques for natural fibres as reinforcement in composites: preparation, processing and characterization. *Carbohydrate Polymers*: 108–121.
8 Kumar, P.S.S. and Allamraju, K.V. (2019). A review of natural fibre composites [Jute, Sisal, Kenaf]. *Materials Today: Proceedings* 18: 2556–2562.
9 Bhoopathi, R., Ramesh, M., Rajaprasanna, R. et al. (2047). Physical properties of glass-hemp-banana hybrid fibre reinforced polymer composites. *Indian Journal of Science and Technology* 10 (7): 1–6.
10 Ramesh, M., Deepa, C., Kumar, L.R. et al. (2020). Life-cycle and environmental impact assessments on processing of plant fibres and its bio-composites: a critical review. *Journal of Industrial Textiles* https://doi.org/10.1177/1528083720924730.
11 Ramesh, M. and Nijanthan, S. (2016). Mechanical property analysis of kenaf–glass fibre reinforced polymer composites using finite element analysis. *Bulletin of Materials Science* 39 (1): 147–157.
12 Ramesh, M., Logesh, R., Manikandan, M. et al. (2017). Mechanical and water intake properties of banana-carbon hybrid fibre reinforced polymer composites. *Materials Research* 20 (2): 365–376.
13 Yazdi, M.F.A., Zakaria, R., Mustaffar, M. et al. (2014). Bio-composite materials potential in enhancing sustainable construction. *Desalination and Water Treatment* 52: 3631–3636.
14 Meints, P.D. and Smith, C.A. (2003). Kenaf seed storage duration on germination, emergence, and yield. *Industrial Crops and Products* 17: 73–79.
15 El-Shekeil, Y.A., Sapuan, S.M., Jawaid, M., and Al-Shujaa, O.M. (2014). Influence of fibre content on mechanical, morphological and thermal properties of kenaf fibres reinforced poly (vinyl chloride)/thermoplastic polyurethane poly-blend composites. *Materials & Design* 58: 130–135.
16 Aji, I.S., Sapuan, S.M., Zainudin, E.S., and Abdan, K. (2009). Kenaf fibres as reinforcement for polymeric composites: a review. *International Journal of Mechanical and Materials Engineering* 4 (3): 239–248.
17 Rashdi, A.A.A., Sapuan, S.M., Ahmad, M.M.H.M., and Khalina, A. (2009). Water absorption and tensile properties of soil buried kenaf fibre reinforced unsaturated polyester composites (KFRUPC). *Journal of Food, Agriculture and Environment* 7: 908–911.
18 Xue, Y., Du, Y., Elder, S. et al. (2009). Temperature and loading rate effects on tensile properties of kenaf bast fibre bundles and composites. *Composites Part B: Engineering* 40: 189–196.
19 Chee, S.S., Jawaid, M., Sultan, M.T.H. et al. (2018). Thermomechanical and dynamic mechanical properties of bamboo/woven kenaf mat reinforced epoxy hybrid composites. *Composites Part B: Engineering* 163: 165–174.

20 Saba, N., Paridah, M.T., and Jawaid, M. (2015). Mechanical properties of kenaf fibre reinforced polymer composite: a review. *Construction and Building Materials* 76: 87–96.

21 Saba, N., Jawaid, M., and Asim, M. (2016). Recent advances in nanoclay/natural fibres hybrid composites. In: *Nanoclay Reinforced Polymer Composites* (eds. J. Mohammad, Q. Abouelkacem and B. Rachid), 1–28. Springer.

22 Saba, N., Alothman, O.Y., Almutairi, Z., and Jawaid, M. (2019). Magnesium hydroxide reinforced kenaf fibres/epoxy hybrid composites: mechanical and thermomechanical properties. *Construction and Building Materials* 201: 138–148.

23 Bhoopathi, R., Deepa, C., Sasikala, G., and Ramesh, M. (2015). Experimental Investigation on mechanical properties of hemp-banana-glass fibre reinforced composites. *Applied Mechanics and Materials* 766: 167–172.

24 Ramesh, M., Rajeshkumar, L., Khan, A., and Asiri, A.M. (2020). Self-healing polymer composites and its chemistry. In: *Self-Healing Composite Materials* (eds. A. Khan, M. Jawaid, S.N. Raveendran and A.M.A. Asiri), 415–427. Woodhead Publishing/Elsevier https://doi.org/10.1016/B978-0-12-817354-1.00022-3.

25 Yusoff, R.B., Takagi, H., and Nakagaito, A.N. (2016). Tensile and flexural properties of polylactic acid-based hybrid green composites reinforced by kenaf, bamboo and coir fibres. *Industrial Crops and Products* 94: 562–573.

26 Maslinda, A.B., Abdul Majid, M.S., Ridzuan, M.J.M. et al. (2017). Effect of water absorption on the mechanical properties of hybrid interwoven cellulosic- cellulosic fibre reinforced epoxy composites. *Composite Structures* 167 (1): 227–237.

27 Jumaidin, R., Sapuan, S.M., Jawaid, M. et al. (2017). Thermal, mechanical, and physical properties of seaweed/sugar palm fibre reinforced thermoplastic sugar palm Starch/Agar hybrid composites. *International Journal of Biological Macromolecules* 97: 606–615.

28 Franc, P.H. and Vega, M.A. (1997). Effect of fibre treatment on the mechanical properties of LDPE–henequen cellulosic fibre composites. *Journal of Applied Polymer Science* 10: 197–207.

29 Asim, M., Paridah, M.T., Saba, N. et al. (2018). Thermal, physical properties and flammability of silane treated kenaf/pineapple leaf fibres phenolic hybrid composites. *Composite Structures* https://doi.org/10.1016/j.compstruct.2018.06.068.

30 Kim, H.S., Yang, H.S., Kim, H.J., and Park, H.J. (2004). Thermogravimetric analysis of rice husk flour filled thermoplastic polymer composites. *Journal of Thermal Analysis and Calorimetry* 76: 395–404.

31 Kumar, S.S., Duraibabu, D., and Subramanian, K. (2014). Studies on mechanical, thermal and dynamic mechanical properties of untreated (raw) and treated coconut sheath fibre reinforced epoxy composites. *Materials & Design* 59: 63–69.

32 Ramesh, M. and Rajesh Kumar, L. (2020, 2020). Bioadhesives. In: *Green Adhesives* (eds. Inamuddin, R. Boddula, M.I. Ahamed and A.M. Asiri), 145–161. Wiley-Scrivener Publisher https://doi.org/10.1002/9781119655053.ch7.

33 Arbelaiz, A., Fernandez, B., Ramos, J., and Mondragon, I. (2006). Thermal and crystallization studies of short flax fibre reinforced polypropylene matrix composites: effect of treatments. *Thermochimica Acta* 440 (2): 111–121.

34 Ridzuan, M., Majid, M.A., Afendi, M. et al. (2016). Thermal behaviour and dynamic mechanical analysis of *Pennisetum purpureum*/glass-reinforced epoxy hybrid composites. *Composite Structures* 152: 850–859.

35 Chee, S.S., Jawaid, M., Sultan, M.T.H. et al. (2019). Accelerated weathering and soil burial effects on colour, biodegradability and thermal properties of bamboo/kenaf/epoxy hybrid composites. *Polymer Testing* 79: 106054.

36 Saba, N., Jawaid, M., Paridah, M., and Alothman, O. (2017). Physical, structural and thermomechanical properties of nano oil palm empty fruit bunch filler based epoxy nanocomposites. *Industrial Crops and Products* 108: 840–843.

37 Dashtizadeh, Z., Abdan, K., Jawaid, M., and Dashtizadeh, M. (2019). Thermal and flammability properties of kenaf/recycled carbon filled with cardanol hybrid composites. *International Journal of Polymer Science* 2019 https://doi.org/10.1155/2019/9168342.

38 Mazlan, A.A., Sultan, M.T.H., Shah, A.U.M., and Safri, S.N.A. (2019). Thermal properties of pineapple leaf/kenaf fibre reinforced vinyl ester hybrid composites. *IOP Conference Series: Materials Science and Engineering* 670 (1): 012030. IOP Publishing,.

39 Mohammed, M., Rozyanty, A.R., Beta, B.O. et al. (2017). Influence of weathering effect in natural environment on thermal properties hybrid kenaf blast/glass fibre and unsaturated polyester composite. *Journal of Physics: Conference Series* 908 (1): 012004.

40 Azammi, A.M., Sapuan, S.M., Ishak, M.R., and Sultan, M.T.H. (2018). Mechanical and thermal properties of kenaf reinforced thermoplastic polyurethane (TPU)-natural rubber (NR) composites. *Fibres and Polymers* 19 (2): 446–451.

41 Samaei, S.E., Mahabadi, H.A., Mousavi, S.M. et al. (2020). The influence of alkaline treatment on acoustical, morphological, tensile and thermal properties of kenaf natural fibres. *Journal of Industrial Textiles* https://doi.org/10.1177/1528083720944240.

42 Idumah, C.I., Hassan, A., and Bourbigot, S. (2018). Synergistic effect of exfoliated graphene nanoplatelets and non-halogen flame retardants on flame retardancy and thermal properties of kenaf flour-PP nanocomposites. *Journal of Thermal Analysis and Calorimetry* 134 (3): 1681–1703.

11

Thermal Properties of Hemp Fiber-Based Hybrid Composites

Hom Nath Dhakal[1] *and Mohini Sain*[2,3]

[1] *University of Portsmouth, School of Mechanical and Design Engineering, Advanced Polymers and Composites (APC) Research Group, Anglesea Road, Anglesea Building, Portsmouth, PO1 3DJ, UK*
[2] *University of Toronto, Center for Biocomposites and Biomaterials Processing, Graduate Department of Forestry, John H. Daniels Faculty of Architecture, Landscape and Design, 33 Willcocks St., Toronto, Ontario, M5S3B3, Canada*
[3] *University of Toronto, Department of Mechanical and Industrial Engineering, 5 King's College Rd., Toronto, Ontario, M5S3G8, Canada*

11.1 Introduction

In the last decades, the unprecedented use of natural resources in an unsustainable way has led to serious environmental concerns indicated by various environmental impacts such as global warming, resource depletion, and ecotoxicity. Due to these concerns, natural fiber-reinforced polymer composites (NFRPCs) have received much attention from research and industrial communities because these materials offer many interesting attributes [1, 2]. These attributes include but not limited to lower density (1.2–1.6 $g\,cm^{-3}$) than that of conventional glass fiber (2.4 $g\,cm^{-3}$), relatively higher specific properties such as strength and modulus in comparison with their conventional counterparts, and renewability and recyclability at the end of their life cycle [1, 2]. Additionally, other advantages include lower energy consumption for production of NFRPCs than glass fiber composites, nonabrasive nature, and lower cost than the conventional synthetic glass and carbon fiber-reinforced composites. Among the NFRPCs, hemp fiber-reinforced composites have a unique attraction [3, 4]. Hemp fiber is lignocellulose fiber, which consists of three key elements, i.e. cellulose, hemicellulose, and some other components such as lignin. Due to these chemical compositions, hemp fiber has high stiffness and strength, making the composite reinforced with it suitable in lightweight applications such as automotive, marine, and constructions [5]. However, these chemical constituents are active polar, which attract moisture. There are inherent associated issues with natural fiber composites such as lower thermal stability, hydrophilic nature leading to lower long-term durability, narrow processing window, and inferior flammability properties. These have restricted the applications of these

Natural Fiber-Reinforced Composites: Thermal Properties and Applications, First Edition.
Edited by Senthilkumar Krishnasamy, Senthil Muthu Kumar Thiagamani,
Chandrasekar Muthukumar, Rajini Nagarajan and Suchart Siengchin.

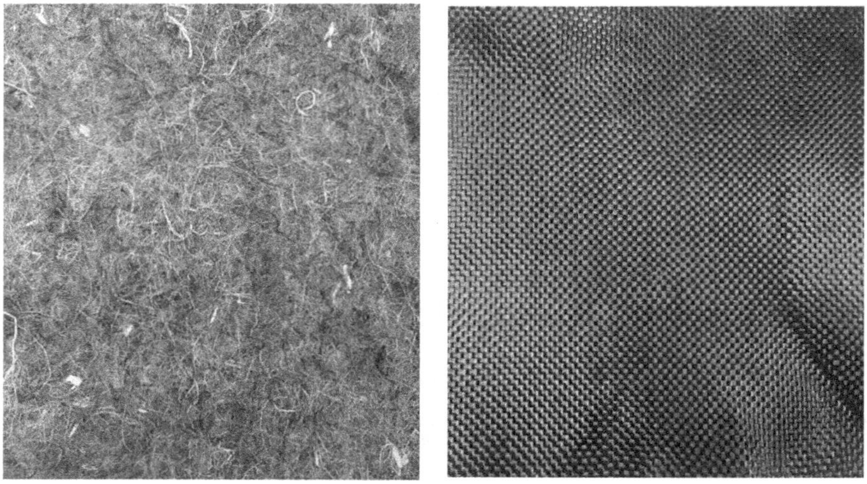

(a) (b)

Figure 11.1 Hemp fiber mat and basalt fabric (a) nonwoven hemp mat and (b) basalt fabric.

composites in low-load-bearing and semi-structural parts in automotive, building and construction, and sports and leisure sectors [6–8].

Moreover, with the hemp being cellulosic fibers, the adhesion between matrix and reinforcement gets affected that eventually leads to poor mechanical and thermal properties. This is one of the key issues relating to the natural fiber such as hemp fiber-reinforced composites. There have been considerable efforts to minimize these effects by introducing several techniques including fiber surface treatments, inclusion of additives and nanoparticles, hybridization of different reinforcements into the natural fibers, and optimization of various processing parameters [9, 10]. Additional parameters that can influence the properties are fiber length and manufacturing processes used. More importantly, various properties (mechanical and thermal) are affected by other aspects including fiber morphology and contents, fiber geometry, and matrix types, among others. Figure 11.1 illustrates the nonwoven hemp mat and basalt fabric often hybridized to improve mechanical and thermal performances of hemp composites.

Hemp fiber (*Cannabis sativa* L.), also known as lignocellulosic fiber, is one of the popular natural bast fibers used in composite reinforcements. The major chemical constituents of mature hemp fiber consist of cellulose (70.2–74.4 wt%) and hemicellulose (17.0–22.4 wt%), while other components are lignin (3.7–5.7 wt%), pectin (0.9 wt%), and wax (0.8 wt%) [11, 12].

This chapter investigates the important characteristics of hemp fiber as reinforcement for composites and their long-term thermal stability. Thermal properties are as important as mechanical properties for the use of hemp composites for various applications. Moreover, the influence of temperature, various environmental conditions, and the influence on thermal behavior are important to understand [13, 14]. Additionally, in this chapter, the influence of various parameters such as surface modification, hybridization, and other processing parameters on the thermal properties such as degradation behavior, thermal stability, and thermal conductivity has been analyzed and discussed. To analyze the thermal stability at different temperatures and environments, various techniques including thermogravimetric analysis (TGA), differential scanning calorimetry (DSC), and thermomechanical analyzer, among others, are reviewed. The correlation between morphological structure, chemical constituents, and other parameters of hemp fiber and its effects on resultant composites in terms of thermal stability are also analyzed [15].

11.2 Thermal Properties Measurements and Importance

11.2.1 Thermogravimetric Analysis (TGA)

Decomposition behavior of polymers and composites is normally evaluated by TGA at nitrogen and air environments. TGA helps to determine the thermal stability by evaluating the weight loss or change in weight. Factors evaluated by this technique comprise onset temperature (T-onset), degradation temperature at different percentages of mass loss, final decomposition, and remaining of residue. TGA behavior or thermal decomposition process of hemp composites, for example, depends on the following parameters:

- The mass loss percentages.
- The initial of decomposition temperature.
- The maximum decomposition temperature.
- Endset temperature (temperature at involution peak).

Information obtained from TGA test can provide how materials perform at various temperatures. It is worth noting that the TGA behavior depends on the chemical composition of hemp fibers. According to various literature, thermogravemetric (TG) curves normally exhibit three phases of weight degradation for hemp composites:

First stage: Evaporation of water (moisture) temperature range between 50 and 100 °C.

Second stage: Decomposition of hemicellulose.

Third stage: Related to the decomposition of lignin and cellulose. Further weight loss can take place between 200 and 350 °C due to the thermal degradability of hemicellulose. The degradations of lignin and hemicellulose normally take place between 300–450 °C and 200–300 °C, respectively. Often, the thermal degradation behavior of naturel fiber such as hemp fiber-reinforced composites is divided into four stages [11, 12, 16]:

(1) Moisture vaporization
(2) Hemicellulose degradation
(3) Cellulose degradation
(4) The lignin degradation

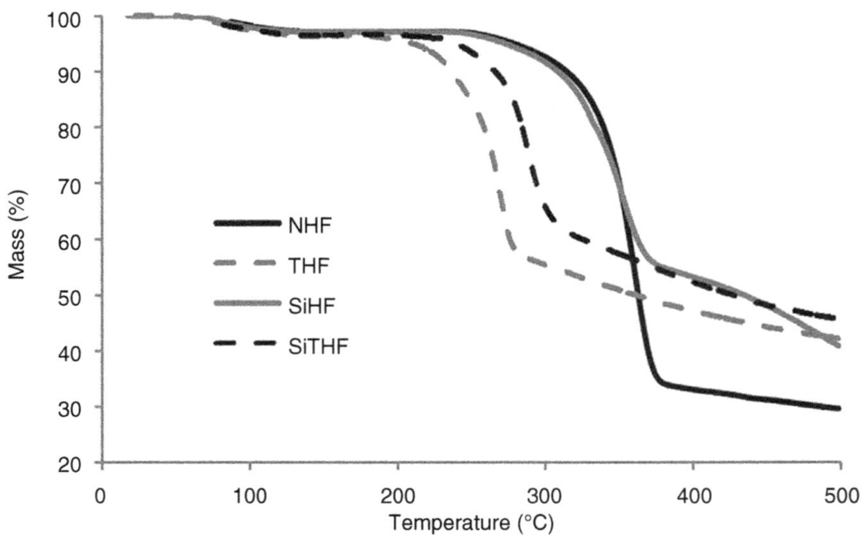

Figure 11.2 TGA curves of the different hemp fabric treated and untreated (NHF, non-treated hemp fabric; THF, Thermotex-treated fabric; SiHF, sol–gel-treated fabric; SiTHF, Thermotex- and sol–gel-treated fabric) [18]. Source: Szolnoki et al. [18].

Often, two key parameters are used to measure the overall thermal stability of composites:

(1) Initial degradation temperature.
(2) Degradation between 200 and 300 °C.

Parameters influencing the thermal stability of hemp composites include heterogeneity of natural fiber constituents that have different thermal reactivities, leading to significant influence on chemical reactions and decompositions [17].

The work presented by Szolnoki et al. on hemp fiber epoxy composites with various surface treatment methods employed has shown thermal decomposition behavior of hemp composites.

The thermal degradation processes of untreated and treated hemp composites displayed different behaviors (Figure 11.2). They showed in their TGA results that treated hemp fabric exhibited improved thermal stability [18].

11.2.2 Hybrid Approach for Thermal Properties Improvement

Hybrid composites are defined as two or more reinforcements in the matrix. Hybridization of some additives and synthetic fibers such as carbon and glass is introduced into natural fibers aiming to offset some of the drawbacks (high moisture uptake, low processing temperature, low fire resistance, and properties variability) of natural fiber composites. A study carried out by Dhakal et al. investigated the influence of glass fiber hybridization on the thermal stability of hemp fiber-reinforced unsaturated polyester composites using TGA tests [11]. Their report highlighted that the decomposition starts close to 240 °C where hemicellulose starts

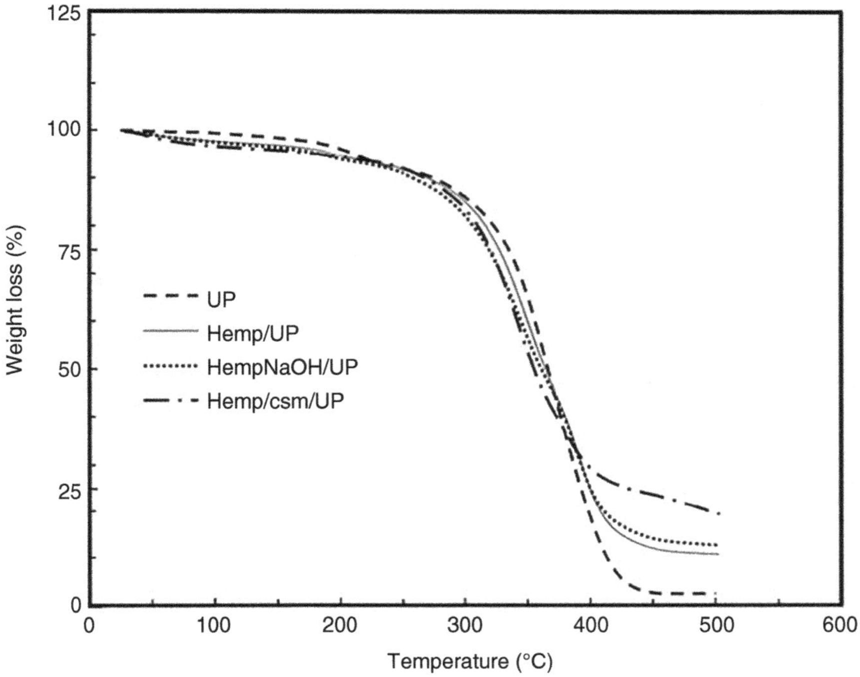

Figure 11.3 TGA curves of neat UP and hemp/UP and treated and glass hybridized hemp composites [11]. Source: Dhakal et al. [11].

the degradation. They highlighted that up to this temperature the composite can be thermally stable. This temperature is important when natural fibers such as hemp are used in composites manufactured using different manufacturing techniques. Their study also highlighted that thermal stability of neat UP was significantly improved by the reinforcement of hemp fiber as illustrated in Figure 11.3.

Moreover, treated hemp/UP composites showed higher thermal stability by showing higher percentage of char yield. Additionally, glass fiber-hybridized specimen displayed highest char yield indicating highest thermal stability among other samples compared.

Their study concluded that thermal properties of hemp/UP composites were significantly improved after the fiber treatment. Moreover, glass fiber hybrid system was also one of the effective approaches in improving the thermal stability of hemp/UP composites.

In the work carried out by Mustapa et al. the properties of cellulose microfibers (CMFs), cellulose nanocrystals (CNCs), and cellulose nanofibrils (CNF) extracted from hemp stalks was investigated in terms of mechanical and thermal properties [17]. Their report highlighted that the incorporation of cellulose helps to enhance the thermal stability. The thermal stability of hemp stalks (R-HS), alkali-treated hemp stalks (A-HS), CMFs, CNC15, CNC30, and CNF was degraded into four different stages as explained earlier.

The recorded onset temperature of R-HS, A-HS, and CMF were at 254, 272, and 290 with maximum degradation temperatures of 328, 329, and 348 °C, respectively. The lower onset temperature of R-HS and A-HS is due to the degradation of non-cellulosic molecules (lignin and hemicellulose).

Mehta et al. reported a similar observation on nonwoven industrial hemp fiber mat-reinforced unsaturated polyester composites. Their work highlighted that hybridization of glass fiber into hemp/UP helped to achieve a balanced thermal and mechanical properties [19]. They observed that alkali-treated hemp fiber exhibited improved maximum rate of decomposition, indicating enhanced thermal stability. Their TGA results exhibited that weight loss of 10% occurred for hemp-untreated fiber at 288 °C. For surface-treated hemp fiber, it was reported at 332 °C, while a weight loss of 20% occurred in the range of 329–366 °C for untreated and surface-treated hemp fibers.

Kandemir et al. compared four key fibers (flax, jute, kenaf, and curaua) in terms of analyzing their mechanical and thermal properties. Their work reported that the first decomposition of these fibers was recorded at 220–230, onset of decomposition temperature and then cellulose degradation at the temperature range of 320–335 °C [20]. Then, a gradual decomposition of lignin requires higher condensation temperature for this behavior. Their analysis highlighted that 80% of weight loss was occurred between the temperature of 150–700 °C for jute and kenaf and curaua where flax exhibited lower mass loss (72%) than the other three fibers. This thermal behavior is illustrated in Figure 11.4 [20]. It is worth noting

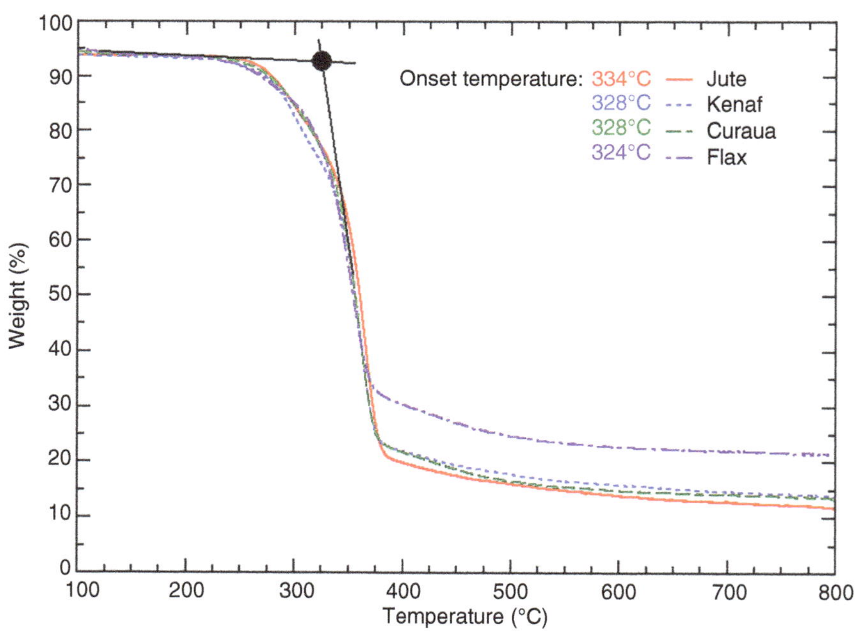

Figure 11.4 TGA traces of flax, jute, kenaf, and curaua fibers [20]. Source: Kandemir et al. [20].

that commonly used natural bast fibers such as flax, jute, kenaf, and curaua show similar thermal stability to hemp fiber composites.

Reported work on TGA behavior of hemp composites from available literature suggests that the thermal stability of hemp composites can be improved using different fiber modifications. The work by Lu and Oza presented the thermal stability and thermomechanical behaviors of hemp fiber-reinforced high density polyethylene (HDPE) composites [21]. They studied the thermal behavior using two different chemical modifications techniques, namely, NaOH and silane treatment. Their work reported that degradation of untreated hemp fiber started about 225 °C. NaOH- and silane-treated hemp started to degrade at about 251 and 247 °C, respectively, showing an improved decomposition. Maximum decomposition for untreated hemp was recorded between 298 and 354 °C. For NaOH- and silane-treated fibers, the maximum decomposition started in the ranges of 318–378 °C and 314–363 °C, respectively. It was highlighted that main chemical constituents of hemp fiber such as hemicellulose degraded around 220–315 °C, whereas degraded cellulose around 370 °C. Their report also suggested that NaOH and silane treatment contributed to higher thermal stability.

11.2.3 Differential Scanning Calorimetry (DSC)

DSC analysis is mainly used to measure energy released or heat flow during heating. Exothermic and endothermic reactions at different temperatures are determined in DSC testing. Thermal phase transformations of hemp fiber can be analyzed by evaluating magnitude and the location of peaks.

Panthapulakkal and Sain developed high-performance polypropylene (PP) composites using short hemp fibers and glass fibers as reinforcement aimed at using these composites in semi-structural or structural applications such as automotive interior parts [22]. Composites were fabricated by melt processing followed by injection molding. Their report highlighted that the incorporation of hemp fiber into PP matrix helped to increase the heat deflection temperature. Furthermore, their work highlighted that the incorporation of glass fiber as hybrid reinforcement helped to increase the heat deflection temperature significantly. Hybridization of glass fiber into hemp/PP composites improved thermal stability of hemp/PP significantly. They concluded that to optimize the effect of hybridization, processing techniques and parameters need to be modified. They further highlighted that hybridization of glass fiber into hemp/PP composites exhibited shifts in the degradation temperature to a higher value confirming increased in thermal stability. This is an illustration of hybrid effects on the improvement in thermal stability of hemp/glass/PP hybrid composites [22].

Panaitescu et al. investigated the effects of different hemp fiber lengths on the mechanical and thermal properties of PP/SEBS/hemp fiber composites [23]. The typical DSC thermograms of PP, PPM, and composites with different hemp fiber lengths are illustrated in Figure 11.5. They reported that the melting temperature/enthalpy for the first and second heating cycles and the crystallization temperature/enthalpy were presented. The fiber length had no influence on the

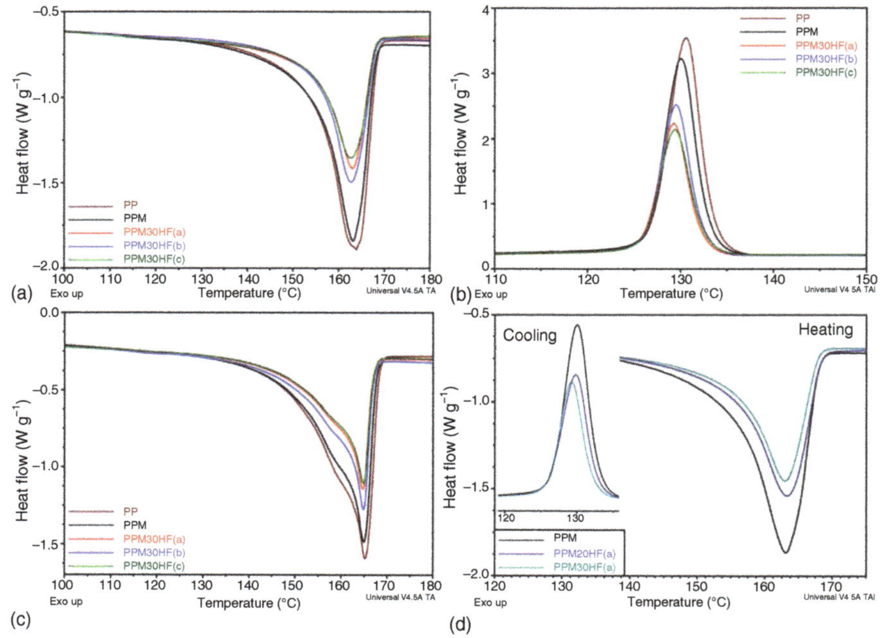

Figure 11.5 DSC thermograms of PP, PPM, and composites with different types of hemp fibers: first heating cycle (a), cooling cycle (b), second heating cycle (c), and melting and crystallization behavior for different concentrations of HF in composites (d) [23]. Source: Panaitescu et al. [23].

melting temperature of composites. However, they observed a slight decrease in crystalline temperature of PPM/hemp fiber composites. This was attributed to the good fiber/matrix interface. They also highlighted that lower amount of fiber has a smaller influence on the crystalline temperature.

11.2.4 Thermal Conductivity

Thermal conductivity (λ) measurement deals with materials' ability to conduct or transport heat. For a homogeneous body, λ and thermal diffusivity (a) are interrelated by specific density ρ and specific heat capacity c_p according to

$$\lambda = ac_p\rho \tag{11.1}$$

where λ is thermal conductivity, a is thermal diffusivity, c_p is specific heat capacity, and *ρ is specific density.*

The thermal conductivity measurement becomes important when hemp fiber composites are used in heat-related applications. The thermal conductivity of natural bast fibers such as hemp depends on various parameters. This is related to heat flow of materials that depends on their morphology and structures. Crystalline and semicrystalline materials are more conductive than amorphous materials. The morphological structure of hemp fiber, for example, plays a significant role in the conduction of heat unlike glass fiber as the hollow structure (lumen) significantly influences the overall thermal conductivity of hemp fibers and its composites. It is not only the fibers but also the matrices that equally influence the thermal conductivity of composites. There are two types of thermal conductivity, namely, in plane and through thickness. Generally, through-thickness or transverse thermal conductivities are far less than in-plane conductivities for natural fiber composites. It is well established that polymers and polymer composites are far less conductive than metals.

Thermal conductivity is an important parameter for natural fiber composites, which measures how fast heat can flow in the materials. Thermal conductivity measured for hemp fiber composites in various literature varies (0.05–0.16 W m^{-1} k^{-1}) since it is influenced by various parameters such as moisture contents.

Thermal conductivity of hemp fiber composites can be altered by incorporating conductive fillers such as graphite, carbon black, and carbon nanotubes. Additionally, thermal conductivity also is influenced by filler and reinforcement content level. The work carried out by Sair et al. measured the thermal conductivity of hemp fiber-reinforced polyurethane (PU) composites using a thermal conductivity analyzer [24]. They reported that with the increase in the contents of hemp fiber, the thermal conductivity of PU matrix was increased, which accordingly indicates that hemp fibers exhibit higher conductivity than pure PU. Often, transverse thermal conductivity can be influenced by the fiber volume fraction. With the increase of hemp fiber content, transverse thermal conductivity can be decreased as hemp fiber has lower thermal conductivity.

Figure 11.6 exhibits in-plane thermal conductivity in relation to fiber aspect ratios of poly(ε-caprolactone)-based hemp composites. It is highlighted that thermal conductivity was influenced by the hemp aspect ratios along with interfacial fiber/matrix interface [25].

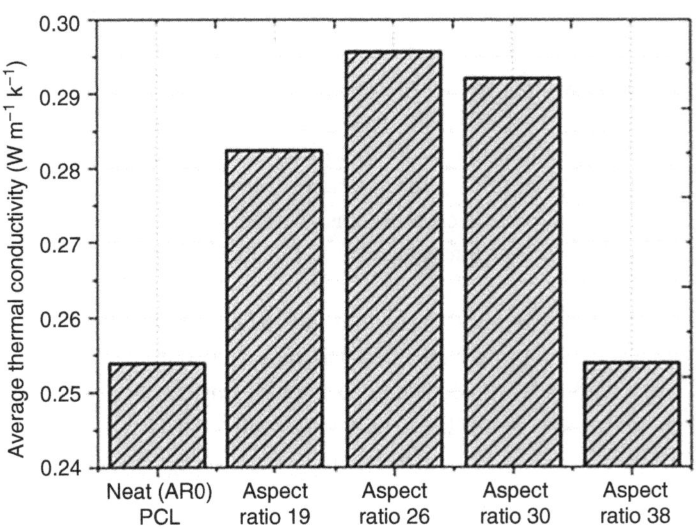

Figure 11.6 Thermal conductivity against aspect ratio of hemp/PCL composites [25]. Source: Dhakal et al. [25].

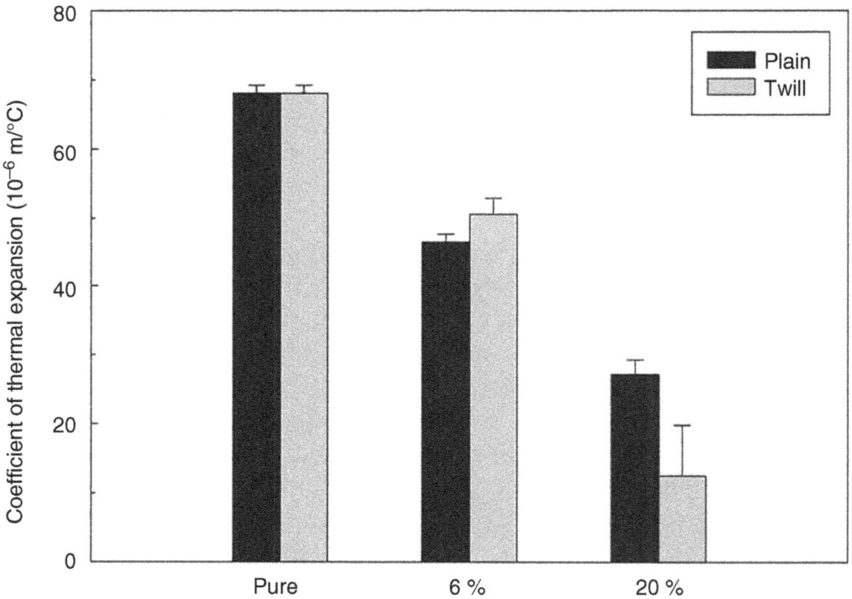

Figure 11.7 Coefficient of thermal expansion (CTE) of neat PLA and hemp/PLA biocomposites [26]. Source: Song et al. [26].

11.2.5 Linear Coefficient of Thermal Expansion

Song et al. investigated the thermal stability of plain and twin hemp fabrics-reinforced polylactic acid (PLA)-based green composites (hemp/PLA) using two different fiber volume fractions (6% and 20%) [26]. Their results indicated that the increment of fiber volume fraction from 6% to 20% was contributing factor in the reduction of the coefficient of thermal expansion (CTE) of the fabricated hemp fabric composites as illustrated in Figure 11.7. They highlighted that the decreased of CTE considerably indicates that hemp/PLA composites have great potential for applications experiencing a wide range of temperatures, such as automobile and aerospace applications. As applicable to mechanical properties, to enhance CTE behavior, good fiber/matrix interface is crucial. This provides the potential of hemp fiber composites for demanding applications such as automotive and building applications.

In the work carried out by Pandey et al. on the influence of hybridisng hemp and flax fibres with glass fiber/PU composites, it is reported that the coefficient of linear thermal expansion (CLTE) of composites containing only natural fiber reinforcements such as flax and hemp fibers showed negative CLTE while hybrid samples exhibited positive values as illustrated in Figure 11.8 [27]. According to their report, this behavior is attributed to shrinking of natural fibers at high temperatures, whereas glass fiber-hybridized composites expand under elevated temperatures. A low CLTE is preferred where high-dimensional stability is required at elevated temperature applications.

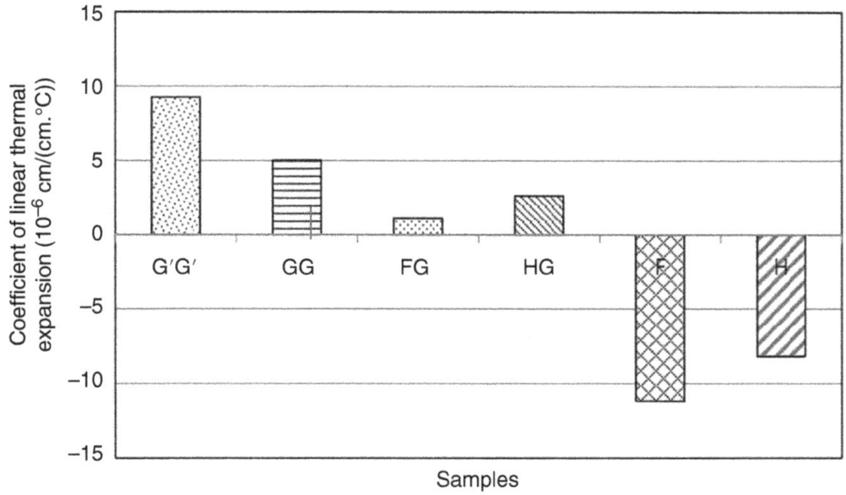

Figure 11.8 Linear coefficient of thermal expansion of different PU-based composites with glass fiber alone (G'G'), glass/flax and glass (FG), hemp–agave and glass fiber (HG), flax fiber alone (F), and hemp fiber alone (H) [27]. Source: Pandey et al. [27].

11.2.6 Dynamic Mechanical Analysis (DMA)

Dynamic mechanical analysis (DMA) is utilized for investigating and analyzing viscoelastic properties of polymers and composite materials such as elastic modulus, loss modulus, and damping coefficient with respect to the function of temperature. The results of DMA indicate the ability of materials to store energy against applied forces without the deformation, leading to the breaking of the materials. With DMA, the glass transition temperature (T_g) can also be calculated based on the responses of elastic modulus and loss modulus with respect to function of temperature.

Ashok et al. investigated thermal stability and dynamic behavior also referred as time-dependent deformation behavior (viscoelastic) that are important parameters in measuring DMA [28]. This is one of the well-established techniques to measure the time-dependent behavior. Key properties measured using this technique are as follows:

- Storage modulus (E').
- Loss modulus (E'').
- Mechanical loss factor tan δ (damping) (E'/E'').

Mehta et al. reported storage modulus and tan δ for neat UP matrix and hemp fiber-reinforced composites [19]. Their results suggested that the storage modulus of surface-treated hemp/UP composites exhibited higher values than that of untreated biocomposites. The results further reiterated that the storage modulus of biocomposites at 40 °C was enhanced to 110–190% compared with neat UP matrix. It was suggested that E-glass/UP showed highest storage modulus. It is worth noting that the storage modulus of neat UP was significantly improved with the hemp fiber reinforcement. The loss modulus of hemp-reinforced composites with surface treatment was recorded higher than that of neat unsaturated polymer matrix. This improvement was attributed to improved adhesion between fiber and matrix due to the fiber surface treatment. It is generally agreed phenomenon that the fiber-reinforced composites exhibit higher loss modulus than neat polymers.

Mehta et al. also highlighted that the tan δ was the highest for the neat unsaturated polyester (UP) resin due to its storage modulus reduction at higher temperatures [19]. The tan δ curves of all composites were reported to shift toward the right as compared with neat matrix. The shift was recorded highest for glass-reinforced composites, indicating higher stiffness.

Mustapa et al. investigated the thermomechanical properties of PLA/hemp composites hybridized with inorganic fillers such as nanosilica with and without the incorporation of tributyl citrate plasticizer [17]. They reported that the thermomechanical properties were significantly improved with the reinforcement of hemp onto PLA matrix. Moreover, the incorporation of nanosilica further improved the thermomechanical properties. Also, the corresponding loss factor tan δ was improved owing to the tributyl incorporation. The mechanism and reasons for properties enhancement of PLA/hemp and nanoparticle-hybridized samples were attributed to the compatibility between PLA and hemp fiber reinforcement. For enhanced storage and loss modulus, the fiber matrix adhesion is important as the applied force is transferred from matrix to fibers via fiber/matrix interface.

The work by Lu and Oza on the untreated and surface-treated hemp HDPE composites reported that surface treatment can play a significant role in improving the storage modulus of composites [21]. They highlighted that the fiber reinforcement helps in obtaining higher storage modulus than neat matrix. By increasing hemp fiber volume fractions from 20% to 40%, the storage modulus increased significantly. However, with hemp fiber volume fraction at 50%, the storage modulus is reduced, which was attributed to less matrix to wet the fiber at high fiber volume fractions. They further highlighted that surface treatment contributed to enhanced storage modulus because of improved wettability of matrix onto fibers as illustrated in Figures 11.9 and 11.10.

Work carried out by Ouajal and Shanks reported that various surface treatment can be instrumental for enhancing thermal properties of hemp fiber-reinforced composites [12]. Figure 11.11 depicts scanning electron microscopy (SEM) images of untreated hemp and treated hemp fibers using various treatment methods. They suggested that non-cellulosic materials need to be removed from the fiber surface using an effective method. To improve thermal stability, the impurities and cellulosic components need to be removed. This will help to enhance the fiber/matrix interface that eventually will enhance the overall mechanical and thermal properties.

An extensive study carried out by Baghaei et al. on mechanical and thermal properties of compression-molded PLA/natural fiber composites hybridized with hemp and lyocell fibers reported that the dynamic mechanical behavior of PLA was significantly enhanced upon the incorporation of hemp and lyocell [29]. Their report highlighted that the storage modulus for example was increased at 30 °C from 2.5 GPa for neat PLA, 3.7 GPa for PLA/lyocell followed by PLA/hemp/lyocell, 3.1 GPa for PLA/hemp, and 2.6 GPa for PLA–PP/hemp hybrid composite. Their

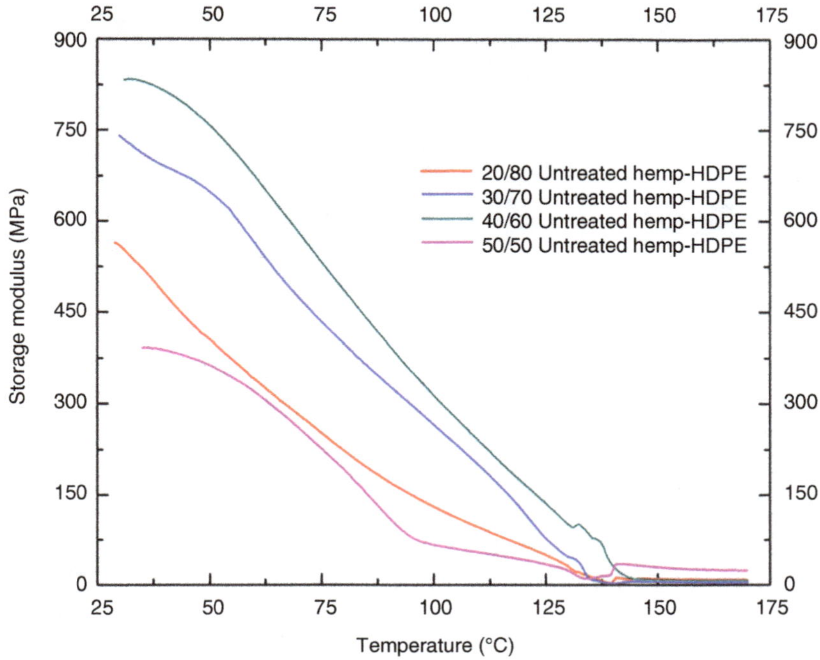

Figure 11.9 Storage modulus of hemp/HDPE composites of untreated hemp fiber with various fiber volume fractions and with NaOH-treated hemp/HDPE composites at various fiber volume fractions [21]. Source: Lu and Oza [21].

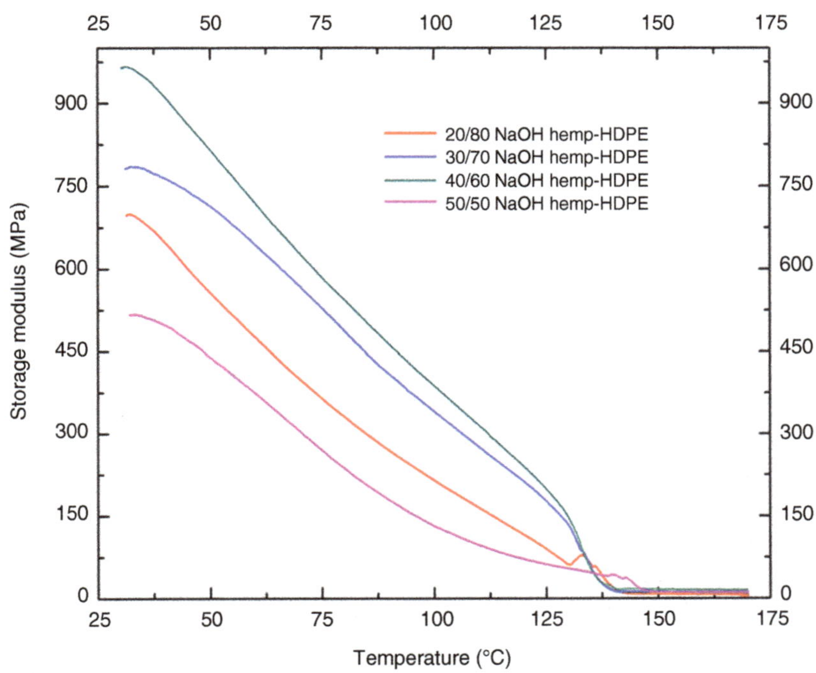

Figure 11.10 Storage modulus of hemp/HDPE composites with NaOH-treated hemp/HDPE composites at various fiber volume fractions [21]. Source: Lu and Oza [21].

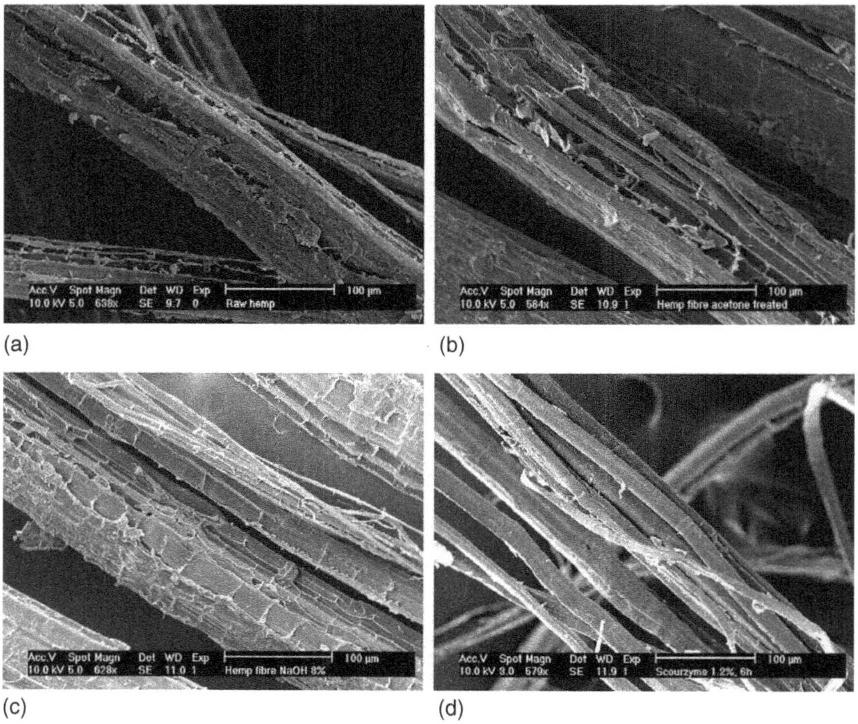

Figure 11.11 SEM micrographs of (a) untreated, (b) acetone-treated, (c) 8% NaOH-treated, and (d) 1.2% Scourzyme-treated hemp fibers [12]. Source: Ouajai and Shanks [12].

work also reported that the storage modulus increased when hemp and lyocell fibers were hybridized into PLA matrix. The improvement in dynamic mechanical behavior is attributed to effective fiber and matrix adhesion. It is clear that fiber/matrix bond strength influences the dynamic mechanical properties such as storage modulus, loss modulus, and damping factor (tan δ). Moreover, these properties are more dependent on the temperature they are tested. At higher temperature, the storage and loss moduli decreased.

The damping or tan δ is related to mobility of molecular in composites that is the ratio of storage modulus and loss modulus, showing the internal friction of the composites and the adhesion of interface. They suggest that the glass transition temperature is often recorded at the maximum of the tan δ, also dependent to the fiber reinforcements. If there is a good compatibility between fiber and the matrix, then tan δ decreases.

With the incorporation of hemp/lyocell also contributed to the increase of glass transition temperature. Their findings reported that the T_g of neat PLA was 65.6 °C, whereas T_g for PLA/lyocell and PLA/hemp/lyocell were 67.5 and 68 °C. For PLA–PP/hemp, T_g was recorded at 69.7 °C. Many reported work suggest that chemical modification between hemp fiber and polymer matrices shows a strong interfacial adhesion. The modification helps achieve the esterification of hemp fibers, leading to the improved interfacial compatibility. It was obvious that in incorporating the reinforcement into PLA and PP, the T_g was increased. It is important to note that an optimal fiber volume fraction and good adhesion between the reinforcements and matrices is crucial for the thermal properties improvement.

There are many literature where basalt fibers are hybridized onto hemp fiber with an aim of improving mechanical and thermal stabilities of hemp composites. Basalt fibers originating from mineral source provides an interesting hybrid reinforcement. Work carried out by Sarasini et al. highlighted that hemp/basalt hybrid composites provided enhanced thermal stability of hemp/HDPE composites [30]. The hybrid composites exhibited improved crystallinity of hemp/HDPE composites. It is evidence from various literature that the improvement on thermal characteristics of hemp composites provides promising sustainable lightweight composites for various applications. The hybrid effects depend on many factors and parameters. The compatibility between hybrid reinforcements and matrices are important. Other critical aspect includes a good fiber/matrix interface so that interfacial properties are enhanced that helps in achieving high mechanical and thermal properties [10].

11.3 Conclusions and Perspectives

In this chapter, thermal performance of hemp composites is presented. It is evidenced from various literature that the variability in its morphological structure and the chemical compositions of hemp fiber and its bundle influence the thermal stability of resultant composites. Thermal property is important as this will provide necessary processing parameters and information on how the hemp fiber-reinforced composites perform at various environments and outdoor exposures. There is a lack of systematic research to bioengineer hemp fiber by biologically tuning the chemical composition of the fiber during plantation. This approach could provide a significant leap in utilizing industrial hemp fiber as a potential alternative of fiberglass even with improved thermal properties.

The surface treatment using various techniques can significantly influence the overall thermal stability of composites. The fiber surface treatment helps in the removal of surface impurities and improvement of fiber/matrix interface adhesion, leading to the improved viscoelastic behaviors and thermal stability. This process helps to improve the thermal decomposition of hemp fiber-reinforced composites by lowering the CTE. Another important approach currently being beneficial is a hybrid approach. Under this approach, the addition of nanoclay and some additives and the incorporation of conventional reinforcements such as glass and carbon into hemp fiber composites have been presented as one of the effective methods in improving the thermal stability. The incorporation of nanofillers has improved important parameters such as storage and loss moduli as a result of improved interfacial interactions, leading to the reduction of molecule mobility. Hybridization of high-strength glass, basalt, and carbon fibers into hemp fiber composites provides improved thermal stability along with other mechanical properties. Moreover, hybridized hemp fiber composites can be cost effective and also provides improved thermal stability. The overall properties improvement is vital for hemp fiber composites, and these composites have the potential to reduce the use of nonrenewable carbon and glass fibers that are used in various semi-structural and structural applications.

References

1 Faruk, O., Bledzki, A.K., Fink, H.P., and Sain, M. (2012). Biocomposites reinforced with natural fibers: 2000–2010. *Progress in Polymer Science* 37: 1552–1596.

2 Dhakal, H.N., Zhang, Z.Y., and Richardson, M.O.W. (2007). Effect of water absorption on the mechanical properties of hemp fibre reinforced unsaturated polyester composites. *Composites Science and Technology* 67: 1674–1683.

3 Mohanty, A.K., Misra, M., and Drzal, L.T. (2002). Sustainable bio-composites from renewable resources: opportunities and challenges in the green materials world. *Journal of Polymers and the Environment* 10: 19–26.

4 Faruk, O., Bledzki, A.K., Fink, H.P., and Sain, M. (2014). Progress report on natural fiber reinforced composites. *Macromolecular Materials and Engineering* 299: 9–26.

5 Li, Y. and Pickering, K.L. (2008). Hemp fibre reinforced composites using chelator and enzyme treatments. *Composites Science and Technology* 68: 3293–3298.

6 Stamboulis, A., Baillie, C.A., and Peijs, T. (2001). Effects of environmental conditions on mechanical and physical properties of flax fibers. *Composites Part A: Applied Science and Manufacturing* 32: 1105–1115.

7 Dhakal, H.N., MacMullen, J., and Zhang, Z. (2015). Moisture measurement and effects on properties of marine composites, Chapter 5. In: *Marine Applications of Advanced Fibre-Reinforced Composites*, 103–122. Cambridge, UK: Woodhead Publishing.

8 Dhakal, H.N., Bourmaud, A., Berzin, F. et al. (2018). Mechanical properties of leaf sheath date palm fibre biomass waste reinforced polycaprolactone (PCL) biocomposites. *Journal of Industrial Crops and Products* 126: 394–402.

9 Khana, A., Asiria, A.M., Jawaid, M. et al. (2020). Effect of cellulose nano fibers and nano clays on the mechanical, morphological, thermal and dynamic mechanical performance of kenaf/epoxy composites. *Carbohydrate Polymers* 239: 116248.

10 Dhakal, H.N., Sarasini, F., Santulli, C. et al. (2015). Effect of basalt fibre hybridisation on post-impact mechanical behaviour of hemp fibre reinforced composites. *Composites Part A: Applied Science and Manufacturing* 75: 54–67.

11 Dhakal, H.N., Zhang, Z.Y., and Bennett, N. (2012). Influence of fibre treatment and glass fibre hybridisation on thermaldegradation and surface energy characteristics of hemp/unsaturated polyester composites. *Composites Part B: Engineering* 43: 2757–2761. http://dx.doi.org/10.1016/j.compositesb.2012.04.036.

12 Ouajai, S. and Shanks, R.A. (2005). Composition, structure and thermal degradation of hemp cellulose after chemical treatments. *Polymer Degradation and Stability* 89: 327–335.

13 Dhakal, H.N., Zhang, Z.Y., Guthrie, R. et al. (2013). Development of flax/carbon fibre hybrid composites for enhanced properties. *Carbohydrate Polymers* 96: 1–8. https://doi.org/10.1016/j.carbpol.2013.03.074.

14 Saba, A., Safwan, M., Sanyang, F. et al. (2017). dynamic mechanical properties of cellulose nanofibers reinforced epoxy composites. *International Journal of Biological Macromolecules* 102: 822–828.

15 Tabari, H.Z., Nourbakhsh, A., and Ashori, A. (2011). Effects of nanoclay and coupling agent on the physico-mechanical, morphological, and thermal properties of wood flour/polypropylene composites. *Polymer Engineering & Science* 51: 272–277.

16 George, M., Mussone, P.G., and Bressler, D.C. (2015). Modification of the cellulosic component of hemp fibers using sulfonic acid derivatives: surface and thermal characterization. *Carbohydrate Polymers* 134: 230–239.

17 Mustapa, I.R., Shanks, R.A., and Daud, N. (2018). Morphological structure and thermomechanical properties of hemp fibre reinforced poly(lactic acid) nanocomposites plasticized with tributyl citrate. *Materials Today: Proceedings* 5: 3211–3218.

18 Szolnoki, B., Bocz, K., Soti, P.L. et al. (2015). Development of fibre reinforced flame retarded epoxy resin composites. *Polymer Degradation and Stability* 119: 68–76.

19 Mehta, G., Drzal, L.T., Mohanty, A.K., and Misra, M. (2006). Effect of fiber surface treatment on the properties of biocomposites from nonwoven industrial hemp fiber mats and unsaturated polyester resin. *Journal of Applied Polymer Science* 99: 1055–1068.

20 Kandemir, A., Pozegic, T.R., Hamerton, I. et al. (2020). Characterisation of natural fibres for sustainable discontinuous fibre composite materials. *Materials* 13: 2129. https://doi.org/10.3390/ma13092129.

21 Lu, N. and Oza, S. (2013). Thermal stability and thermo-mechanical properties of hemp-high density polyethylene composites: effect of two different chemical modifications. *Composites Part B: Engineering* 44: 484–490.

22 Panthapulakkal, S. and Sain, M. (2007). Injection-molded short hemp fiber/glass fiber-reinforced polypropylene hybrid composites: mechanical, water absorption and thermal properties. *Journal of Applied Polymer Science* 103: 2432–2441.

23 Panaitescu, D.M., Fierascu, R.C., Gabor, A.R., and Nicolae, C.A. (2020). Effect of hemp fibre length on the mechanical and thermal properties of polypropylene/SEBS/hemp fibre composites. *Journal of Materials Research and Technology* 9: 10768–10781.

24 Sair, S., Oushabi, A., Kammouni, A. et al. (2018). Mechanical and thermal conductivity properties of hemp fiber reinforced polyurethane composites. *Case Studies in Construction Materials* 8: 203–212.

25 Dhakal, H.N., Ismail, O.S., Beaugrand, J. et al. (2020). Characterisation of nano-mechanical, surface and thermal properties of hemp fibre-reinforced polycaprolactone (HF/PCL) biocomposites. *Applied Sciences* 10: 2636. https://doi.org/10.3390/app10072636.

26 Song, Y.S., Lee, J.T., Ji, D.S. et al. (2012). Viscoelastic and thermal behavior of woven hemp fiber reinforced poly(lactic acid) composites. *Composites Part B: Engineering* 43: 856–860.

27 Pandey, P., Bajwa, D., Ulven, C., and Bajwa, S. (2016). Influence of hybridizing flax and hemp-agave fibers with glass fiber as reinforcement in a polyurethane composite. *Materials* 9: 390. https://doi.org/10.3390/ma9050390.

28 Ashok, R.B., Srinivasa, C.V., and Basavaraju, B. (2019). Dynamic mechanical properties of natural fibre composites – a review. *Advanced Composites and Hybrid Materials* 2: 586–607.

29 Baghaei, B., Skrifvars, M., Rissanen, M., and Ramamoorthy, S.K. (2014). Mechanical and thermal characterization of compression moulded polylactic acid natural fiber composites reinforced with hemp and lyocell fibers. *Journal of Applied Polymer Science* 131: 40534.

30 Sarasini, F., Tirillo, J., Sergi, C. et al. (2018). Effect of basalt fibre hybridisation and sizing removal on mechanical and thermal properties of hemp fibre reinforced HDPE composites. *Composite Structures* 188: 394–406.

12

Thermal Properties of Cellulose Nanofibers and Their Composites

Sadia Khalid[1], Tania Ali[1,2], Asiya Gul[1,2], and Sara Qaisar[1]

[1]*National Centre for Physics, Nanosciences and Technology Department, Quaid-i-Azam University Campus, Shahdra Valley Road, Islamabad, 45320, Pakistan*
[2]*Hazara University, Department of Chemistry, Mansehra, 21300, Khyber Pakhtunkhwa, Pakistan*

12.1 Introduction

Nanocellulose posseses high aspect ratio, larger surface area, and abundant surface hydroxyl groups. It exhibits unique physiochemical, mechanical, thermal, and optical properties owing to its structural attributes [1]. Nanocellulose-based hybrid materials are the demand of flexible, greener, and sustainable socioeconomic needs. The surface modification strategies of nanocellulose have further paved the way for advanced applications such as industries, agriculture, biomedical, energy, and environment.

Surface functionalization of cellulose nanofibers (CNFs) using chemical and physical approaches such as blending, polymerization, covalent modification, and cross-linking via hydroxyl groups within structure of CNF has been a wide area of research. Surface modification of CNF with carbon materials, inorganic nanoparticles, two-dimensional layered materials, and particularly polymers has been remained a vast research area for future applications, particularly aerospace and flexible electronics [2].

With the increasing energy demands, thermally insulating materials synthesized using natural available resources are preferred for energy-efficient buildings. CNF aerogels have potential for thermal insulating applications but have flammability issues. Thermomechanical properties of the pure CNFs need to be improved for such applications by material designing. CNF-based composites have been extensively studied by researchers for their thermal and mechanical properties. However, CNF-based hybrid materials exhibit constraints and limitations, owing to their low thermal stability and design incompatibility. Therefore, the current challenge is to provide stable functionalities and rational design strategies for the development and commercialization of innovative electronic and energy-related devices. In this chapter, we represent the main features of CNFs and its emerging applications as thermal insulator or flame-retardant material.

Natural Fiber-Reinforced Composites: Thermal Properties and Applications, First Edition.
Edited by Senthilkumar Krishnasamy, Senthil Muthu Kumar Thiagamani, Chandrasekar Muthukumar, Rajini Nagarajan and Suchart Siengchin.

Table 12.1 Forms of nanocellulose with general values of dimensions (diameter and length).

Cellulose	Cellulose nanocrystals (CNCs)	Cellulose nanofibres (CNFs)	Bacterial nanocellulose (BNC)
Synthesis	Acid hydrolysis	Mechanical treatment	Biological
Morphology	Whiskers	Fibrils	Pellicles
Diameter length	5–70 nm 50–500 nm	2–50 nm >2000 nm	20–100 nm 100 to >1000 nm
Source	Biomass	Biomass	Bacteria

Source: Based on Chakrabarty and Teramoto [3].

12.2 Nanocellulose

Plants, algae, fungi, and bacteria are the main source of nanocellulose. Nanocellulose has been prepared from physical (mechanical treatment), chemical (tempo-oxidation), and biological (bacteria) methods. Natural cellulose $(C_6H_{10}O_5)_n$ fibres are composed of crystalline and amorphous parts. Crystalline structure of cellulose consists of β-D-glucose monomer repetitions linked via β-(1,4) glycoside linkage. Disintegration of cellulose at nanolevel gives rise to nanocellulose, cellulose nanocrystals (CNCs) (also known as nanowhiskers) and CNFs. Bacterial nanocellulose or BNC is the purest one among all the forms of cellulose that is extracted from cell-free systems. Different forms of nanocellulose are summarized in (Table 12.1) [3].

12.3 Cellulose Nanofibers (CNFs)

CNFs are renewable, biocompatible, and low-density form of nanocellulose. These have been extensively applied for environment clean and control systems because of their thermal stability, mechanical strength, and biodegradability. Unlike CNCs, crystallinity of CNFs varies from 40 to 90% (Figure 12.1) [4]. CNFs differ in their dimensions (aspect ratio) according to the source of extraction or fabrication process. Typical diameter of CNFs varies from 2 to 50 nm. CNF exhibit high aspect ratio (>150), high thermal stability (devolatilization temperature of 250–350 °C) [5], low thermal expansion coefficient (TEC) (0.1 ppm K^{-1}) [6], and high tensile modulus (23.9 GPa) with high tensile strength (383.3 MPa) [7]. High transparency

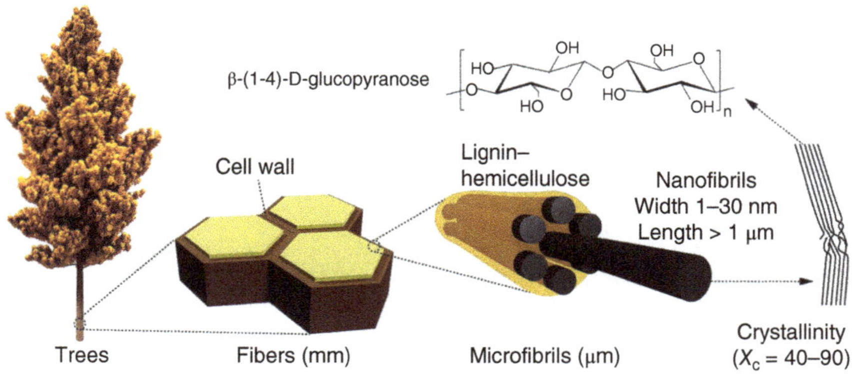

Figure 12.1 Schematic illustration of the origin of cellulose nanofibrils (CNFs) in trees. Natural plant cell walls are composed of nanofibrils embedded in a lignin–hemicellulose matrix. The nanofibrils contain crystalline and amorphous domains of linear polymer chains of β-(1-4)-D-glucopyranose [4]. Source: Reproduced from Benítez and Walther [4] Copyright (2017), with permission from The Royal Society of Chemistry.

and chemical and thermal durability of CNFs make them potential candidates for a number of applications such as optoelectronics, energy storage and conversion devices, sensors, catalysts, food packaging, and electronics [8].

12.4 CNF Preparation

CNFs have been extracted from plants (wood, cotton, rice husks, wheat straws, etc.), algae (*Rhizoclonium, Cladophora, Microdyction*, etc.), and bacteria (Gram positive such as *Gluconacetobacter hansenii* and Gram negative such as *Aerobacter*, *Sarcina*, *Salmonella*, etc.) [9]. Bottom-up (biosynthesis via bacteria) and top-down (disintegration via plants) strategies have been applied for the production of CNFs. CNFs have been produced by physical, chemical, and enzymatic methods and are prepared by much simpler physical methods (mechanical treatments) such as ball milling [10], steam explosion, ultrasonication, grinding, homogenization, microfluidization, and electrospinning with ~70% yield. Chemical or enzymatic pretreatments have also been utilized to facilitate the disintegration of cellulose to extract CNFs. Posttreatments such as chemical modification or surface functionalization of CNFs have also been used (Figure 12.2) [11]. However, CNF is an inexpensive ecofriendly nanoreinforcement material; therefore, the aspect ratio, crystallinity, and degree of polymerization should be maintained during extraction process [12].

12.5 Surface Functionalization of CNFs

The surface of the nanocellulose is usually hydrophilic due to surface hydroxyl groups that favors the surface modification via physical or chemical interaction [8]. The functionalization of nanocellulose can be designed and controlled according

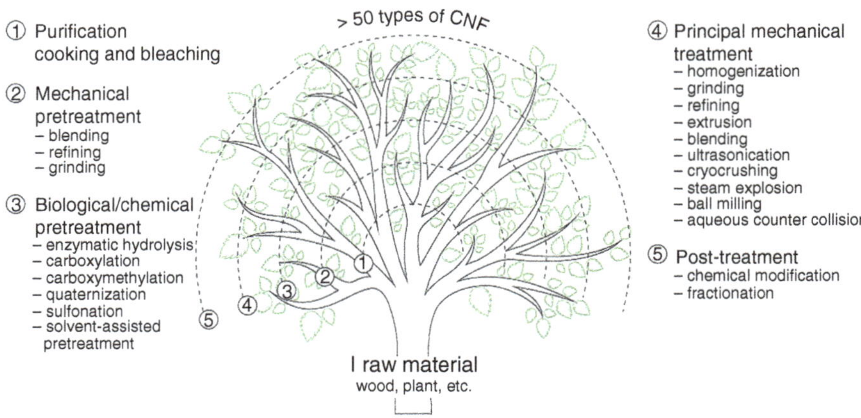

Figure 12.2 Schematic diagram of CNF production "tree" [11]. Source: Reproduced from Nechyporchuk et al. [11] Copyright (2016), with permission from Elsevier.

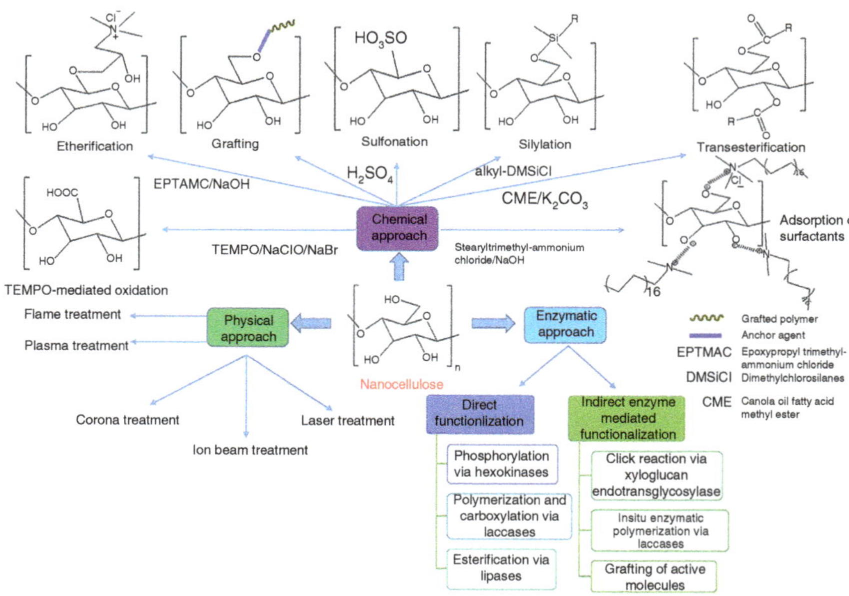

Figure 12.3 Schematic representation of the most commonly used surface modification routes of nanocellulose [13]. Source: Reproduced from Trache et al. [13] Copyright (2020), with permission from Frontiers in Chemistry under Creative Commons CC BY license.

to the desired application by various approaches (Figure 12.3) [13]. Replacement of hydroxyl groups CNCs with carboxylic group (R–COOH) is a well-known TEMPO (2,2,6,6-tetramethylpiperidine 1-oxyl)-mediated oxidation method. Several other chemical modification methods have also been used including acid treatment, surfactant treatment, polymer grafting (grafting onto and grafting from), etc.

Generally, thermal decomposition temperature of nanocellulose lies between 200 and 300 °C. Thermal stability of the nanocellulose varies with surface hydroxyl

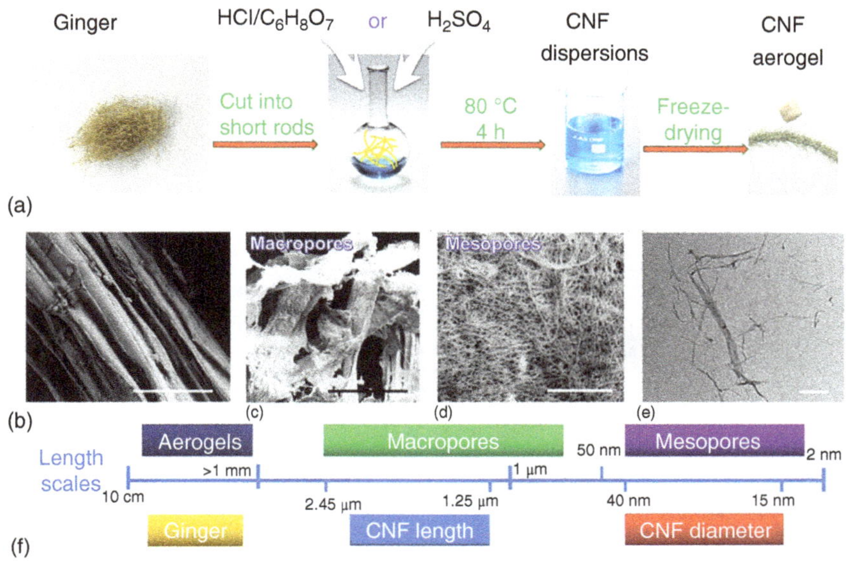

Figure 12.4 (a) Schematic illustration of experimental processes, (b) Scanning electron microscopy (SEM) image of ginger fiber, (c) macroscaled 7-CNF aerogel, (d) nanoscaled 7-CNF, (e) Transmission electron microscopy (TEM) image of 9-CNF, and (f) schematic illustrations of relevant structure dimensions. Scale bars of (b) 100 μm, (c) 100 μm, (d) 1 μm, and (e) 200 nm [15]. Source: Reproduced from Wang et al. [15] Copyright (2018), with permission from American Chemical Society.

groups that leads to the degradation of polymer by water absorption through hydrolysis. Above and beyond, preparation methods, extraction techniques, drying procedures, surface functional groups, and crystallinity index define the thermal behaviour of the cellulose-based nanocomposites. The higher sulfate groups on the nanocellulose extracted by sulfuric acid hydrolysis lead to lower thermal stability of the nanocellulose, hence rapid degradation. Surface modification of the nanocellulose with hydrophobic functional group improves the dispersion by reducing agglomeration [14].

For example, wasted ginger was used by Wang et al. to extract carboxylated CNF (aspect ratio of 144) through acid hydrolysis (Figure 12.4). Mixed acid system (hydrochloric acid/citric acid in varying ratios: v/v = 9/1, 7/3, and 5/5 designated as 9-CNF, 7-CNF, and 5-CNF, respectively) was used to prepare CNF aerogels. CNF aerogels showed enhanced mechanical properties and excellent shape recovery. Thermal stability (T_{max} = 345.7 °C) of 7-CNF was greater than other carboxylated CNF samples, owing to the lower degree of damage to the crystalline region [15].

12.6 CNF-Based Composites

Recent research and development (R&D) projects aim flexible and facile CNF preparation methods at large scales and renewable sources. One of the commonly applied

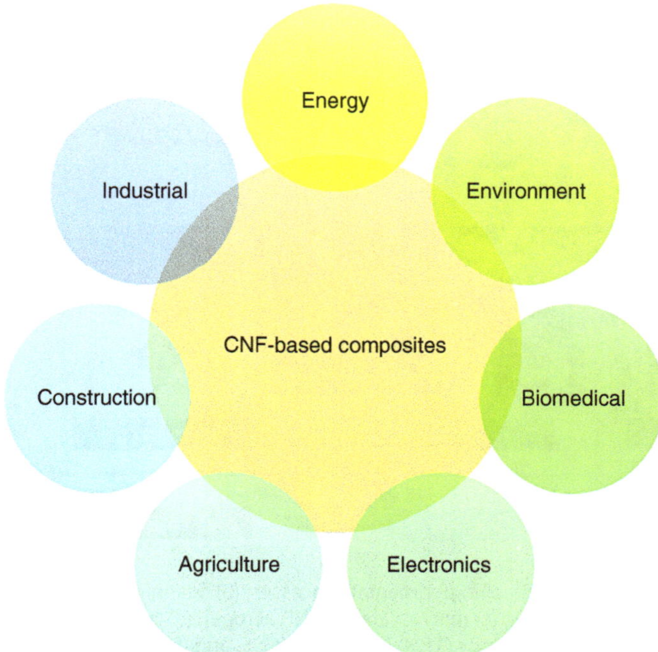

Figure 12.5 Applications of cellulose nanofiber-based composites.

strategies is the modification or surface functionalization of CNFs to make valuable composites. The better interaction between filler and matrix leads to the higher activation energy required to break the interface. CNF-based composites have been employed in diverse fields. Some of CNF-based composites are comprehensively described in this section (Figure 12.5).

For example, the aerogels of CNFs/silica have been prepared by one step sol–gel polymerization process by aging of silica precursor in the presence of CNFs. Polymerization and hydrolyzation of sodium silicate (Na_2SiO_3) are done in the presence of CNFs with changed ratios for the sack of hydrogel production, whose young modulus and storage verified CNF as a function of structural skeleton. At the most favorable 8 : 2 ratio of CNF/Na_2SiO_3 composite, the homogeneous dispersion of silica/CNF-based hydrogel was freeze-dried for the preparation of mesoporous aerogel (specific surface area, $342\,m^2\,g^{-1}$; pore volume, $0.86\,cm^3\,g^{-1}$; density, $7.7\,mg\,cm^{-3}$). By using sol–gel method, the naturally occurring bio-based cellulose and silica-based aerogel have high tensile strength and young modulus in the range of 177 to 28.5 kPa. The adsorption capacity for capturing functionalized CNF/silica aerogels was $1.49\,mmol\,g^{-1}$ [16]. Therefore, CNF/silica-based aerogels have potential to be used as CO_2 adsorbent and thermal insulators.

Natural bacterial cellulose (BC) and polymerizable deep eutectic solvent (PDES) (BC–PDES) composites were prepared by a facile in situ photopolymerization method. The multichannel networks of BC–PDES polymernanofiber composite exhibited improved mechanical properties such as stretchability and compression resistance. The stretchability was observed with 290% of strain and 0.8 MPa of stress, while compression resistance was observed with 75% of strain and 6.8 MPa of stress. The electrical conductivity of BC–PDES polymernanofiber composite was 0.18 S m^{-1} with 90% of optical transmittance. Notably, BC–PDES composites could retain a long-term stable mechanical property even in a high humidity environment. The sensitivity to strain was estimated as 1.29 of gauge factor (GF) with ~35 ms of response time for pressure. These multifunctional BC–PDES composites might be promising candidates for flexible electronics [17].

The other example is the tetragonal barium titanate (BTO) and CNF-based nanocomposite films prepared by solution casting method. They showed high dielectric constant in aqueous media. Thermal properties of CNF/BTO composite were improved due to the reinforcement of BTO; consequently, it can be a potential candidate for energy storage applications [18]. The addition of few layers of MXene (Ti_3C_2) nanosheets in CNFs using vacuum-assisted filtration (CNF)/Ti_3C_2 improved the thermal conductivity of flexible composite films to 11.57 (in plane) and 1.57 (out of plane) W m^{-1} K^{-1}. Ti_3C_2/CNF flexible composite films have potential to be used in flexible electronic devices for heat dissipation [19].

Reinforcement of elastomers by nanocellulose has been widely investigated to enhance the thermomechanical aspects of rubbers. A mixture of aqueous hydrogenated acrylonitrile–butadiene rubber (H-NBR) latex and aqueous dispersion of TEMPO-oxidized cellulose nanofiber (TOCN) was prepared with or without applying mechanical treatments. The 0–10.0% volume ratios of TOCN were used to prepare the mixtures of TOCN/H-NBR. Hot pressing was applied to cross-linked the TOCN/H-NBR composite sheets with peroxide of H-NBR molecules. Excellent mechanical properties such as dimensional stability, tensile strength, tensile modulus, ductility, and storage modulus are achieved by applying shear forces toward high temperature [20]. CNF exhibits aggregation issues. Zinc oxide submicron particles have been used for blocking the aggregation of CNFs through intercalation between fibers. NR/CNF was obtained by dry blending of CNF powder with natural rubber (NR). Mechanical properties of the composite were improved [21].

Zeolitic imidazolate framework-67 (ZIF-67) with particle size of 50–200 nm was used for the preparation of ZIF-67@CNF membrane with pore size of 150–350 nm to overcome the issues related to inhomogeneity in pore size and pore distribution. ZIF-67@CNF membrane exhibited enhanced thermal behaviour with better thermal stability and surface wettability with low value of thermal shrinkage. Li-ion batteries (LIBs) with ZIF-67@CNF separator membrane showed better rate capability and cyclic and electrochemical stabilities [22].

PPy@CNF–PEG hybrid composite was prepared by coating CNF sponge with polypyrrole (PPy@CNF) and polyethylene glycol (PEG). This composite exhibited 169.7 J g^{-1} melting enthalpy. PPy@CNF-PEG converted 1.75–1.9 V of electrical enrgy to thermal energy with 85.1% efficiency. CNF-based composites with promising electrothermal conversion efficiency can be the potential candidates for energy conversion devices [23].

12.7 Thermal Properties of CNF Composites

Thermomechanical behaviour of a polymer-based composite is the backbone of the stability and reliability of energy storage and electronic devices. CNF-based composites have superb thermal retardant properties; nevertheless, high flammability limits their application. The mechanical strength and thermal conductivity of nanocomposites of CNFs and nanodiamond (ND) were studied by Kato et al. [24] ND have high modulus as compared with graphene oxides that are conventionally used as nanocarbon fillers in an aqueous medium. It has topmost thermal conductivity as compared with other elemental substances. Other than carbon, ND contain hydrophilic heteroatom like oxygen on its surface that enables it for good dispersion in water and fix rigid structure at the center. It was evident that ND enhanced the mechanical strength and improved CNF's thermal conductivity. Particularly, TEMPO-oxidized CNFs were strengthened efficiently as compared with modified quaternary ammonium CNFs due to the robust coordination with ND and high dispersion of ND.

Using an alkaline treatment technique, thermal stability of nanocellulose CNF has been enhanced by converting cellulose crystalline structure to cellulose. Conventional method required 20 wt% sodium hydroxide solutions and interdigitation and aggregation of cellulose, which results in difficulties in the production of CNF. A pretreatment method was build up by using a low concentration of alkaline solution for the making of good dispersion of CNF with better thermal stability. Through a French pressure cell method, CNF with 90 nm diameter was prepared by treatment of 2 wt% sodium hydroxide solutions at a temperature less than 0 °C, followed by homogenization. The resulted CNF had comparatively high thermal stability as compared with CNF prepared without NaOH pretreatment. The improved thermal stability suggests a new method for the growth of bio-based composites [25].

The electrodes of flexible solar cells (FSC) are composed of polymer substrates. The polymer substrates are very important in determining the performance of electrode and device credibility. Mostly, FSC electrodes made from polymer substrates have a limitation of high TECs and are thermally unstable while their exposure to thermal cycling impacts. Modified CNF were used to develop a nanocellulose/epoxy hybrid substrate. These substrates showed enhanced thermal behaviour with improved optical transparency. The nanoscale structure and a large number of functional groups at the surface of CNFs enabled it for compatible coordination with epoxy matrix (Figure 12.6). Thermal properties of composite substrate were enhanced, having 19 ppm K^{-1} of TEC with 71.8 °C glass transition temperature, and half-life thermal decomposition temperature ($T_{d,50\%}$) is increased by 376 °C. Ultimate strength of 41% and ultimate strain of 121.5% were observed with 89% transmittance at 600 nm. CNF-based hybrid substrates are interesting candidates for stable, efficient, and economic FSCs [26].

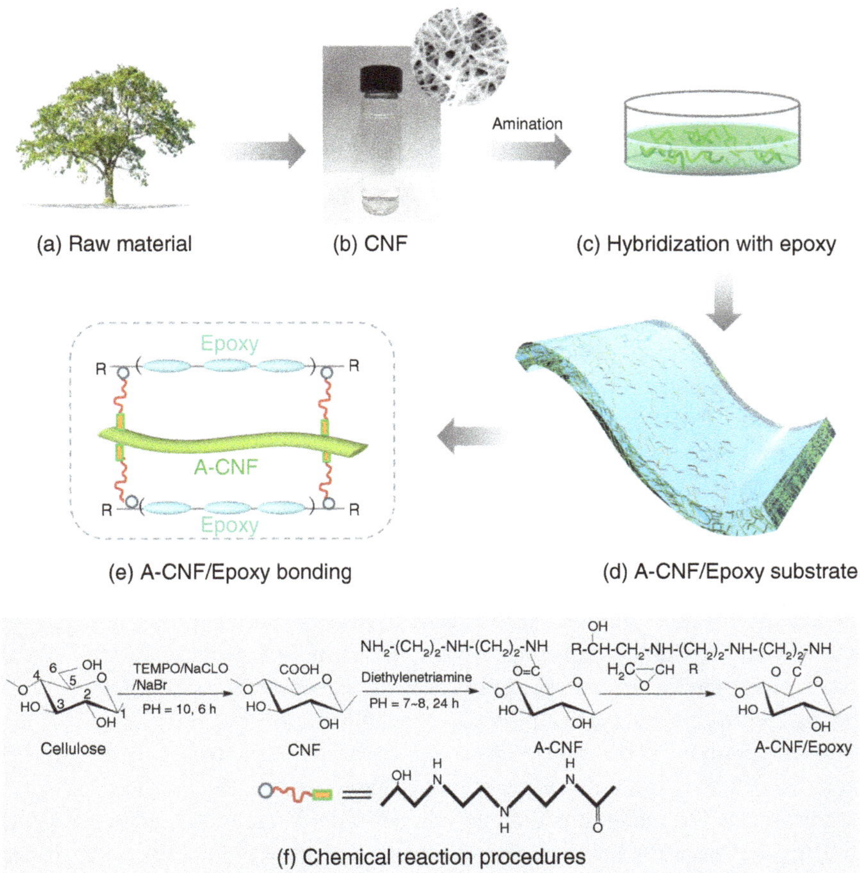

Figure 12.6 Schematic of the preparation of transparent and thermally stable hybrid substrates for flexible solar cell (FSC) electrodes with CNFs and epoxy resin. (a) Starting material of natural wood. (b) Cellulose nanofibrillated fibers (CNFs) oxidized with TEMPO. (c) Aminated CNFs (ACNF) blended with epoxy resin. (d) A-CNF/epoxy hybrid substrate for FSC devices. (e) Improved thermal and mechanical properties ascribed to the desirable interfacial interaction between amino groups of A-CNF and epoxide groups of epoxy. (f) Brief illustration of chemical reactions. The inset of (b) is an atomic force microscopy (AFM) image of CNFs [26]. Source: Reproduced from Wang et al. [26] Copyright (2020), with permission from The Royal Society of Chemistry under Creative Commons Attribution-NonCommercial 3.0 Unported Licence.

12.8 Current Status: CNF-Based Composites

The heat removal in early devices has become a foremost issue due to the challenging advances in electronic devices. High thermal conductivities are required for thermal management of electrically insulating materials. Nowadays nanocellulose and related materials have been emphasized and investigated due to their outstanding properties and their renewability [27].

Fabrication of ecofriendly green plastics is very important to replace the conventional synthetic plastics for a sustainable ecological environment and better economy. The green plastic (cellulose/functionalized graphene composite plastics [CGPs]) was prepared using cellulose and amino-functionalized graphene (GPs). The CGPs exhibited improved thermomechanical behaviour as compared with synthetic plastics. The incorporation of 6 wt% functionalized graphene into cellulose showed 9.0 W m^{-1} K^{-1} in-plane thermal conductivity. The homogeneously dispersed GP and CGP alignments have resulted in the enhanced mechanical properties that might be arised from the hydrogen bonding between cellulose and amino-functionalized graphene. CGPs have potential to be applied as green plastics in thermal management devices [28].

CNF-based aerogels were prepared using *N*-methylol dimethylphosphonopropionamide (MDPA) and 1,2,3,4-butanetetracarboxylic acid as coadditives through environmentally friendly post-cross-linking and freeze-drying techniques. CNF/MDPA/BTCA (1,2,3,4-butanetetracarboxylic acid) aerogel has shown a flame-resistant performance, extraordinary self-smothering conduct, and fundamentally expended char yield (268%) than pure CNF aerogel. Then, flexibility of the aerogel samples was improved as the resilience of aerogel somewhat diminished. Besides, the aerogel samples have excellent heat-insulating properties, having 0.03258 W m^{-1} K^{-1} of thermal conductivity. These properties ensure the CNF-based aerogel as potential insulating candidates for heat-defensive equipment like clothing and fire protection of advanced spacesuit components in the future [29].

The fire-retardant performance of a material can be evaluated by its fire growth index (FGI) that is the ratio of peak heat release rate (PHRR) to the peak time. The burning of the material is directly related to the value of FGI. The higher its value, higher the risk of fire to spread. While the fire propagation index (FPI) of the material is the ratio of time to ignition (TTI) and PHRR. FPI is associated with the flame-retardant performance of the material. The higher the FPI, the lower the flame hazard. Flammability of the material decreases with the higher value of FPI and by lowering FGI. Cellulose itself is a flammable material; therefore, cellulose-based hybrid materials have been extensively investigated as sustainable insulating green material. For example, fire-resistant behaviour has been observed in CNF-based composites with the incorporation of aluminum hydroxide nanoparticles (AH NPs). The composite aerogels indicated remarkable heat retardant properties such as PHRR of 280 to 22 W g^{-1} with 13.2 to 1.6 kJ g^{-1} of heat discharge [30].

Nano-wrapping of CNFs with nanosheets of molybdenum disulfide (MoS_2) via chemical cross-linking revealed the thermal conductivity of 28.09 mW m^{-1} K^{-1}. Limiting oxygen index (LOI) is the measure (vol%) of relative flammability. The minimum oxygen concentration in oxygen and nitrogen mixture is required to support the material's flaming combustion. LOI for CNF and MoS_2 composites was observed as 34.7% with the total heat release of 0.4 MJ m^{-2} [31].

The graphene and cellulose nanofiber composite (CNFG) flexible film prepared by vacuum-assisted filtration exhibited improved thermal conductivity and heat spreading behaviour. The of the CNFG film (graphene loading: 50 wt%) showed 164.7 W m^{-1} K^{-1} of thermal conductivity in plane and 5.0 W m^{-1} K^{-1} through

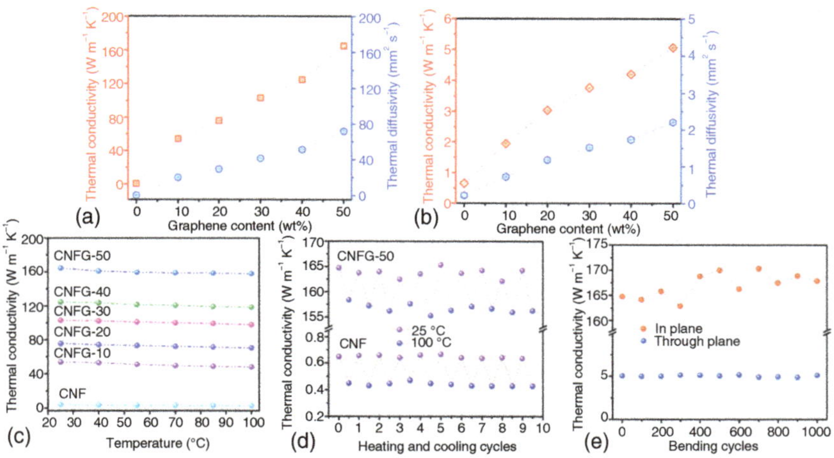

Figure 12.7 Thermal conductive properties of CNFG composite films: thermal conductivities and thermal diffusivities of neat CNF and CNFG composite films with different graphene loadings in the (a) in-plane and (b) through-plane directions. (c) Thermal conductivities of CNF and CNFG composite films as functions of temperature. (d) Thermal conductivities as functions of heating and cooling cycles for CNFG-50 and neat CNF. (e) In-plane and through-plane thermal conductivities of CNFG-50 after different numbers of bending cycles [32]. Source: Reproduced from Chen et al. [32] Copyright (2018), with permission from The Royal Society of Chemistry.

plane and thermal stability in radial and axial directions (Figure 12.7). CNFG film (graphene loading: 50 wt%) exhibited a tensile strength of 72.3 ± 5.4 MPa. Accordingly, CNFG flexible films can be applied as heat spreaders in delicate electronic devices [32].

Layer-by-layer technique is the most suitable and rapid method that modifies the surface properties of aerogel to give extraordinary flame-resistant characteristics. The cationic chitosan, anionic poly(vinylphosphonic) acid, and anionic montmorillonite (MMT) clay were deposited using layer-by-layer method on highly porous CNF aerogel (CH/PVPA/CH/MMT) showing thermally stable and flame-retardant properties [33].

CNF-based fluorinated CNT (NFCs/FCNTs [fluorinated carbon nanotubes]) composite film prepared by vacuum-assisted filtration showed well-arranged layered structure along planar direction. NFCs/FCNTs exhibited low value of interfacial thermal resistance of NFCs/FCNTs with 14.1 $W m^{-1} K^{-1}$ of in-plane thermal conductivity and excellent electrical insulation and mechanical properties (FCNT loading: 35 wt%) [34].

The incorporation of sodium bicarbonate (40 wt%) in CNFs (sodium bicarbonate [SBC]) enhanced the fire retardancy of the CNF-based aerogel with 0.20 cm^{-1} of burning velocity as compared with pure CNF-based aerogel (5.84 cm^{-1}). Besides, SBC provides control over the morphology and porosity because of the dependent on concentration lamellar layer formation during aerogel preparation. A suitable and efficient preparation method has been suggested for flame-retardant CNF-based aerogel. This could set organization for the next-generation bio-based material and insulation [35].

Table 12.2 CNF-based flame retardant composites and their thermal properties.

Sr. #	Composites	Limited oxygen index (%)	Thermal conductivity ($W\,m^{-1}\,K^{-1}$)	References
1.	PLA/APP@CNF	27.5%	—	[38]
2.	MTMS@CNF	—	26.1×10^{-3} to 39.6×10^{-3}	[39]
3.	PVA/BNNS/APP	29.5 ± 0.5 %	—	[40]
4.	CNF/ZB	29.5–40%	0.0276–0.0298	[41]
5.	CNF/BP	19.5–23.9 %	0.275–0.448	[42]
6.	CNF/BNNS Silylated CNF/BNNS	—	0.50 0.57	[43]

Cellulose nanofibers (CNF), poly(lactic acid) (PLA), ammonium polyphosphate (APP), methyltrimethoxysilane (MTMS), polyvinyl alcohol (PVA), zinc borate (ZB), black phosphorus (BP), and boron nitride nanosheets (BNNSs).

CNF-based aerogels would be potential candidates as absorber of kinetic energy, heat insulator, and insulating materials for buildings [36]. Flame-retardant CNF and sodium MMT-based aerogel with enhanced mechanical properties are cross-linked to melamine-formaldehyde resins into suspension of precursor using freeze-drying method. The effect of melamine-formaldehyde (MF) on CNF and CNF/MMT was confirmed by scanning electron microscopy. It was confirmed that CNF cross-linking does not change the structure of CNF and CNF/MMT aerogel. Although cross-linking enhances mechanical and flame properties of materials. The production of CNF-based aerogel from sustainable and bio-based materials improved the mechanical properties and high flame conductivity [37]. Recent CNF-based flame-retardant materials and their thermal properties are summarized in Table 12.2.

The reduced graphene nanosheets confined with zirconium phosphate (ZrP/RGO [reduced graphene oxide]) into CNF by building porous ZrP/RGO/CNF lamella-bridge architecture via unidirectional freeze-casing method are introduced (Figure 12.8). The lamellar alignments of the fibers exhibited 18 $mW\,m^{-1}\,K^{-1}$ of thermal conductivity with 104 $kN\,m\,kg^{-1}$ of maximal specific Young's modulus. LOI was 33.5% and PHRR was 14.1 $kW\,m^{-2}$. The porous structure with lamellar alignment suggested better insulating properties by inhibiting thermal radiations, oxygen infusion, and volatile outflow. Nanoscale engineering offers a rational designing of natural architectures for high-performance materials [44].

12.9 Outlook and Future Perspective

Thermal management in the flexible electronic devices has been remained one of the major issues for the sustainable applications. CNF-based green composites have a great potential in dealing with such thermal management issues. Cellulose is the

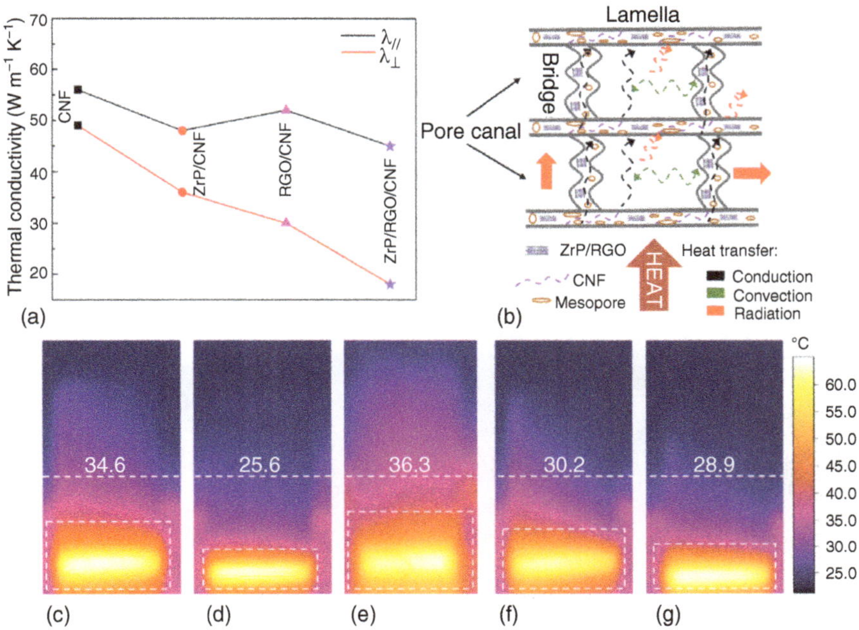

Figure 12.8 Thermal insulating performance of biomimetic structural ZrP/RGO/CNF aerogel. (a) Thermal conductivities parallel (λ//) and perpendicular (λ⊥) to lamella alignments of CNF, RGO/CNF, ZrP/CNF, and ZrP/RGO/CNF aerogels. (b) Schematic illustration of the contributions of convection, conduction, and radiation to thermal conductivities perpendicular to lamella alignments of ZrP/RGO/CNF aerogel. Temperature profiles of heat transfer (c) parallel and (d) perpendicular to the lamella alignments in ZrP/RGO/CNF aerogel. Temperature profiles of heat transfer perpendicular to the lamella alignments of (e) CNF, (f) ZrP/CNF, and (g) RGO/CNF aerogels. Heated volumes are marked by white frames [44]. Source: Reproduced from Wang et al. [44] Copyright (2020), with permission from The Royal Society of Chemistry under Creative Commons Attribution-NonCommercial 3.0 Unported Licence.

renewable and smart green biocompatible natural polymer for economic and sustainable large-scale applications (Figure 12.9).

However, there are some challenges while working with CNFs such as aggregation, biodegradation, and multistep processing methods. For example, the conversion of CNF suspension into aerogels or films is an additional critical step. CNFs exhibit high flammability that limits its applications as insulating material.

Nanoengineering has provided unique path to modify thermal properties of CNFs. Self-fire-extinguishing and flame-retardant properties of the CNFs have been improved by numerous strategies such as the incorporation of inorganic compounds (e.g. carbon materials, metal oxides, metal sulfides, and two-dimensional layered materials). CNF-based hybrid aerogels provide an alternative way of improving the thermal properties. In summary, CNF-based composites are smart and safe sustainable green materials for heat-resistant and flame-retardant applications.

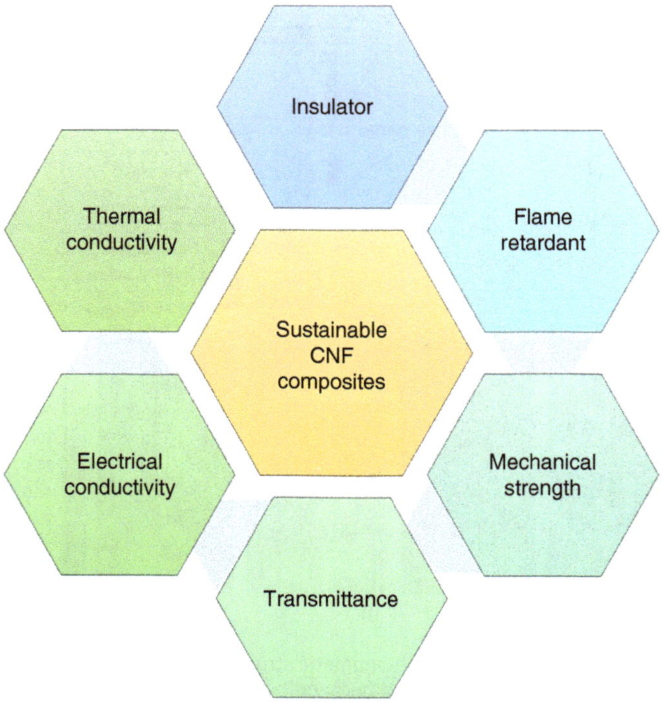

Figure 12.9 Schematic summary of sustainable CNF-based composites.

References

1 Nogi, M., Kim, C., Sugahara, T. et al. (2013). High thermal stability of optical transparency in cellulose nanofiber paper. *Applied Physics Letters* 102: 181911.

2 Thomas, P., Duolikun, T., Rumjit, N.P. et al. (2020). Comprehensive review on nanocellulose: recent developments, challenges and future prospects. *Journal of the Mechanical Behavior of Biomedical Materials* 110: 103884.

3 Chakrabarty, A. and Teramoto, Y. (2018). Recent advances in nanocellulose composites with polymers: A guide for choosing partners and how to incorporate them. *Polymers* 10: 517.

4 Benítez, A.J. and Walther, A. (2017). Cellulose nanofibril nanopapers and bioinspired nanocomposites: a review to understand the mechanical property space. *Journal of Materials Chemistry A* 5: 16003–16024.

5 Borsoi, C., Zimmernnam, M.V.G., Zattera, A.J. et al. (2016). Thermal degradation behavior of cellulose nanofibers and nanowhiskers. *Journal of Thermal Analysis and Calorimetry* 126: 1867–1878.

6 Guan, Q.-F., Yang, H.-B., Han, Z.-M. et al. (2020). Lightweight, tough, and sustainable cellulose nanofiber-derived bulk structural materials with low thermal expansion coefficient. *Science Advances* 6: eaaz1114.

7 Kafy, A., Kim, H.C., Zhai, L. et al. (2017). Cellulose long fibers fabricated from cellulose nanofibers and its strong and tough characteristics. *Scientific Reports* 7: 17683.

8 Abdul Khalil, H.P.S., Bhat, A.H., and Ireana Yusra, A.F. (2012). Green composites from sustainable cellulose nanofibrils: a review. *Carbohydrate Polymers* 87: 963–979.

9 Prakash Menon, M., Selvakumar, R., Suresh kumar, P., and Ramakrishna, S. (2017). Extraction and modification of cellulose nanofibers derived from biomass for environmental application. *RSC Advances* 7: 42750–42773.

10 Piras, C.C., Fernández-Prieto, S., and De Borggraeve, W.M. (2019). Ball milling: a green technology for the preparation and functionalisation of nanocellulose derivatives. *Nanoscale Advances* 1: 937–947.

11 Nechyporchuk, O., Belgacem, M.N., and Bras, J. (2016). Production of cellulose nanofibrils: a review of recent advances. *Industrial Crops and Products* 93: 2–25.

12 Hussin, M.H., Trache, D., Chuin, C.T.H. et al. (2019). Extraction of cellulose nanofibers and their eco-friendly polymer composites. In: *Sustainable Polymer Composites and Nanocomposites*, 653–691. Springer.

13 Trache, D., Tarchoun, A.F., Derradji, M. et al. (2020). Nanocellulose: from fundamentals to advanced applications. *Frontiers in Chemistry* 8: 392.

14 Gan, P.G., Sam, S.T., Abdullah, M.F.b., and Omar, M.F. (2020). Thermal properties of nanocellulose-reinforced composites: a review. *Journal of Applied Polymer Science* 137: 48544.

15 Wang, D., Yu, H., Fan, X. et al. (2018). High aspect ratio carboxylated cellulose nanofibers cross-linked to robust aerogels for superabsorption–flocculants: paving way from nanoscale to macroscale. *ACS Applied Materials & Interfaces* 10: 20755–20766.

16 Jiang, F., Hu, S., and Hsieh, Y.-l. (2018). Aqueous synthesis of compressible and thermally stable cellulose nanofibril–silica aerogel for CO2 adsorption. *ACS Applied Nano Materials* 1: 6701–6710.

17 Wang, M., Li, R., Feng, X. et al. (2020). Cellulose nanofiber-reinforced ionic conductors for multifunctional sensors and devices. *ACS Applied Materials & Interfaces* 12: 27545–27554.

18 Tao, J., Cao, S.-a., Feng, R., and Deng, Y. (2020). High dielectric thin films based on barium titanate and cellulose nanofibrils. *RSC Advances* 10: 5758–5765.

19 Song, G., Kang, R., Guo, L. et al. (2020). Highly flexible few-layer Ti3C2 MXene/cellulose nanofiber heat-spreader films with enhanced thermal conductivity. *New Journal of Chemistry* 44: 7186–7193.

20 Noguchi, T., Endo, M., Niihara, K. et al. (2020). Cellulose nanofiber/elastomer composites with high tensile strength, modulus, toughness, and thermal stability prepared by high-shear kneading. *Composites Science and Technology* 188: 108005.

21 Hirase, R., Nagatani, A., and Yuguchi, Y. (2020). Development of powdering method for cellulose nanofibers assisted by zinc oxide for compounding reinforced natural rubber composite. *Current Research in Green and Sustainable Chemistry* 3: 100005.

22 Sun, X., Xu, W., Zhang, X. et al. (2021). ZIF-67@cellulose nanofiber hybrid membrane with controlled porosity for use as Li-ion battery separator. *Journal of Energy Chemistry* 52: 170–180.

23 Chen, L., Lv, J., Ding, L. et al. (2020). A shape-stable phase change composite prepared from cellulose nanofiber/polypyrrole/polyethylene glycol for electric-thermal energy conversion and storage. *Chemical Engineering Journal* 400: 125950.

24 Kato, T., Matsumoto, T., Hongo, C., and Nishino, T. (2018). Mechanical and thermal properties of cellulose nanofiber composites with nanodiamond as nanocarbon filler. *Nanocomposites* 4: 127–136.

25 Lee, H., Sundaram, J., Zhu, L. et al. (2018). Improved thermal stability of cellulose nanofibrils using low-concentration alkaline pretreatment. *Carbohydrate Polymers* 181: 506–513.

26 Wang, R., Yu, H., Dirican, M. et al. (2020). Highly transparent, thermally stable, and mechanically robust hybrid cellulose-nanofiber/polymer substrates for the electrodes of flexible solar cells. *ACS Applied Energy Materials* 3: 785–793.

27 Sato, K., Tominaga, Y., and Imai, Y. (2020). Nanocelluloses and related materials applicable in thermal management of electronic devices: a review. *Nanomaterials* 10: 448.

28 Song, N., Hou, X., Chen, L. et al. (2017). A green plastic constructed from cellulose and functionalized graphene with high thermal conductivity. *ACS Applied Materials & Interfaces* 9: 17914–17922.

29 Guo, L., Chen, Z., Lyu, S. et al. (2018). Highly flexible cross-linked cellulose nanofibril sponge-like aerogels with improved mechanical property and enhanced flame retardancy. *Carbohydrate Polymers* 179: 333–340.

30 He, C., Huang, J., Li, S. et al. (2018). Mechanically resistant and sustainable cellulose-based composite aerogels with excellent flame retardant, sound-absorption, and superantiwetting ability for advanced engineering materials. *ACS Sustainable Chemistry & Engineering* 6: 927–936.

31 Yang, L., Mukhopadhyay, A., Jiao, Y. et al. (2017). Ultralight, highly thermally insulating and fire resistant aerogel by encapsulating cellulose nanofibers with two-dimensional MoS_2. *Nanoscale* 9: 11452–11462.

32 Chen, Y., Hou, X., Kang, R. et al. (2018). Highly flexible biodegradable cellulose nanofiber/graphene heat-spreader films with improved mechanical properties and enhanced thermal conductivity. *Journal of Materials Chemistry C* 6: 12739–12745.

33 Köklükaya, O., Carosio, F., and Wågberg, L. (2017). Superior flame-resistant cellulose nanofibril aerogels modified with hybrid layer-by-layer coatings. *ACS Applied Materials & Interfaces* 9: 29082–29092.

34 Wang, X. and Wu, P. (2018). Fluorinated carbon nanotube/nanofibrillated cellulose composite film with enhanced toughness, superior thermal conductivity, and electrical insulation. *ACS Applied Materials & Interfaces* 10: 34311–34321.

35 Farooq, M., Sipponen, M.H., Seppälä, A., and Österberg, M. (2018). Eco-friendly flame-retardant cellulose nanofibril aerogels by incorporating sodium bicarbonate. *ACS Applied Materials & Interfaces* 10: 27407–27415.

36 Gupta, P., Singh, B., Agrawal, A.K., and Maji, P.K. (2018). Low density and high strength nanofibrillated cellulose aerogel for thermal insulation application. *Materials & Design* 158: 224–236.

37 Wang, L., Cui, L., Sánchez-Soto, M. et al. (2018). Highly flame retardant melamine-formaldehyde cross-linked cellulose nanofibrils/sodium montmorillonite aerogels with improved mechanical properties. *Macromolecular Materials and Engineering* 303: 1800379.

38 Yin, W., Chen, L., Lu, F. et al. (2018). Mechanically robust, flame-retardant poly(lactic acid) biocomposites via combining cellulose nanofibers and ammonium polyphosphate. *ACS Omega* 3: 5615–5626.

39 Gupta, P., Verma, C., and Maji, P.K. (2019). Flame retardant and thermally insulating clay based aerogel facilitated by cellulose nanofibers. *The Journal of Supercritical Fluids* 152: 104537.

40 Zhang, Q., Wang, X., Tao, X. et al. (2019). Polyvinyl alcohol composite aerogel with remarkable flame retardancy, chemical durability and self-cleaning property. *Composites Communications* 15: 96–102.

41 Cheng, X., Zhu, S., Pan, Y. et al. (2020). Fire retardancy and thermal behaviors of cellulose nanofiber/zinc borate aerogel. *Cellulose* 27: 7463–7474.

42 Du, X., Qiu, J., Deng, S. et al. (2020). Flame-retardant and form-stable phase change composites based on black phosphorus nanosheets/cellulose nanofiber aerogels with extremely high energy storage density and superior solar-thermal conversion efficiency. *Journal of Materials Chemistry A* 8: 14126–14134.

43 Liu, Y., Zhang, Y., Liao, T. et al. (2020). Boron nitride–nanosheet enhanced cellulose nanofiber aerogel with excellent thermal management properties. *Carbohydrate Polymers* 241: 116425.

44 Wang, D., Peng, H., Yu, B. et al. (2020). Biomimetic structural cellulose nanofiber aerogels with exceptional mechanical, flame-retardant and thermal-insulating properties. *Chemical Engineering Journal* 389: 124449.

13

Influence of Graphene Nanoparticles on Thermal Properties of the Natural Fiber-Based Hybrid Composites

Theivasanthi Thirugnanasambandan

Kalasalingam Academy of Research and Education (Deemed to be university), International Research Centre, Krishnankoil, 626126, Tamil Nadu, India. ORCID: https://orcid.org/0000-0002-2280-9316

13.1 Introduction

Natural fibers are the recyclable products that are obtained from the renewable resources. Their attractive properties include their ability to replace the synthetic fibers in composites owing to their lower cost and more environmental sustainability, biodegradability, lightweight, easy to fabricate, and lower density. They reduce the CO_2 emission and fossil fuel utilizations as well. The natural fibers such as hemp, flax, jute, kenaf, and sisal are utilized to replace synthetic fibers in many applications. The properties of the natural fibers that make them suitable for many applications include low weight, high specific stiffness, and specific strength.

Polymer composites find many applications in the industries due to their properties such as good processing ability, high production rates, low manufacturing cost, and flexibility in making complex geometries. Low-level mechanical and interfacial characters are some drawbacks of the natural fiber composites. These properties can be improved by some methods: (i) nanomaterials (such as graphene oxide [GO] and graphene) coating and (ii) addition of natural fibers with synthetic fibers (like carbon fibers). Natural fibers are also subjected to plasma treatment, heat treatment, and corona treatment to improve their mechanical properties. Acetylation, various treatments (like silane, peroxide, and plasma treatment), and etherification are some of the processes employed to modify natural fibers. However, these techniques are costly and more time consuming.

With regard to the arrangement of fibers, longitudinally laid fibers are possessing improved mechanical properties compared with the transversally laid fibers. High-performance natural fibers can replace synthetic fibers by reducing carbon footprint applied in automobile industries, ship constructions, civil construction works, and windmills. Jute fiber is a natural fiber (biodegradable and totally safe to environment) and produced in large scale. Its cost is lower than flax fibers. There is high demand for jute fibers (100 000 ton year^{-1}) to be used in the automobile

Natural Fiber-Reinforced Composites: Thermal Properties and Applications, First Edition.
Edited by Senthilkumar Krishnasamy, Senthil Muthu Kumar Thiagamani, Chandrasekar Muthukumar, Rajini Nagarajan and Suchart Siengchin.

industry. These fibers can be added with a small volume fraction of graphene and can be applied for various lightweight structural applications [1].

13.2 Graphene

Graphene possesses single-atom thickness and 2D honeycomb structure. It has also excellent mechanical, electrical, and thermal characters and is used as filler material in epoxy resin composites and in natural fiber-reinforced polymer composites due to its high strength, high surface area (2600 m^2 g^{-1}), unique graphitized plane structure, and low manufacturing cost. GO sheets possess a high chemical reactivity and contain more functional groups on its surface. GO has high hydrophilicity and can form stable colloidal suspensions. Functionalized graphene is superior to carbon nanotubes (CNTs) and expanded graphite because it can interact well with the polymers.

GO is functionalized using 2-amino-4,6-didodecylamino-1,3,5-triazine (ADDT). GO has several functional groups (like epoxide, hydroxyl, carboxyl, and carbonyl) at the edges. The thermal stability of GO is confirmed through thermogravimetric analysis (from the results of enhanced residue formation and reduced mass loss rate). The solubility of GO–ADDT is also increased in the organic solvents, and the compatibility is high with polymers [2].

A high cross-plane thermal conductivity K is achieved when the graphene and multilayer graphene mixture is utilized in a composite. Thermal conductivity measurements (done in laser flash apparatus) show the increase of K by 2300% for the 10 vol% of filler loading. The thermal conductivity is increased from 5.8 to 14 W m^{-1} K^{-1} in thermal grease with 2% graphene loading without losing the mechanical properties [3].

Graphene has high thermal conductivity value, i.e. 3000–5000 W m^{-1} K^{-1} at room temperature. It is higher than the thermal conductivity value of pyrolytic graphite (2000 W m^{-1} K^{-1}) at room temperature. As per the report of Shahil et al. graphene shows 100 times anisotropy of heat flow between in-plane and out-of-plane directions. Thermal conductivity in the in-plane direction is high due to the covalent sp^2 bonding between carbon atoms. The out-of-plane heat flow is low due to the weak van der Waals coupling [4]. Electromechanical resonators in nanoelectromechanical systems (NEMSs) working in extreme environments made of graphene are reported to vibrate at high temperature of over 900 °C, the highest operating temperature ever reported [5].

13.3 Polymer/Graphene Composites

The performance of a composite depends on the uniform dispersion of the graphene. When carbon nanomaterials are used to improve the properties of a composite, many difficulties arise. CNTs are costly because some catalysts are employed during its synthesis. Graphene can be prepared in large scale with high purity for commercial applications. Functionalization of graphene improves the heat transport properties. The filler and matrix interface will support to increase the thermal conductivity of graphene-based polymer composites.

The thermal conductivity of polymers is in the range of 0.2–0.5 $W\,m^{-1}\,K^{-1}$ at room temperature. Various parameters of 2D graphene sheets such as thickness of the sheets, surface area, and length of the graphene nanoribbons affect the thermal conductivity of the composite materials. The thermal characters of the composites vary due to the concentration of filler material (graphene). A proper content of graphene can facilitate more effective phonon transmission between the layers of graphene. As per the literatures, thermal conductivity of graphene nanoribbons is close to stainless steel [6].

Graphene foam (GF) is added with pliable polydimethylsiloxane (PDMS) to increase the thermal properties. This possesses unique interconnected structure. It can improve the thermal conductivity of the composite of up to 0.56 $W\,m^{-1}\,K^{-1}$. This enhancement is 300% of pure PDMS and even higher than the composite made with graphene sheets. The coefficient of thermal expansion is also very low when compared with pure PDMS and the composite made with graphene sheets [7].

The in-plane thermal conductivity of composites of up to 1.98 $W\,m^{-1}\,K^{-1}$ can be achieved with the addition of 20 wt% graphene nanoplatelets (GNPs). The high-level in-plane thermal conductivity of graphene supports for this. Mixing the graphene in the polymer matrices increases the viscosity of polymer resin. Hence, it is difficult to achieve uniform graphene dispersion. This results in agglomeration of graphene and formation of cracks, pores, or pinholes in the composites [8].

Amine functionalization (2 wt% loading) of the expanded GNPs with epoxy resins (done in the high-pressure processor and three-roll milling) can increase the thermal conductivity of the composite of up to 36%. A conductive network is formed in this epoxy matrix that is responsible to increase the thermal conductivity in epoxy. The limitation in heat propagation is due to the acoustic phonons [9].

The thermal properties of polypropylene (PP) composites filled with GNPs are analyzed using differential scanning calorimetry (DSC) and a thermal gravimetric analyzer (TGA). It is found that the melting temperature is shifted to higher values when filled with graphene. An increase in the weight fraction and lateral dimension of GNPs causes an increase in the thermal stability of the composite. The decomposition temperature increases when the GNP weight fraction is increased and the GNP lateral dimension is decreased. It is due to the sheet barrier function of the GNPs [10].

polyvinyl chloride (PVC) and graphene are mixed in *N,N*-di-methyl formamide (DMF) to get a clear dispersion. Stabled graphene dispersion in DMF is prepared by sonication of graphite flakes with 0.5 M cetyl trimethylammonium bromide (CTAB) solution in glacial acetic acid. Graphene/PVC composite films are made by sonication followed by drop casting. Graphene enhances the thermal stability of poly(vinyl chloride) by rising the glass transition temperature of the polymer [11].

Diglycidyl ether of bisphenol-A (DGEBA) molecules are grafted onto reduced graphene oxide (RGO). This functionalized RGO and epoxy sonicated in acetone for 30 minutes followed by the ball milling. Then, the curing agent is added by mechanical stirring. After that, the prepared composite mixture is poured into the previously heated steel molds to form the composite. Furthermore, 1.00 wt% of DGEBA-functionalized RGO addition with epoxy increases the decomposition temperature by 28 °C compared with the neat epoxy [12].

The epoxy/graphene nanocomposite is prepared by sonicating graphene nanoparticle (GNP) and DER 331 for 30 minutes at the temperature of 60–70 °C. The curing agent is added and poured into a silicon mold (for curing process at the temperature 120 °C for the period of one hour). The properties of graphene such as fraction, surface roughness, high aspect ratio, intrinsic crystallinity of the filler, degree of dispersion, orientation on the filler/matrix interface, and thermal contact resistance produce some effects on the thermal conductivity characters of the polymer composites. The thermal conducting path or network is more vital in increasing the thermal conductivity of the epoxy nanocomposites. GNP in the epoxy matrix acts as a thermal bridge, and the thermal chains improve the heat flow. This enhancement in the thermal conductivity is due to the thermal conductivity of the GNP. In GNP, the phonon is able to move in both parallel and perpendicular directions to the surface. Increasing the filler loading in the nanocomposites decreases the distance between fillers. It results in the formation of thermal chains. For 0.5 wt% of GNP loading, the thermal conductivity value of the epoxy nanocomposites is increased from 0.21 to 0.27 $W\,m^{-1}\,K^{-1}$ [13].

To get a neat dispersion of graphene without damaging its structure in epoxy resin, graphene is functionalized using a silane coupling agent, (3-glycidyloxypropyl) trimethoxysilane. GNPs are modified with nitric acid (200 ml) for four hours at 80 °C, thus forming GO. Silane coupling agent (15.0 g), methanol (15.0 g), glacial acetic acid (0.15 ml), and distilled water (0.75 ml) are mixed with GO (1.0 g) in toluene (200 ml). The nanocomposite is prepared by dispersing of graphene in epoxy resin by a three-roll mill. This dispersion is put in molds (after addition of hardener) and cured at 120 °C the temperature for eight hours [14].

GO is superior to the filler materials (like CNTs and clay) used as a filler material in polymer composites. Linear low-density polyethylene (LLDPE)/GO nanocomposites are made by melt compounding, extrusion, and injection molding. During DSC analysis of the prepare composites, the melting temperature and crystallinity of the composite remain unchanged. The melting peaks are broad confirming that the composite is partially crystalline. There is a shift in the crystallization peak toward the positive direction of temperature when the content of GO is increased. The thermal parameters such as melting temperature, heat of fusion, heat of transition, and degree of crystallinity are analyzed. The variation in crystallinity is due to the filler materials that prevent the polymer chains to form the crystalline phase. The melting temperature of the composite appears at about 127 °C for 1.00 wt% of GO of LLDPE. Moreover, the crystallization temperature appears at 116 °C for 2 wt% of GO. It increases by the addition of GO. The increase of crystallization temperature is due to the heterogeneous nucleation of GO nanosheets during crystallization of LLDPE [15].

Graphene is used as filler material to modified epoxy resin for thermal interface applications. The thermal conductivity value of graphene is 4.9 $W\,m^{-1}\,K^{-1}$ (30 wt% loading at room temperature shows 1900% enhancement) with a high thermal stability even at high temperatures. There exists a strong coupling of phonon modes between graphene and epoxy resin. Graphene shows a high influence on the thermal conductivity property of polymer composites. Large graphene sheets and surface functional groups decrease the Kapitza thermal resistance at the interface region of graphene and epoxy resin. SiC, Al_2O_3, and boron nitride (BN) are used as fillers to epoxy resin for thermal management issues. However, the filler loading more than 60 vol% results in a high-density and low-level mechanical performance.

A high thermal conductivity can be achieved in thermal interface materials even for low loading of carbon nanomaterials. The thermal conductivity of monolayer graphene (supported on silicon substrate) decreases because of phonon leaking and scattering at the interface. The thermal conductivity of graphene (prepared in the methods like chemical vapor deposition and mechanical exfoliation) is higher than the chemically RGO. Heat flow in condensed matter happens due to acoustic phonons and free electrons. Particularly the contribution from phonon is more in graphene (even though it behaves like metals) due its high electrical conductivity and mobility of charge carriers.

Graphene exhibits a strong sp^2 covalent bonding. This bond permits the heat conduction due to the in-plane lattice vibrations. Also, it offers a high thermal conductivity when graphene is added as thermal conducting filler to the composites. Thermal conductivity is classified into intrinsic (only for pure crystals) and extrinsic (applicable to crystals with impurities and defects that results in scattering of phonons). Graphene has high intrinsic thermal conductivity and can offer more heat transport channels. It also provides a high thermal conductivity to the composite because of the lower level thermal boundary resistance present in between graphene and matrix. In 3D graphene nanosheet (3DGNs), when graphene is added more than 55%, percolation occurs in the composite, i.e. the thermal conductivity increases dramatically. The epoxy resin is not able to fill the voids of the 3DGNs because of their highly porous nature [16].

13.4 Polymer/Natural Fiber Composites

When natural fibers are reinforced in polymeric composites, an important issue to be considered is the damage from heat. The composition of natural fibers is mostly carbon, oxygen, silicon, and hydrogen. Mixing the curaua fiber (CF) (a Brazilian fiber) with epoxy improves the adhesion. This fiber can be functionalized with GO. Natural fibers contain large hemicellulose and lignin that reduce the mechanical characters and interfacial shear strength in epoxy matrix.

Both hemicellulose and lignin decrease the crystalline and hydrophilic nature of the natural fibers. They can be removed by treating jute fibers using alkali substances. Silane and plasma treatment on jute fibers increases the interfacial shear strength. However, these treatments are considered to be expensive. When the temperature is increased in natural fibers, various changes (like chemical and/or physical) occur. The thermal analysis of natural fibers is usually performed in the temperature range between 290 and 490 °C. TGA is the technique used to analyze the thermal stability and related characters of the materials.

Cured epoxy resins possess low thermal and viscoelastic properties. To develop the dynamic mechanical characters of the epoxy resins, nano-oil palm empty fruit bunch (OPEFB) fibers are added at various loadings of 1%, 3%, and 5%. The addition of nano-OPEFB filler makes the composites lightweight with high thermal stability. The improved characters can be confirmed through the dynamic mechanical analysis [17].

Treated bamboo fibers (TBF) by using 6 wt% concentration of sodium hydroxide (NaOH) solution are dispersed in the matrix of epoxy resin. TBF show the thermal stability character better than the untreated bamboo fibers (UBF). Thermal analysis of UBF, TBF, neat epoxy, and bamboo fiber/epoxy composites is shown in Figure 13.1a,b. It is observed that these composites have been decomposed at lower temperatures. However, they have good thermal stability compared to neat epoxy. This property leads to potential engineering applications. Figure 13.1b shows the highest differential thermogravimetry (DTG) peak for UBF at 368 °C and for TBF at 374 °C. However, the TBF exhibits higher rate of DTG (1.335 wt%/°C) than the UBF (1.082 wt%/°C). P(epoxy)/UBF has Y_c (char yield) value 9.7%, and P(epoxy)/TBF has less Y_c value, i.e. 8.9%. Likewise, TBF exhibits Y value 13.1% that is less than the Y_c value (16.7%) of UBF [18].

Dissolution of the lignin (present in the fibers) by NaOH is the major reason for less Y_c value. Decomposing of lignin is difficult [19]. The weight loss of bamboo fibers (BF) in different temperature ranges, i.e. 200–330 °C, 330–356 °C, and 356–450 °C, is due to the decomposition of hemicellulose, cellulose, and lignin, respectively [20, 21]. Like bamboo fibers, reinforcing date palm fibers into epoxy makes the composites with improved thermal and mechanical properties. As per the report of Gheith et al., reinforcing the date palm fibers at loadings of 40%, 50%, and 60% by wt. enhance the thermal stability and dynamic mechanical characters of epoxy composites like storage modulus (E'), loss modulus (E''), and damping factor as well [22].

Natural fibers (like alfa, kenaf, hemp, oil palm, coir and sisal fibers [SFs]) possess good mechanical characters, but they have low plasticity. Also, fibers and matrix compatibility is less because the fibers are hydrophilic nature and the matrix is hydrophobic nature. The fibers are able to reduce the onset temperature in the thermal degradation process of low-density polyethylene (LDPE). Reducing the onset temperature leads to the reduction of the general thermal stability character of the material. The composites are found to be less viscous during the extrusion and molding than the neat LDPE.

The high shear and frictional forces present in between the Doum fibers and LDPE matrix break the LDPE chains. It makes changes in the matrix properties. Crystallites are produced in LDPE/5 wt% Doum fiber composites. However, the natural fibers act as nucleating agent that leads to the germination, spherolites growth as well. Fiber content (10–30 wt%) provides a delay in crystallites growth because of the restricted alignment in the polymer network caused by the natural fibers. Hence, during melting of the polymer, the molecules have low mobility in lamellae surfaces [23].

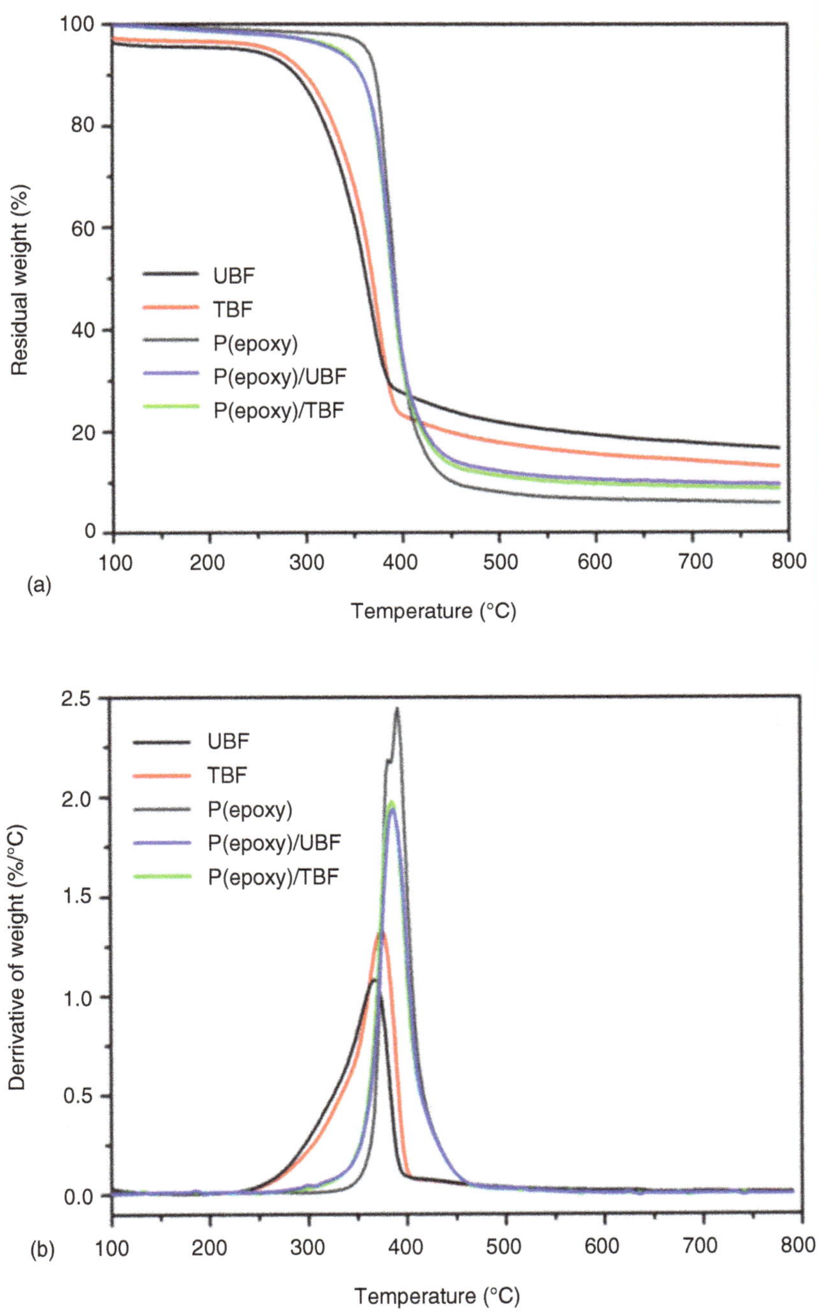

Figure 13.1 Thermal analysis of untreated bamboo fibers (UBF) and 6 wt% NaOH-treated bamboo fibers (TBF), neat epoxy, and bamboo fiber/epoxy composites. (a) Thermogravimetry curves. (b) Differential thermogravimetry curves. Source: Zhang et al. [18]. P(epoxy): neat epoxy. P(epoxy)/UBF: UBF-reinforced epoxy composites. P(epoxy)/TBF: TBF-reinforced epoxy composites.

13.5 Polymer/Natural Fiber/Graphene Composites

Graphene is added with fibers by solvent mixing, melt processing, and in situ polymerization. It can also be mixed with the fiber by spinning and dispersion. It increases the surface area of the composite. This will improve the thermal characters of the composite such as thermal stability, polymer crystallization, and glass transition.

Thermal characters of the GO-coated piassava fiber/epoxy composites (GO/PF/epoxy) are analyzed by TGA and DSC. The neat PF is stable up to 200 °C. When PF is coated with GO, the hemicellulose and lignin degrade only at higher temperatures. In DSC analysis, the neat PF got an endothermic peak at the temperature of 125 °C. It is attributed to the molecular bond breakage. In the case of DSC analysis of GO-coated PF, there is no such peak. The enhancement in the properties like damping capacity and viscous stiffness related to the epoxy composite coated with GO is confirmed by the dynamic mechanical analysis (DMA). TGA analysis of neat PF shows mass loss of 7.8%, 26.2%, and 32.4% at three different temperatures. The first mass loss occurs below 150 °C due to the evaporation of moisture and volatile compounds in the fiber. The hemicellulose present in the PF decomposes at 288 °C temperature, and the lignin decomposes at 359 °C. When the PF is coated with GO, the peaks are shifted to higher temperatures of 317 and 479 °C. These results confirm that GO offers thermal protective coating for the PF [24].

Polypropylene (PP) is functionalized with GO to improve the thermal stability. In this composite, the shift in the weight loss peaks occurs from 375 °C (for neat PP) to 469 °C (for PP/1.0 wt% of GO). This shift is due to the fact that GO restrict the movement of polymer chains and can prevent the emission of degradation products ([25]). The DSC analysis shows endothermic peaks at 125 and 61 °C with enthalpy of 71.3 and 138.9 $J g^{-1}$ for the neat PF and GO-coated PF, respectively. The improvement in the thermal stability of the PF by coating with GO can be demonstrated by DMA [24].

The storage modulus (E') curves infer the information regarding viscoelastic deformation and exhibit three regions such as glassy, glass transition, and rubbery plateau region [26]. Their values are reduced when the PF is coated with GO. The GO coating increases the stiffness of the material that leads to the reduction of storage modulus values. The loss modulus (E'') is a viscous response to dissipate the applied energy. The neat PF and GO-coated PF have same loss modulus values of 500 MPa at 75 °C temperature [27]. Curves of variation of tangent delta with temperature give information on the mechanical damping factor of the material [28]. The damping factor of GO/PF/epoxy matrix is higher than the neat PF composite. It is due to the strong interfacial bond of GO with fiber/matrix. The glass transition temperature of 10 vol% GO-coated PF is found to be 80.7 °C. These results suggest the possibility of using GO/PF/epoxy composites for applications above room temperature [24].

Jute fibers can replace glass fibers because their specific gravity is ~1.3 (specific modulus of ~40 $GN m^{-2}$). The glass fiber has specific gravity of ~2.5 (specific modulus of ~40 $GN m^{-2}$). A strong bonding or mechanical interlocking in between the jute fibers and GO occurs. It increases the interfacial shear strength to 236% and tensile strength to 96% approximately. The untreated jute fiber surface is smooth and featureless. Its surface roughness increases when it is treated with heat and alkali because of the removal of noncellulosic materials. The bonding between GO and jute fibers offers uniform surface that can be observed in scanning electron microscope (SEM) analysis. The alkaline treatment reduces the diameter of the jute fiber. However, the treatment of the jute fiber with GO increases its diameter. The thermal stability of jute fibers increases by at least 11 °C by GO coating that is confirmed in TGA analysis [29].

Random fiber orientation and low-level fiber/matrix interface are the major reasons for the deprived mechanical characters of natural fibers. Jute fibers are arranged in parallel direction with graphene to improve the interfacial adhesion. GO and G flakes are coated over jute fiber by hot pressing method and added with epoxy by a vacuum resin infusion process. Lignin, waxes, and hemicellulose present in the jute fibers pack the fibrils inside the polymer matrix and prevent them to come out from the matrix.

The combing of jute fibers results in fiber separation and hot pressing on combed fiber. This will increase the fiber volume fraction V_f. Graphene and GO coating further enhances V_f. Thirty centimeter-long pieces of jute fibers are treated with 0.5 wt% NaOH solutions to remove hemicelluloses. Graphene/GO-coated jute fibers are hand-sprayed with 1 wt% of epoxy that is then hot pressed at 120 °C for 30 minutes [30].

Physical absorption method is employed for the coating of GO onto SF. Polypropylene is modified with maleic anhydride-*grafted* polypropylene (MAPP). The composites are fabricated by injection molding and extrusion compounding methods. The crystallization behavior and the thermal property of the prepared composite are analyzed using DSC. The composite is showing a very good adhesion of SFs with the polypropylene matrix. The initial decomposition temperature (IDT) and the maximal decomposition temperature (MLT) of the composites are increased due to the synergistic effect of GO and MAPP. The melting temperatures are also increased that in turn increases the crystallinity of the composite, which is in well agreement with the results of X-ray diffraction (XRD) analysis [31].

Epoxy composites are prepared by adding with natural and synthetic fibers to improve the thermal and dynamic mechanical properties. Hybrid fillers reduce the usage of synthetic fibers. GNP are added with Kevlar/*Cocos nucifera* sheath (CS)/epoxy composites to increase the thermal properties at various wt% of 0.25, 0.5, and 0.75 by ultrasonication process. The different weight ratios of Kevlar/CS samples (S1), (S2), (S3), (S4), and (S5) are 100 : 0, 75 : 25, 50 : 50, 25 : 75, and 0 : 100. Moreover, the fiber/matrix weight ratio is 45 : 55 [32].

Figure 13.2a demonstrates the incorporation process of GNP into epoxy [33]. In this process, GNP are dispersed in acetone (15 mg ml^{-1}) for the period of one hour using ultrasonicator. In the next step, the dispersed solution is mixed with epoxy using a magnetic stirrer kept on hot plate (60 °C) for 20 minutes. This step is useful to remove the acetone present in the solution. Also, uniform dispersion is achieved by sonication for 30 minutes at 100 W, making uniformly dispersed GNP/epoxy

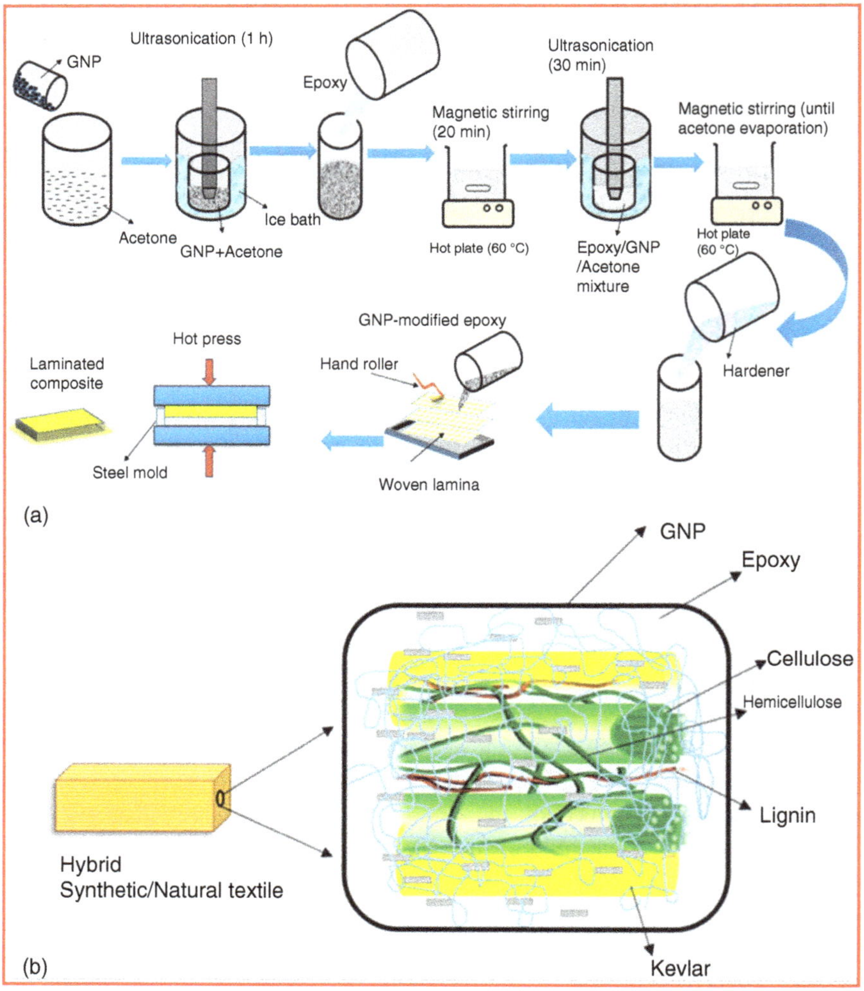

Figure 13.2 Schematic diagram shows the fabrication and materials of hybrid composites. (a) Dispersion of graphene nanoplatelets (GNPs) into epoxy matrix to fabricate laminated composites. Source: Jesuarockiam et al. [33]. (b) Various synthetic and hybrid materials of hybrid composites. Source: Jesuarockiam et al. [32].

solution [34]. This solution is cooled for one hour at room temperature. Finally, joint amine hardener is mixed in this solution to obtain the end product. Figure 13.2b shows the different synthetic and hybrid materials present in the hybrid composites. From this, a strong interface created in between the reinforcement and matrix by the GNP can be seen [32].

The stiffness and rigidity of the composite can be analyzed from the storage modulus curve [35, 36]. Figure 13.3 exhibits the impact of adding GNP on storage modulus (E') of Kevlar/epoxy (Figure 13.3a), hybrid composites (Figure 13.3b–d), and CS/epoxy composites (Figure 13.3e). When the temperature is increased, the

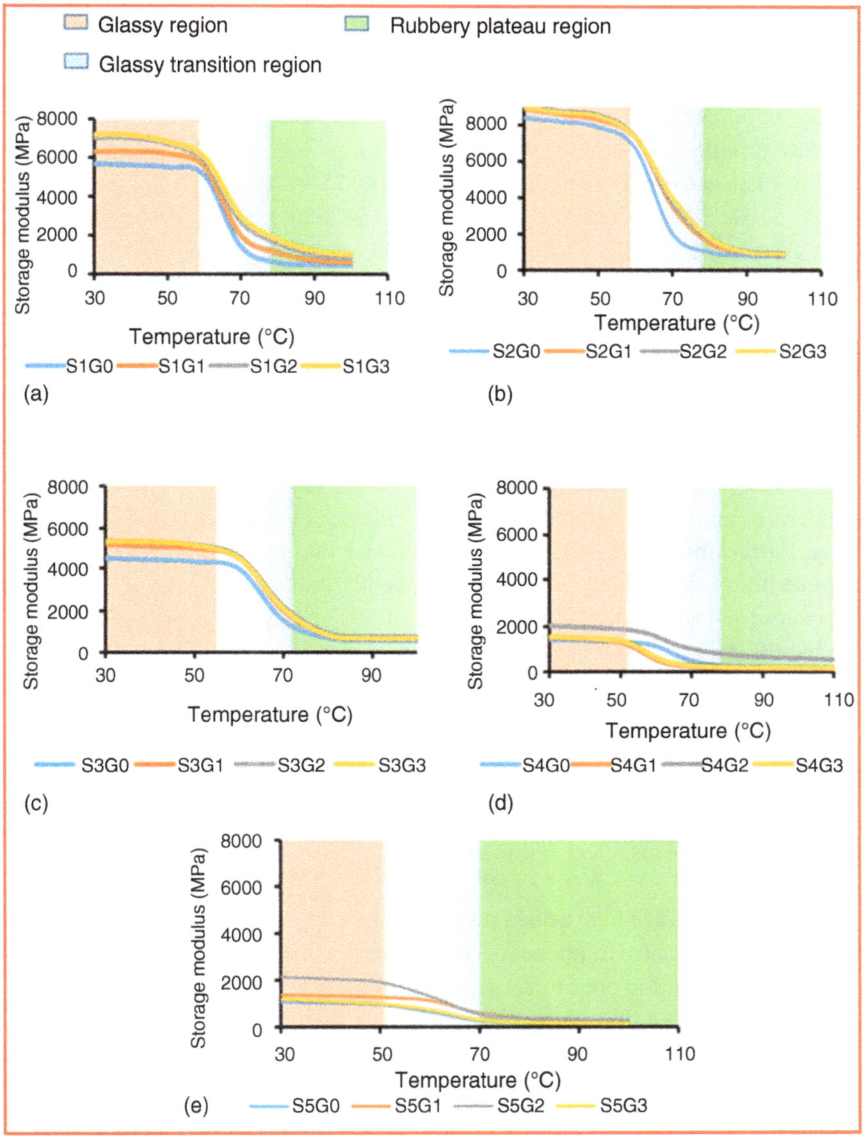

Figure 13.3 Effect of Graphene nanoplatelets (GNP) on the storage modulus (*E′*) of Kevlar/*Cocos nucifera* sheath (CS)/epoxy composites with different weight ratios. (a) S1, (b) S2, (c) S3, (d) S4, and (e) S5. Source: Jesuarockiam et al. [32]. Weight ratios of Kevlar/CS samples: S1 (100 : 0), S2 (75 : 25), S3 (50 : 50), S4 (25 : 75), and S5 (0 : 100). Addition of GNP: G0 is 0 wt%, G1 is 0.25 wt%, G2 is 0.5 wt%, and G3 is 0.75 wt%.

storage modulus plot (Figure 13.3a–e) undergone three phases. It is observed that the GNP addition makes changes the E' values of rigid and T_g regions only.

In the first phase (glossy region), the composite is stiff and rigid because of the presence of rigid polymeric chain (which has tightly packed molecules). In the second phase (glass transition region), E' is decreased above T_g due to the movement of polymeric chain. This movement affects stiffness and fiber/matrix adhesion [37]. In the third phase (rubbery plateau region), due to accelerated mobility in polymeric chain (at high level temperature), no variation in E' occurs [26]. Overlapping the curves of E' plots confirms that the addition of GNP does not make any variation in E' value [38].

Figure 13.3a–e exhibit the addition of GNP at 0.25 wt% (S1G1, S2G1, and S3G1), 0.5 wt% (S1G2, S2G2, and S3G2), and 0.5 wt% (S4G2 and S5G2) enhancing the E' values except at 0.75 wt% of GNP (S1G3, S2G3, and S3G3) and 0.75 wt% of GNP (S4G3 and S5G3). Synergistic effect of GNP aggregation and stress concentration (at 0.75 wt% of GNP) prevents the enhancing of E' value. This can be seen in Figure 13.3d–e. S3G3 (50 wt% CS at 0.75 wt% of GNP) exhibits E' value just like S1G0 (100 wt% of Kevlar/epoxy composites). Hence, it can be applied to replace the S1G0 Kevlar/epoxy composites (from [32]).

Reduced level in the polymeric chain movement (due to stiffening effect) enhances E' value. Also, higher level loading of nanofiller leads to decline E' value [39]. Reinforcement of nanofiller in the epoxy matrix enhances the stress transfer rate from matrix. Furthermore, the enhanced stress transfer rate increases E' value [39, 40]. Synthetic fiber-based composites have E' value higher than the natural fiber-based composites [41]. However, the hybridization of Kevlar and *C. nucifera* sheath in the GNP-modified epoxy results in superior performance [42].

GO is coated on the SF by physical absorption method. Then the fiber is reinforced with polypropylene that is modified by MAPP. This composite is fabricated by injection molding and extrusion compounding techniques. The synergistic effect of GO and MAPP shows an enhancement in the IDT and the MLT and a high thermal stability. The melting temperature and the crystallinity of the composite are also increased. This effect is due to the more interface adhesion of the SF with the matrix [31].

Hemp fibers are added in various proportions such as 0.3, 0.5, and 0.7 wt% of the matrix L-12 epoxy resin. Graphene (0.3 wt%) is mixed to improve the thermal properties. Hemp fibers exhibit poor interfacial adhesion with the polymer matrix that can be enhanced by the addition of graphene. This method harvests the best qualities of natural fiber composites with the help of graphene [43].

The CFs derived from the Amazonian curaua (*Ananas erectifolius*) plant leaves are able to replace aramid fibers in multilayered armor system. Graphene oxide coated fibers (GOCF) to enhance the interfacial bond in polymer composites. Thermogravimetric analysis (TGA) explains the evaporation of moisture, degradation (hemicellulose, cellulose, and lignin), and residue generated by the neat fiber and the GO-coated fiber. In differential thermal analysis (DTA), the decomposition temperatures are shifted to higher temperatures for GOCF than for CF. This confirms the improvement in the thermal stability of the fibers when coated with GO [44].

Images of the as-received, non-coated CFs and GOCFs are shown in Figure 13.4a,b, respectively. Thermogravimetry analysis (TGA) of these fibers is presented in Figure 13.4c, and DTA is presented in Figure 13.4d. It is observed from Figure 13.4c that the onset of the degradation is at temperature approximately 64–150 °C in these fibers. It indicates the moisture evaporation from these fibers. The major mass degradation step starting temperature is at 293 and 300 °C for CF and GOCF fibers, respectively, and the residue produced by the CF fibers and by GOCF is 15% and 14%, respectively. Some literatures have reported that the said mass degradation temperatures are attributed to the degradation of hemicellulose, cellulose, and lignin [45–48].

In the Figure 13.4d, three stages are observed: the first stage is in between 250 and 300 °C. It refers to the decomposition of the hemicelluloses (degradation temperature at 272 °C). The second process occurs in between 293 and 350 °C (maximum degradation temperature at 327 °C), indicating the cellulose decomposition. In the third stage, lignin decomposition occurs in between the temperature 400 and 450 °C (maximum degradation temperature at around 422 °C). However, in the case of GOCF fibers, the degradation has been shifted to higher temperatures. It indicates about the improved thermal stability of GOCF fibers [45–48]. GO coating forms insulation to heat propagation that delays the degradation and improves the thermal stability [29, 44].

The thermal stability of the waste of OPEFB is analyzed by adding with GO. This GO/cellulose aerogel bio-nanocomposite degrades slowly even for low GO loading that is confirmed by TGA [49]. Conventionally GO is synthesized by a modified Hummers method. Graphite (2 g) is added in the 50 ml of concentrated H_2SO_4. $NaNO_3$ (2 g) and $KMnO_4$ (4 g) are then added. In the next step, the prepared mixture is mixed in 100 ml of deionized water and stirred at the temperature of 96 °C for the period of one hour. Hydrogen peroxide is finally added and the resultant is washed in deionized water. Yao et al. have reported that carboxylic-functionalized GO is prepared by refluxing 0.1 g of GO in 25 ml of concentrated H_2SO_4 and 75 ml of concentrated HNO_3 and stirred at 98 °C temperature for the period of four hours. Then, the prepared mixture is deposited on the carbon fiber surface. It enhances the interfacial characters of the polymer composites [50].

Coir fibers are reinforced in the composite of graphene/epoxy/carbon fiber. The carbon fibers and the coir fibers are placed in an intercalated manner in this composite. Graphene is added in various proportions, i.e. 1, 2, 4, and 6 wt% of the holding matrix. Coir fiber is treated with the solutions of 5% NaOH and 0.3% $KMnO_4$. Graphene is treated with CTAB and distributed in the epoxy by sonication process. After the NaOH alkaline treatment, there is no pectin residue in the fiber. However, the breakage of the ether linkages created by the alkaline cleavage occurs. The surface roughness gives rise to an easy bonding between the fibers. The oxide groups of the GO enhance the interfacial rigidity of carbon fibers. CTAB involves in proper mixing of graphene with epoxy [51].

GO is coated on carbon fiber by electrophoretic deposition to improve the surface roughness and surface energy. GO sheets increase the wettability of resin on carbon fiber, and interlocking is obtained between CFs and epoxy resin (Wang et al. [52]). GO is added with glass fiber/epoxy composites to study the enhancement in thermal properties [53]. GO sheets is decorated onto the surface of carbon fibers to enhance the interfacial characters in carbon fiber/epoxy composites [54].

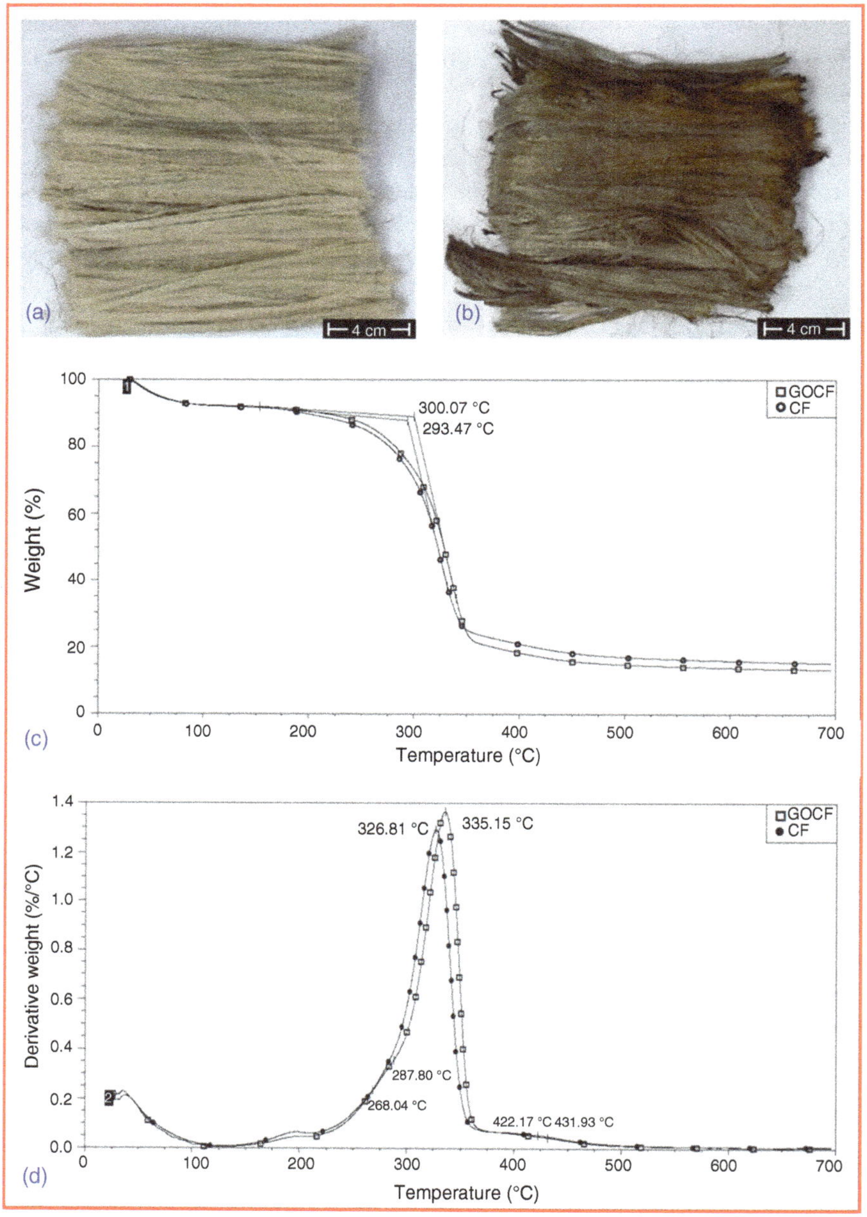

Figure 13.4 Thermal analysis of curaua fibers (CF) and graphene oxide-coated curaua fibers (GOCF). (a and b) Photographs of CF and GOCF. (c and d) show their thermogravimetry analysis and differential thermal analysis curves, respectively. Source: Costa et al. [44].

13.6 Conclusion

Natural fibers are ideal replacement for synthetic fibers in polymer composites. Modification of natural fibers through various methods enhances the characteristics of the composites. Decorating natural fibers with nanoparticles is one such method. High thermal conductivity of graphene is useful to develop the thermal properties of natural fiber-reinforced polymer composites. Thermal properties such as melting temperature, crystallinity, and degradation of the composites are well studied using DSC, TGA, and DMA analysis. The results confirm that graphene can be employed in polymer composites for several applications.

Acknowledgments

The author expresses her thanks to her husband Mr. G. Sankar for his help in this work. Also, she acknowledges the assistances of International Research Center, Kalasalingam Academy of Research and Education (Deemed to be University), Krishnankoil-626126 (India).

References

1 Phys.org. 2018. *Adding Graphene To Jute Fibres Could Give Natural Alternative To Man-Made Materials.* [online] Available at: https://phys.org/news/2018-11-adding-graphene-jute-fibres-natural.html. [Accessed 3 December 2020].

2 Tang, X.-Z., Li, W., Yu, Z.-Z. et al. (2011). Enhanced thermal stability in graphene oxide covalently functionalized with 2-amino-4, 6-didodecylamino-1, 3, 5-triazine. *Carbon* 49 (4): 1258–1265.

3 Shahil, K.M. and Balandin, A.A. (2011). Graphene-based nanocomposites as highly efficient thermal interface materials. *Graphene Based Thermal Interface Materials*: 1–18.

4 Shahil, K.M.F. and Balandin, A.A. (2012). Thermal properties of graphene and multilayer graphene: Applications in thermal interface materials. *Solid State Communications* 152 (15): 1331–1340.

5 Ye, F., Lee, J., and Feng, P.X.-L. (2018). Electrothermally tunable graphene resonators operating at very high temperature up to 1200 K. *Nano Letters* 18 (3): 1678–1685.

6 Chen, J., Liu, B., and Gao, X. (2020). Thermal properties of graphene-based polymer composite materials: A molecular dynamics study. *Results in Physics* 16: 102974.

7 Zhao, Y.-H., Wu, Z.-K., and Bai, S.-L. (2015). Study on thermal properties of graphene foam/graphene sheets filled polymer composites. *Composites Part A: Applied Science and Manufacturing* 72: 200–206.

8 Kim, S.Y., Noh, Y.J., and Yu, J. (2015). Thermal conductivity of graphene nanoplatelets filled composites fabricated by solvent-free processing for the

excellent filler dispersion and a theoretical approach for the composites containing the geometrized fillers. *Composites Part A: Applied Science and Manufacturing* 69: 219–225.

9 Chatterjee, S., Wang, J.W., Kuo, W.S. et al. (2012). Mechanical reinforcement and thermal conductivity in expanded graphene nanoplatelets reinforced epoxy composites. *Chemical Physics Letters* 531: 6–10.

10 Liang, J.Z., Wang, J.Z., Tsui, G.C.P., and Tang, C.Y. (2018). Thermal properties and thermal stability of polypropylene composites filled with graphene nanoplatelets. *Journal of Thermoplastic Composite Materials* 31 (2): 246–264.

11 Vadukumpully, S., Paul, J., Mahanta, N., and Valiyaveettil, S. (2011). Flexible conductive graphene/poly (vinyl chloride) composite thin films with high mechanical strength and thermal stability. *Carbon* 49 (1): 198–205.

12 Wan, Y.-J., Yang, W.-H., Yu, S.-H. et al. (2016). Covalent polymer functionalization of graphene for improved dielectric properties and thermal stability of epoxy composites. *Composites Science and Technology* 122: 27–35.

13 Zakaria, M.R., Abdul Kudus, M.H., Md Akil, H., and Mohd Thirmizir, M.Z. (2017). Comparative study of graphene nanoparticle and multiwall carbon nanotube filled epoxy nanocomposites based on mechanical, thermal and dielectric properties. *Composites Part B: Engineering* 119: 57–66.

14 Ahmadi-Moghadam, B., Sharafimasooleh, M., Shadlou, S., and Taheri, F. (2015). Effect of functionalization of graphene nanoplatelets on the mechanical response of graphene/epoxy composites. *Materials & Design (1980-2015)* 66: 142–149.

15 Li, J., Gunister, E., and Barsoum, I. (2019). Effect of graphene oxide as a filler material on the mechanical properties of LLDPE nanocomposites. *Journal of Composite Materials* 53 (19): 2761–2773.

16 Tang, B., Hu, G., Gao, H., and Hai, L. (2015). Application of graphene as filler to improve thermal transport property of epoxy resin for thermal interface materials. *International Journal of Heat and Mass Transfer* 85: 420–429.

17 Saba, N., Jawaid, M., Alothman, O.Y., and Almutairi, Z. (2019). Evaluation of dynamic properties of nano oil palm empty fruit bunch filler/epoxy composites. *Journal of Materials Research and Technology* 8 (1): 1470–1475.

18 Zhang, K., Wang, F., Liang, W. et al. (2018). Thermal and mechanical properties of bamboo fiber reinforced epoxy composites. *Polymers* 10 (6): 608.

19 De Rosa, I.M., Santulli, C., and Sarasini, F. (2010). Mechanical and thermal characterization of epoxy composites reinforced with random and quasi-unidirectional untreated Phormium tenax leaf fibers. *Mater. Des.* 31: 2397–2405.

20 Chen, H., Yu, Y., Zhong, T. et al. (2017). Effect of alkali treatment on microstructure and mechanical properties of individual bamboo fibers. *Cellulose* 24: 333–347.

21 Wang, F., Zhou, S., Yang, M. et al. (2018). Thermo-Mechanical Performance of Polylactide Composites Reinforced with Alkali-Treated Bamboo Fibers. *Polymers* 10: 401.

22 Gheith, M.H., Aziz, M.A., Ghori, W. et al. (2019). Flexural, thermal and dynamic mechanical properties of date palm fibres reinforced epoxy composites. *Journal of Materials Research and Technology* 8 (1): 853–860.

23 Arrakhiz, F.Z., El Achaby, M., Malha, M. et al. (2013). Mechanical and thermal properties of natural fibers reinforced polymer composites: doum/low density polyethylene. *Materials & Design* 43: 200–205.

24 Da Costa Garcia Filho Fabio, Fernanda Santos da Luz, Michelle Souza Oliveira, Artur Camposo Pereira, Ulisses Oliveira Costa, and Sergio Neves Monteiro. "Thermal behavior of graphene oxide-coated piassava fiber and their epoxy composites." Journal of Materials Research and Technology (2020).

25 Yuan, B., Bao, C., Song, L. et al. (2014). Preparation of functionalized graphene oxide/polypropylene nanocomposite with significantly improved thermal stability and studies on the crystallization behavior and mechanical properties. *Chem Eng J* 237: 411–420.

26 Saba, N., Jawaid, M., Alothman, O.Y., and Paridah, M. (2016). A review on dynamic mechanical properties of natural fibre reinforced polymer composites. *Constr. Build. Mater.* 106: 149–159.

27 Mohanty, S., Verma, S.K., and Nayak, S.K. (2006). Dynamical mechanical and thermal properties of mape treated jute/HDPE composites. *Compos Sci Technol* 66: 538–547.

28 Zhang, Z., Wang, P., and Wu, J. (2012). Dynamic mechanical properties of EVA polymer modified cement paste at early age. *Phys Procedia* 25: 305–310.

29 Sarker, F., Karim, N., Afroj, S. et al. (2018). High-performance graphene-based natural fiber composites. *ACS applied materials & interfaces* 10 (40): 34502–34512.

30 Sarker, F., Potluri, P., Afroj, S. et al. (2019). Ultrahigh Performance of Nanoengineered Graphene-Based Natural Jute Fiber Composites. *ACS applied materials & interfaces* 11 (23): 21166–21176.

31 Chen, J., Huang, Z., Lv, W., and Wang, C.C. (2018). Graphene oxide decorated sisal fiber/MAPP modified PP composites: Toward high-performance biocomposites. *Polymer Composites* 39: E113–E121.

32 Jesuarockiam, N., Jawaid, M., Zainudin, E.S. et al. (2019a). Enhanced Thermal and Dynamic Mechanical Properties of Synthetic/Natural Hybrid Composites with Graphene Nanoplateletes. *Polymers* 11 (7): 1085. https://doi.org/10.3390/polym11071085.

33 Jesuarockiam, N., Jawaid, M., Zainudin, E.S. et al. (2019). Improved Mechanical and Moisture-Resistant Properties of Woven Hybrid Epoxy Composites by Graphene Nanoplatelets (GNP). *Materials* 12 (8): 1249. https://doi.org/10.3390/ma12081249.

34 Wang, F., Drzal, L.T., Qin, Y., and Huang, Z. (2016). Size effect of graphene nanoplatelets on the morphology and mechanical behavior of glass fiber/epoxy composites. *J. Mater. Sci.* 51: 3337–3348.

35 Asim, M., Jawaid, M., Nasir, M., and Saba, N. (2018). Effect of fiber loadings and treatment on dynamic mechanical, thermal and flammability properties of pineapple leaf fiber and kenaf phenolic composites. *J. Renew. Mater.* 6: 383–393.

36 Joseph, S., Appukuttan, S.P., Kenny, J.M. et al. (2010). Dynamic mechanical properties of oil palm microfibril-reinforced natural rubber composites. *J. Appl. Polym. Sci.* 117: 1298–1308.

37 Hameed, N., Sreekumar, P., Francis, B. et al. (2007). Morphology, dynamic mechanical and thermal studies on poly (styrene-co-acrylonitrile) modified epoxy resin/glass fibre composites. *Compos. A Appl. Sci. Manuf.* 38: 2422–2432.

38 Md Shah, A.U., Sultan, M.T., and Jawaid, M. (2019). Sandwich-structured bamboo powder/glass fibre-reinforced epoxy hybrid composites–Mechanical performance in static and dynamic evaluations. *J. Sandw. Struct. Mater.*

39 Rasana, N., Jayanarayanan, K., Deeraj, B., and Joseph, K. (2019). The thermal degradation and dynamic mechanical properties modeling of MWCNT/glass fiber multiscale filler reinforced polypropylene composites. composites. *Compos. Sci. Technol.* 169: 249–259.

40 Samal, S.K., Mohanty, S., and Nayak, S.K. (2009). Polypropylene—bamboo/glass fiber hybrid composites: Fabrication and analysis of mechanical, morphological, thermal, and dynamic mechanical behavior. *Reinf. Plast. Compos.* 28: 2729–2747.

41 Braga, R. and Magalhaes, P. Jr., (2015). Analysis of the mechanical and thermal properties of jute and glass fiber as reinforcement epoxy hybrid composites. *Mater. Sci. Eng. C* 56: 269–273.

42 Sathishkumar, T., Naveen, J.A., and Satheeshkumar, S. (2014). Hybrid fiber reinforced polymer composites—A review. *J. Reinf. Plast. Compos.* 33: 454–471.

43 Hallad, S.A., Banapurmath, N.R., Patil, V. et al. (2018). Graphene reinforced natural fiber nanocomposites for structural applications. In: *IOP Conf. Ser. Mater. Sci. Eng. IOP Publ., vol. 376, p. 012072.*

44 Costa, U.O., Nascimento, L.F.C., Garcia, J.M. et al. (2019). Effect of graphene oxide coating on natural fiber composite for multilayered ballistic armor. *Polymers* 11 (8): 1356.

45 Caraschi, J.C. and Leãto, A.L. (2000). Characterization of curaua fiber. Mol. Cryst. Liq. Cryst. Sci. Technol. Sect. A. *Mol. Cryst. Liq. Cryst.* 353: 149–152.

46 Correa, A.C., de Morais Teixeira, E., Pessan, L.A., and Mattoso, L.H.C. (2009). Cellulose nanofibers from curaua fibers. *Cellulose* 17: 1183–1192.

47 Sharma, P.R., Joshi, R., Sharma, S.K., and Hsiao, B.S. (2017). A simple approach to prepare carboxycellulose nanofibers from untreated biomass. *Biomacromolecules* 18: 2333–2342.

48 Spinace, M.A., Lambert, C.S., Fermoselli, K.K., and De Paoli, M.A. (2008). Characterization of lignocellulosic curaua fibres. *Carbohydr. Polym.* 77: 47–53.

49 Gan, S., Zakaria, S., Chia, C.H., and Kaco, H. (2018). Effect of graphene oxide on thermal stability of aerogel bio-nanocomposite from cellulose-based waste biomass. *Cellulose* 25 (9): 5099–5112.

50 Yao, X., Gao, X., Jiang, J. et al. (2018). Comparison of carbon nanotubes and graphene oxide coated carbon fiber for improving the interfacial properties of carbon fiber/epoxy composites. *Composites Part B: Engineering* 132: 170–177.

51 Kuclourya, T., Jain, M.K., Mudliar, S., and Thamba, N.B. (2020). Statistical analysis and investigation of tensile test data of coir composites reinforced with graphene, epoxy and carbon fibre. *Proceedings of the Institution of*

Mechanical Engineers, Part L: Journal of Materials: Design and Applications: 1464420720939998.

52 Wang, C., Zhao, M., Li, J. et al. (2017). Silver nanoparticles/graphene oxide decorated carbon fiber synergistic reinforcement in epoxy-based composites. *Polymer* 131: 263–271.

53 Prusty, R.K., Ghosh, S.K., Rathore, D.K., and Ray, B.C. (2017). Reinforcement effect of graphene oxide in glass fibre/epoxy composites at in-situ elevated temperature environments: An emphasis on graphene oxide content. *Composites Part A: Applied Science and Manufacturing* 95: 40–53.

54 Zhang, X., Fan, X., Yan, C. et al. (2012). Interfacial microstructure and properties of carbon fiber composites modified with graphene oxide. *ACS applied materials & interfaces* 4 (3): 1543–1552.

14

Influence of Nanoclay on the Thermal Properties of the Natural Fiber-Based Hybrid Composites

Sabarish Radoor[1], Jasila Karayil[2], Reshma Soman[3], Aswathy Jayakumar[4], Edayileveettil K. Radhakrishnan[4], Jyotishkumar Parameswaranpillai[1], and Suchart Siengchin[1]

[1] *King Mongkut's University of Technology North Bangkok, The Sirindhorn International Thai-German Graduate School of Engineering (TGGS), Department of Mechanical and Process Engineering, 1518 Wongsawang Road, Bangsue, Bangkok, 10800, Thailand*
[2] *Directorate of Technical Education, Government Women's Polytechnic College, Department of Chemistry, Malaparamba, Calicut, Kerala, 673009, India*
[3] *CSIR - National Institute of Oceanography, Regional Centre Kochi, Dr. Salim Ali Road, Ernakulam, 682018, Kerala, India*
[4] *Mahatma Gandhi University, School of Biosciences, P.D. Hills (P.O.), Kottayam, 686560, Kerala, India*

14.1 Introduction

The fiber-reinforced composites are lightweight composite with excellent performance in fields such as automotive, marine, aerospace, civil engineering, etc. [1–3]. Synthetic fiber such as glass fiber, carbon fiber, etc. has been widely used to reinforce composites. However, the high-cost, nonavailability, and nonbiodegradable nature of synthetic fiber urges researchers to find an alternative [4–6]. Natural fibers are abundantly available and possess interesting features such as environment friendly, biodegradability, renewability, and nontoxicity [7, 8]. Several researchers have employed natural fiber to reinforce polymer composites and suggested it as a promising alternative for synthetic fiber. Compared with synthetic fibers, natural fiber-based composites are lightweight and possess good thermal and acoustical insulation [9–11]. In spite of its advantage over synthetic fiber, nature fiber suffers from poor compatibility with hydrophobic polymer matrix, which limits its application. The interfacial interaction between the polymer and the natural fibers is generally improved by introducing different filler such as graphite, montmorillonite (MMT), mica, carbon, and nanoclay [12–16].

In the recent years, there have been increasing utilizations of renewable material as filler in the polymer composites. Nanoclay is one of the best examples for renewable filler. It has a layered aluminosilicate structure with interesting features such as high surface area, high aspect ratio, ion-exchange ability, and easy dispersibility in the polymer matrix. Owing to its nontoxicity, easy procurement, and inexpensive nature, it is considered as an alternative for expensive filler such as carbon

Natural Fiber-Reinforced Composites: Thermal Properties and Applications, First Edition.
Edited by Senthilkumar Krishnasamy, Senthil Muthu Kumar Thiagamani,
Chandrasekar Muthukumar, Rajini Nagarajan and Suchart Siengchin.

nanotube (CNT) [17–19]. Among different type of clay, MMT is widely used as filler as it is easily available and environmentally safe and has distinguishable chemistry [20]. Halloysite is another example of nanotubes that have been used as fillers for composites. It is a multiwalled aluminosilicate consisting of octahedral alumina and tetrahedral silica sheets [21]. Organo clays are another important class of clay that received great attention from researchers, owing to its greater compatibility with hydrophobic polymer [22]. The incorporation of nanoclay will affect the thermal, mechanical, and barrier properties of composite. The effect of nanoclay on the physical properties of natural fiber-based polymer composite is displayed in Table 14.1.

14.2 Effect of Nanoclay on the Thermal Stability of Natural Fiber-Based Hybrid Composites

Thermal property is one of the major restrictions in the application of natural fiber-reinforced composites. However, thermal properties (thermal stability, thermal expansion, decomposition temperature, glass transition temperature, etc.) of composite is obtained by thermogravimetric analysis (TGA) and differential scanning calorimetry (DSC) analysis. TGA is widely employed to study the thermal behavior of composites [39]. The thermal degradation with an increasing temperature along with numerical calculation of degradation can be analyzed. The thermal stability of natural fibers depends on their chemical constituents. As a result, different type of natural fibers has different decomposition temperature. The degradation of lignin, hemicellulose, and cellulose content in fiber occurs at higher temperature [40]. The aromatic ring present in lignin content of natural fibers can increase the thermal decomposition temperature [41]. The thermal degradation of natural fibers such as sisal, reed, and water hyacinth was reported to occur between 290 and 490 °C [42]. The first weight loss from 50 to 100 °C has been reported due to the evaporation of water molecules from natural fibers. The degradation of hemicelluloses occurs at 200–350 °C, and lignin and cellulose decomposition appears between 300 and 450 °C.

The poor thermal stability of natural fibers is due to the presence of low-molecular-weight components such as cellulose, hemicellulose, and lignin. For instance, the thermal degradation of jute fibers mainly appeared at temperature 280 and 360 °C and is attributed to the decomposition of the three components [43]. The low thermal stability of natural fiber-reinforced polymer composites restricts their application. Therefore, natural fibers are often reinforced with several polymers and nanomaterials (nanoclay, CNTs, silicate nanoparticle, graphite, etc.). The addition of nanomaterials has significant performance as it can improve the physical, thermal, and mechanical properties of composites [44]. Owing to its low cost and high thermal stability (stable up to 800 °C), nanoclay is one of the best choices as a filler for natural fiber-based polymer composites [45].

There are several studies regarding the nanoclay-mediated enhancement of thermal properties of composites. Khan et al. [23] studied the effect of different filler such as nanoclay MMT, organoclay montmorillonite (OMMT), and cellulose nanofibers

Table 14.1 Effect of nanoclay on the properties of natural fiber polymer composite.

Samples	Properties	References
Cellulose nanofibers, nanoclay/kenaf epoxy composite	With addition of 0.95% of CNF nanoclay, a significant improvement in the mechanical and thermal stability of composite was noted	[23]
Nanoclay/bamboo/kenaf/epoxy hybrid nanocomposite	Nan clay addition enhances the thermal stability, water adsorption, and swelling property of composite	[24]
Nanoclay/flax fiber-reinforced geopolymer composite	A significant improvement in the durability and mechanical property of composite was noted	[25]
Bamboo fiber/nanoclay/HDPE	Nanoclay addition enhances the modulus properties of the HDPE–clay hybrid composite	[26]
Nanoclay/polypropylene	Storage modulus of polymer is remarkably improved with the addition of nanoclay	[27]
Nanoclay/starch composite	The incorporation of nanoclay improves the tensile strength and rigidity of the composite The addition of nanoclay also improves the water vapor barrier property of the composite	[28]
Pine cone fiber/clay hybrid	Tensile modulus of hybrid composite increases with nanoclay content	[29]
Polyvinyl alcohol (PVA)/clay/wood nanocomposite	The addition of nanoclay improves the tensile modulus and linear dimensional stability of composite	[30]
HDPE/wood/nanoclay	The tensile modulus and elongation at break increase with nanoclay (1%) addition	[31]
Jute/nanoclay/polyester	Addition of nanoclay reduces the water absorption property and also improves the mechanical property	[32]
Sisal fiber/nanoclay/metakaolin/ Portland cement	The incorporation of nanoclay significantly enhances the flexural properties	[33]
Wood fiber/nanoclay/HDPE	The mechanical properties and flame retardancy of wood fiber/HDPE are found to improve with the addition of nanoclay	[34]
Bamboo fiber/nanoclay/epoxy/polyester	The thermal, water absorption, and mechanical properties of the composite are found to be superior to pristine composite	[35]
Nanoclay/sisal fiber/epoxy composite	Mechanical properties such as tensile, dynamic, and wear properties of composite are found to significantly improve with nanoclay addition	[36]
Wheat straw flour (WSF)/nanoclay/foamed HDPE	Even small percentage of nanoclay (2%) addition improves the mechanical properties of composite	[37]
Wood fiber/nanoclay/polypropylene	Nanoclay addition improves the mechanical, physical, and water absorption properties of composite	[38]

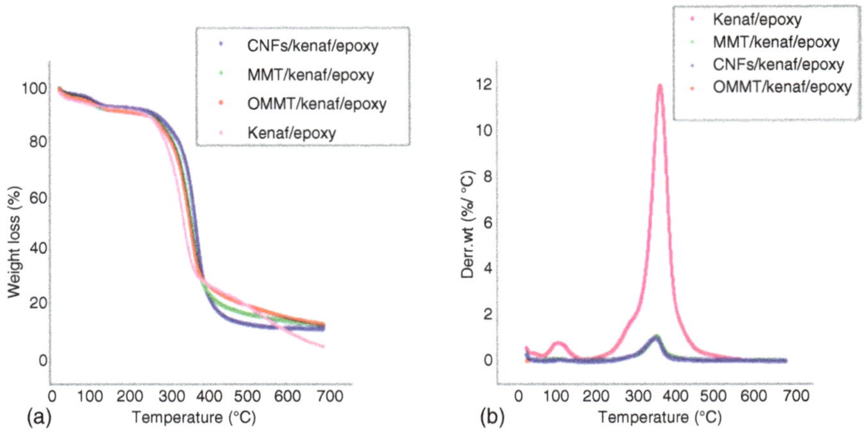

Figure 14.1 (a) TGA and (b) DTG plot of different hybrid composites. Source: Khan et al. [23].

(CNFs) on the properties of kenaf/epoxy hybrid composites. Here, all the developed composites lost their first weight at 100–150 °C due to the loss of water and moisture. The decomposition of cellulose content of kenaf fibers and aromatic compounds of epoxy occurred at 300–400 °C. The weight loss observed at 150–500 °C implies the decomposition of organic content adsorbed over the organically modified MMT surfaces. The dehydroxylation of MMT layers were reported to be observed between 500 and 800 °C. Their studies showed that the filler-incorporated composites exhibit better thermal properties than that of pure composite. This is attributed to the fact that filler particle can restrict the diffusion of volatile products (Figure 14.1).

Sajna et al. [46] studied the effect of nanoclay on the thermal stability of poly(lactic acid)/banana fiber biocomposites. The TGA results showed that upon the addition of nanoclay, the char residue (CR) and the degradation temperature of composite increase. Hasan et al. [47] reported that the thermal stability of composite is sensitive to clay content. The authors reported that the addition of 5% nanoclay to jute/kenaf-based polyester composite significantly improved its thermal stability. Similarly, Shahroze et al. [48] varied the clay (organically modified MMT) content from 0% to 5% and investigated its effect on the thermal properties of sugar palm fiber/polyester composite. The increased thermal properties of the composites have been reported due to the transfer of heat from matrix to thermally stable nanoclay. Ramesh and coworkers [49] studied the effect of MMT clay on the thermal properties of polylactic acid-treated aloe vera fiber (PLA–TAF) biocomposite. The thermal stability of the composites was analyzed by noting the T10 (10% weight loss temperatures) and T75% (75% weight loss temperatures). The T10 and T75% of PLA–TAF appeared at 295 and 338 °C, respectively. In incorporating 3 wt% of nanoclay, the T10 and T75% of composites change to 299 and 350 °C, respectively. The study led by Mohan and Kanny [50] suggested that nanoclay-infused banana fiber improves the thermal properties of epoxy composites. Here, the nanoclay acts as a barrier medium and enhances the decomposition temperature of the composites that was confirmed by TGA analysis. The TGA result shows that the thermal decomposition temperatures (T_d) of nanoclay-treated composite is higher than that of untreated composites. Hossen et al. [51] reported that the high thermal stability of the nanoclay-incorporated jute fiber/polyethylene (PE) composites could be due to the increased interfacial interaction and reduced microvoids in the system. These results were in agreement with the previous reports. For instance, Eng et al. [52] reported that the introduction of nanoclay in polylactic acid/polycaprolactone/oil palm mesocarp fiber biocomposites significantly improves the thermal decomposition temperature of the system. Previous studies show that the thermal properties of composites are highly sensitive to the type of nanoclay. Delhom et al. [53] studied the effect of different type of nanoclay (MMT and Cloisite clay) on the degradation temperature and char yield of natural fiber-reinforced (cotton, ramie, kenaf, and wood pulp) nanocomposite. The result indicated that the polymer composite incorporated with MMT displayed high degradation temperature and char yield. This is attributed to the different intercalation behavior of MMT and Cloisite. The concentration of nanoclay also found to play a major role in controlling the properties of composites (Figure 14.2). Kumar and Kumar [54] investigated the effect of the addition of different weight percentage (1, 2, 3, 5, and 10 wt%) of MMT Cloisite 30B nanoclay on bamboo/epoxy hybrid composites. Their studies reported that the first, second, and

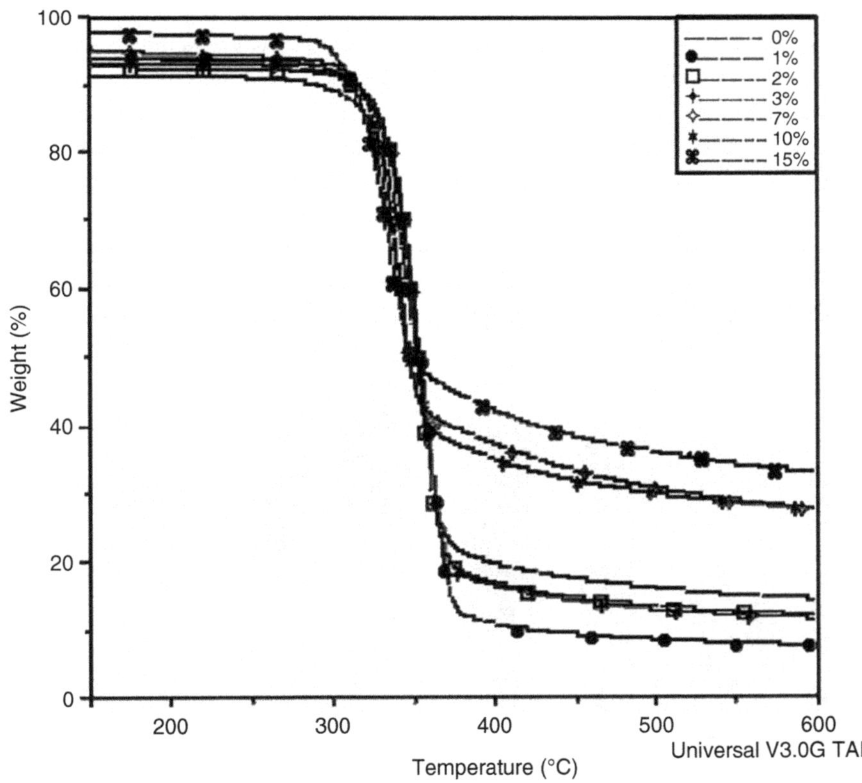

Figure 14.2 TGA thermograms of different percentage of nanocomposites. Source: Delhom et al. [53].

final weight losses were observed at 150, 250–450, and above 450 °C, respectively. The addition of 3% nanoclay on hybrid composite resulted in good thermal performance. The fine dispersion of nanoclay and the good interfacial adhesion between clay nanoparticles and the polymer matrix could have restricted the thermal motion of epoxy fiber and thus improved the thermal stability of composite. The reinforcement of banana fiber and Cloisite 30B nanoclay on poly(lactic acid) has been studied [46]. Here, nanoclay acts gas barrier and enhances thermal stability. Mandal et al. [55] has been reported the reinforcement of nanoclay on wood/plant oil composites. Their report suggests that the addition of MMT resulted in the increase in decomposition temperature to 405, 427, and 415 °C, whereas that of control composite was 389 °C. In light of the reports above, it is evident that nanoclay plays a significant role in improving the thermal properties of hybrid composites.

14.3 Effect of Nanoclay on the Inflammability of Natural Fiber-Based Hybrid Composites

Flammability, in general, refers to the capability of a substance to ignite and spread the flame and to generate heat. The flammability properties of natural fiber-based composites could be improved by introducing a flame retardant into the composite system. However, many types of fire retardants were used in the market, of which halogenated fire retardant like hexabromocyclododecane are proficient and had governed the market until the 1990s. Even so, due to their hazardous nature and bioaccumulation on the environment, its usage has been restricted (Figure 14.3) [56]. Therefore, alternative inorganic fillers or metal hydroxides that are more ecofriendly and non-halogenated in nature are often been employed in the preparation of flame-retardant composite. Still, to fulfill the adequate fire-retardant proficiency, a higher amount of filler loading is required in the conventional fire retardants that might lead to a secondary issue in the stagnation of mechanical

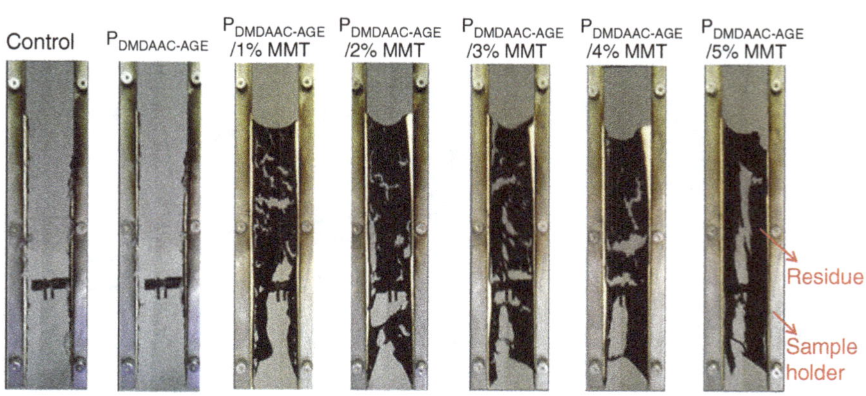

Figure 14.3 Images of control fabric and fabrics finished with polymer dimethyl diallyl ammonium chloride-allyl glycidyl ether (PDMDAAC-AGE/MMT), following the vertical flame test. Source: Gao et al. [56].

properties. To rectify this problem, clay-based minerals constituting the nanoscale particles are used as fire retardant. Nanoclay gains attention as it enhances the fire retardation capacity of the polymer composite without dealing with its physical and mechanical properties. One of the attractive features of these types of filler is that even at small loading, it displays sufficient flame retardancy. The fire retardancy of clay-based nanocomposites is due to the protective shielding of clay palettes, which tend to hinder the diffusion of volatile product generated during decomposition [57].

The flammability analysis is assessed by different parameters including CR, the total time to ignite (TTI), time to peak heat release rate (TPHRR), smoke production rate (SPR), limiting oxygen index (LOI), average heat release rate, total heat release, cone calorimetry, TGA, DSC, dynamic mechanical analysis (DMA), and vertical and horizontal flame tests [58–60]. A group of scientists from Egypt and Poland introduce newly modified MMT into polyurethane polymer and explored its flame-retardant properties. The clay-modified polymer system exhibits high thermal stability, low heat release rate, and high char yield production. Therefore, authors claimed that the MMT-modified polyurethane could act as an effective flame-retardant system [59]. A crucial role is also played by degree of exfoliation of nanoclay. Lee and coworkers [34] reported that utilization of maleated (high density polyethylene) HDPE (PE-g-MAn) will improve the exfoliation of nanoclay. Their studies show that exfoliated nanoclay/wood fiber/HDPE composite displayed high flame retardancy. The flame retardancy of nanoclay-based composite could be also improved by controlling its dispersion in polymer matrix. Gilman [61] reported that flame retardancy of the composites could be enhanced by increasing its dispersion. Flammability properties of intumescent flame-retardant polypropylene (PP)/kenaf nanocomposites were reported. The authors concluded that the introduction of nanoparticles (MMT and halloysite nanotube [HNT]) will alter the flammability of composites. For the same weight percent, MMT exhibits better flame-retardant performance than HNT [62]. Another study showed that the incorporation of nanoclay into the cotton fabrics improved the flame-retardant properties of cotton fabrics [63].

Chee and his colleagues [57] studied the thermo-oxidative stability and flammability of nanoclay hybrid composites. In their work, three different nanoclays such as MMT, HNTs, and organically modified MMT were used. The flammability properties of the composites were studied by different parameters like UL94 horizontal burning, limited oxygen index, cone calorimetry, and smoke density tester. The result showed that the addition of nanoclay to the hybrid fiber composites will improve the flame retardancy and thermo-oxidative stability of the composites. The introduction of nanoclay increases its LOI value, decrease peak heat release rate (PHRR), and decrease burning rate. The nanoclay acts as an insulation barrier and hinders the exchange of heat and mass within the solid and gas phase during the combustion process. Consequently, the charring effects increase.

In short, the flame-retardant mechanisms include the evolution of barrier against heat and gaseous volatile compounds by the transfer of nanoclay toward the surface, char formation in between the flame and the surface, and suppression of bubble formation during combustion.

14.4 Effect of Nanoclay on the Melting and Crystallization (DSC) of Natural Fiber-Based Hybrid Composites

Composites possess different types of crystallization processes. One of the techniques used to measure the isothermal crystallization behavior of the composite is thermoanalytical technique called DSC. There are several factors that control the crystalline process that includes the annealing time, cooling rate, and temperature of crystallization [64]. Delhom and his colleagues [53] improved the flame retardancy of cellulose fiber by incorporating it with nanoclay. The authors reported that the addition of nanoclay will alter the crystalline structure of cellulose, and therefore the melt temperature of the composite decreases from 186 to 172 °C.

Chen and Yang [65] investigated the effect of organo-nanoclay-treated kraft fiber on the crystalline behavior of HDPE and HDPE/maleic anhydride polyethylene (MAPE) composite. DSC experiment revealed that clay-treated HDPE–MAPE composite possesses high melting point (126.67 °C), crystallization peak temperature (116.95 °C), and crystallinity level (59.3). The result thus suggests that the presence of clay and MAPE increases the fiber nucleating rate and also improved the interfacial interaction between fiber and polymer matrix (Figure 14.4). In their study, Biswal and his colleagues [66] reported that the combined presence of nanoclay (3%) and banana fiber (30%) increases the T_m (melting temperature) of composite from 162 to 166.37 °C while T_c (crystallization peak temperature) increases from

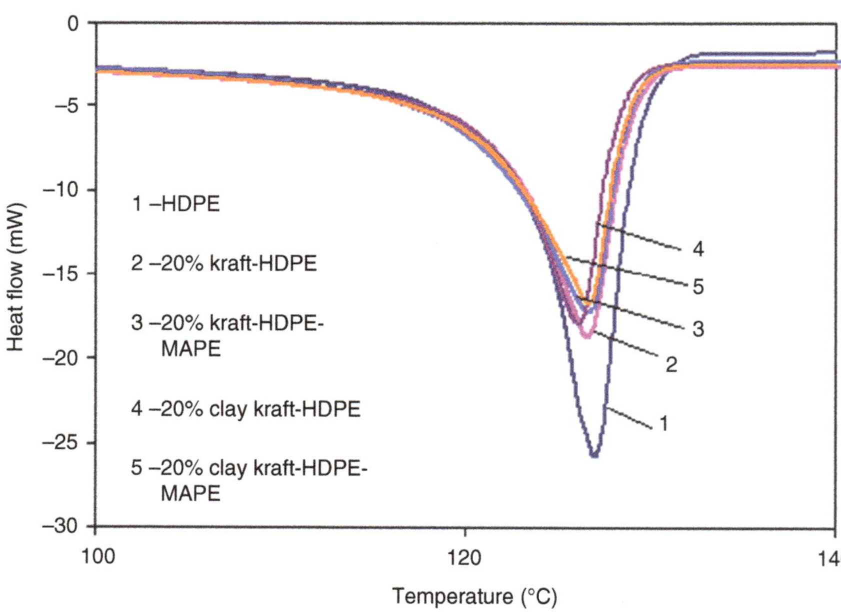

Figure 14.4 DSC curves of 20% organo-nanoclay-treated or organo-nanoclay-untreated fiber–HDPE and HDPE–MAPE composites including neat HDPE as a control. Source: Chen and Yan [65].

116.4 to 120.99 °C. This is attributed to the fact that nanoclay and banana fiber act as nucleating agent and promote the crystallization rate. This was supported by the work of Sahoo et al. [67]. The authors reported that clay and fiber act as heterogeneous nucleating agent and promote the crystallization rate. The influence of organically modified nanoclay on the performance of pineapple leaf fiber-reinforced polypropylene nanocomposites was reported. The authors confirmed their behavior by optical microscopic technique [68]. However, high percentage of clay had a negative effect on the crystallization of low density polyethylene (LDPE) composite [69]. Similarly, Lei et al. [31] also reported that the addition of clay has a negative effect on the crystallization rate. The incorporation of 2% nanoclay decreases the crystallization peak temperature of HDPE pine composite from 114.4 to 113.1 °C, and crystallinity level decreases from 59% to 54%. This report was well supported by Rahman and coworkers [70]. They also noted a reduction in the crystallization peak temperature and crystalline level with the addition of nanoclay.

14.5 Effect of Nanoclay on the Glass Transition Temperature of Natural Fiber-Based Hybrid Composites

Glass transition temperature (T_g) is one of the characteristic features of polymeric material. The temperature range over which the polymer changes from a glassy state into a rubbery state is considered as the T_g of the polymer. The glass transition is related to energy dissipation and mobility of the free amorphous phase. The lower the T_g, the easier the mobility of the free amorphous phase in the composites. T_g also depends on polymer chain length, branching, cross-linking, loading time, and plasticizer [71]. For most of the polymers, T_g is above the room temperature and is generally determined from thermal analysis technique such as DSC and DMA. The temperature in the middle of the inclined region of DSC curve is taken as the T_g of the polymer material. While in DMA measurement, temperature of the maximum loss modulus is considered as the T_g. Previous studies indicate that the incorporation of either filler or fiber will modify the T_g of polymer. Nanoclay is one of the frequently used fillers to alter this [72]. Arulmurugan and Venkateshwaran [73] introduced MMT nanoclay as filler in jute fiber-reinforced polyester composite. The effect of nanoclay on the storage modulus, tensile strength, interlaminar shear strength (ILSS), and T_g is discussed. The authors observed a considerable increment in the T_g of composite with the addition of nanoclay and jute fiber. This is because of the positive effect of nanoclay and the fiber on the stiffness of the composite. Also, nanoclay- and fiber-incorporated composite has better polymer/matrix interface and therefore will enhance the stiffness of the composite. Mohan and Kanny [50] also observed an enhancement in the glass transition temperature of composite with the addition of nanoclay. They compared the T_g of untreated, treated, and nanoclay-loaded composites. DMA revealed that nanoclay-incorporated sample displayed high T_g (72 °C). Authors justified the results by stating that high thermal property of nanoclay could have slow down the transition from glassy phase into rubbery phase. Khan et al. [23] studied the effect of different nanoclay (MMT and OMMT) on the mechanical

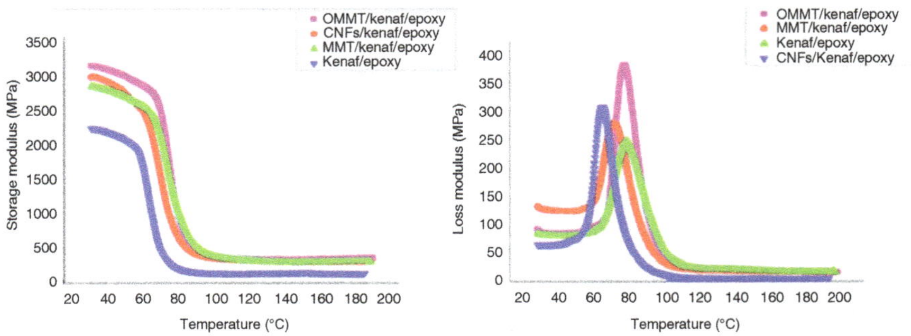

Figure 14.5 Storage modulus (E') and loss modulus (E'') plot of kenaf/epoxy and hybrid nanocomposites. Source: Khan et al. [23].

performance of kenaf fiber-reinforced epoxy nanocomposites. The DMA result indicates that OMMT is a better filler than MMT to enhance the T_g of the material as it offers greater interfacial adhesion between fiber and polymer (Figure 14.5). Paluvai et al. [74] reported that the inclusion of nanoclay and chemically treated fiber has a positive impact on the T_g of epoxy/unsaturated polyester (UP) composite. In the absence of MMT clay and fiber, the T_g of composite is found to be 101 °C. However, with the addition of clays and silane-treated fibers, the T_g increases to 118 °C. This is attributed to the enhancement in the cross-linking in the composite that is perhaps triggered by nanoclay. Lin and Renneckar [75] reported that crystallization rate plays an important role in controlling T_g. According to their study, with the increase in crystallization rate, a decline in T_g was observed. Biswal et al. [76] investigated the effect of nanoclay on the T_g of treated banana fiber/polypropylene (PP) composites. A 5 °C shift in T_g peak is observed for nanoclay-incorporated polymer composite. The authors claimed that the addition of nanoclay will decrease the mobility of polymer matrix and thus could influence the glass transition temperature.

14.6 Conclusion

Nanoclays are one of the attractive fillers for natural fiber-based polymer composites. Studies have demonstrated that the incorporation of nanoclay highly influences the properties of composites, especially thermal and mechanical properties. The clay-modified polymer system exhibits high thermal stability, low heat release rate, and high char yield production. At low concentration, nanoclay act acts as nucleating agent and promote crystallization. It also improves the interfacial addition between the fiber and polymer matrix. However, high clay concentration is found to have a depression effect on crystallization. Thus, melting point, crystallization peak temperature, and crystallinity level of polymer composite could be tuned with the addition of clay.

Acknowledgments

The authors gratefully thank the King Mongkut's University of Technology North Bangkok (KMUTNB), Thailand, through the Post-Doctoral Program (Grant No. KMUTNB-63-Post-01 and KMUTNB-64-Post-03 to SR and Grant No. KMUTNB-BasicR-64-16) for their financial support in this study.

References

1 Norkhairunnisa, M. and Majid, D.L. (2016). A review on thermal performance of hybrid natural fiber/nanoclay polymer composites. In: *Nanoclay Reinforced Polymer Composites* (eds. M. Jawaid, A. Qaiss and R. Bouhfid), 151–174. Springer.
2 Al-Oqla, F.M. and Sapuan, S.M. (2014). Natural fiber reinforced polymer composites in industrial applications: feasibility of date palm fibers for sustainable automotive industry. *Journal of Cleaner Production* 66: 347–354.
3 Mohammed, L., Ansari, M.N.M., Pua, G. et al. (2015). A review on natural fiber reinforced polymer composite and its applications. *International Journal of Polymer Science* 2015: 1–15.
4 Rahman, R. and Putra, S.Z.F.S. (2019). Tensile properties of natural and synthetic fiber-reinforced polymer composites. In: *Mechanical and Physical Testing of Biocomposites, Fibre-Reinforced Composites and Hybrid Composites* (eds. M. Jawaid, M. Thariq and N. Saba), 81–102. Woodhead Publisher.
5 do Nascimento Santos, M.J., Delgado, J.M.P.Q., and Barbosa de Lima, A.G. (2017). Synthetic fiber-reinforced polymer composite manufactured by resin transfer molding technique: foundations and engineering applications. *Diffusion Foundations* 14: 21–42.
6 Shahzad, A. and Nasir, S.U. (2017). *Mechanical properties of natural fiber/synthetic fiber reinforced polymer hybrid composites.* In: *Green Biocomposites* (eds. M. Jawaid, S.M. Sapuan and O.Y. Alothman), 355–396. Springer.
7 Abhemanyu, P.C., Prassanth, E., Kumar, T.N. et al. (2019). *Characterization of natural fiber reinforced polymer composites. Emerging Trends in Mechanical Engineering, AIP Conference Proceedings* 2080, 020005-1–020005-7.
8 Radoor, S., Karayil, J., Rangappa, S.M. et al. (2020). A review on the extraction of pineapple, sisal and abaca fibers and their use as reinforcement in polymer matrix. *Express Polymer Letters* 14 (4): 309–335.
9 Vijayan, R. and Krishnamoorthy, A. (2019). Review on natural fiber reinforced composites. *Materials Today: Proceedings* 16: 897–906.
10 Pickering, K.L., Efendy, M.G.A., and Le, T.M. (2016). A review of recent developments in natural fibre composites and their mechanical performance. *Composites Part A: Applied Science and Manufacturing* 83: 98–112.

11 Gurunathan, T., Mohanty, S., and Nayak, S.K. (2015). A review of the recent developments in biocomposites based on natural fibres and their application perspectives. *Composites Part A: Applied Science and Manufacturing* 77: 1–25.

12 Khalili, P., Tshai, K.Y., and Kong, I. (2017). Natural fiber reinforced expandable graphite filled composites: Evaluation of the flame retardancy, thermal and mechanical performances. *Composites Part A: Applied Science and Manufacturing* 100: 194–205.

13 Rajini, N., Jappes, J.T.W., Rajakarunakaran, S., and Jeyaraj, P. (2012). Mechanical and free vibration properties of montmorillonite clay dispersed with naturally woven coconut sheath composite. *Journal of Reinforced Plastics and Composites* 31 (20): 1364–1376.

14 Zhao, Y., Dang, W., Lu, Z. et al. (2018). A novel mica-based composite with hybrid aramid fibers for electrical insulating applications: largely improved mechanical properties and moisture resistance. *Polymer International* 67 (2): 204–211.

15 Santulli, C. (2019). Mechanical and Impact damage analysis on carbon/natural fibers hybrid composites: a review. *Materials* 12 (3): 517–529.

16 Deepak, K., Reddy, N.S., and Naidu, T.V.S. (2015). Thermosetting polymer and nano clay based natural fiber bio-composites. *Procedia Materials Science* 10: 626–631.

17 Bordes, P., Pollet, E., and Averous, L. (2009). Nano-biocomposites: biodegradable polyester/nanoclay systems. *Progress in Polymer Science* 34 (2): 125–155.

18 You, Z., Mills-Beale, J., Foley, J.M. et al. (2011). Nanoclay-modified asphalt materials: preparation and characterization. *Construction and Building Materials* 25 (2): 1072–1078.

19 Lau, K.-t., Gu, C., and Hui, D. (2006). A critical review on nanotube and nanotube/nanoclay related polymer composite materials. *Composites Part B: Engineering* 37 (6): 425–436.

20 Uddin, F. (2008). Clays, nanoclays, and montmorillonite minerals. *Metallurgical and Materials Transactions A* 39 (12): 2804–2814.

21 Joussein, E., Petit, S., Churchman, J. et al. (2018). Halloysite clay minerals – a review. *Clay Minerals* 40 (4): 383–426.

22 Zeng, Q.H., Yu, A.B., Lu, G.Q., and Paul, D.R. (2005). Clay-based polymer nanocomposites: research and commercial development. *Journal of Nanoscience and Nanotechnology* 5 (10): 1574–1592.

23 Khan, A., Asiri, A.M., Jawaid, M. et al. (2020). Effect of cellulose nano fibers and nano clays on the mechanical, morphological, thermal and dynamic mechanical performance of kenaf/epoxy composites. *Carbohydrate Polymers* 239: 1–30.

24 Chee, S.S., Jawaid, M., Sultan, M.T.H. et al. (2020). Effects of nanoclay on physical and dimensional stability of bamboo/Kenaf/nanoclay reinforced epoxy hybrid nanocomposites. *Journal of Materials Research and Technology* 9 (3): 5871–5880.

25 Assaedi, H., Shaikh, F.U.A., and Low, I.M. (2018). Effect of nanoclay on durability and mechanical properties of flax fabric reinforced geopolymer composites. *Journal of Asian Ceramic Societies* 5 (1): 62–70.

26 Han, G., Lei, Y., Wu, Q. et al. (2008). Bamboo–fiber filled high density polyethylene composites: effect of coupling treatment and nanoclay. *Journal of Polymers and the Environment* 16 (2): 123–130.

27 Venkatesh, G.S., Deb, A., Karmarkar, A., and Chauhan, S.S. (2012). Effect of nanoclay content and compatibilizer on viscoelastic properties of montmorillonite/polypropylene nanocomposites. *Materials & Design* 37: 285–291.

28 Müller, C.M.O., Laurindo, J.B., and Yamashita, F. (2011). Effect of nanoclay incorporation method on mechanical and water vapor barrier properties of starch-based films. *Industrial Crops and Products* 33 (3): 605–610.

29 Arrakhiz, F.Z., Benmoussa, K., Bouhfid, R., and Qaiss, A. (2013). Pine cone fiber/clay hybrid composite: mechanical and thermal properties. *Materials & Design* 50: 376–381.

30 Meng, Q.K., Hetzer, M., and De Kee, D. (2010). PLA/clay/wood nanocomposites: nanoclay effects on mechanical and thermal properties. *Journal of Composite Materials* 45 (10): 1145–1158.

31 Lei, Y., Wu, Q., Clemons, C.M. et al. (2007). Influence of nanoclay on properties of HDPE/wood composites. *Journal of Applied Polymer Science* 106 (6): 3958–3966.

32 Dewan, M.W., Hossain, M.K., Hosur, M., and Jeelani, S. (2013). Thermomechanical properties of alkali treated jute-polyester/nanoclay biocomposites fabricated by VARTM process. *Journal of Applied Polymer Science* 128 (6): 4110–4123.

33 Wei, J. and Meyer, C. (2014). Sisal fiber-reinforced cement composite with Portland cement substitution by a combination of metakaolin and nanoclay. *Journal of Materials Science* 49 (21): 7604–7619.

34 Lee, Y.H., Kuboki, T., Park, C.B. et al. (2010). The effects of clay dispersion on the mechanical, physical, and flame-retarding properties of wood fiber/polyethylene/clay nanocomposites. *Journal of Applied Polymer Science* 118 (1): 452–461.

35 Kushwaha, P.K. and Kumar, R. (2011). Reinforcing effect of nanoclay in bamboo-reinforced thermosetting resin composites. *Polymer-Plastics Technology and Engineering* 50 (2): 127–135.

36 Mohan, T.P. and Kanny, K. (2011). Water barrier properties of nanoclay filled sisal fibre reinforced epoxy composites. *Composites Part A: Applied Science and Manufacturing* 42 (4): 385–393.

37 Babaei, I., Madanipour, M., Farsi, M., and Farajpoor, A. (2014). Physical and mechanical properties of foamed HDPE/wheat straw flour/nanoclay hybrid composite. *Composites Part B: Engineering* 56: 163–170.

38 Tabari, H.Z., Nourbakhsh, A., and Ashori, A. (2011). Effects of nanoclay and coupling agent on the physico-mechanical, morphological, and thermal properties of wood flour/polypropylene composites. *Polymer Engineering & Science* 51 (2): 272–277.

39 Abraham, J., Mohammed, A.P., Ajith Kumar, M.P. et al. (2018). Thermoanalytical techniques of nanomaterials. In: *Characterization of Nanomaterials* (eds. S.M. Bhagyaraj, O.S. Oluwafemi, N. Kalarikkal and S. Thomas), 213–236. Woodhead Publisher.

40 Yang, H., Yan, R., Chen, H. et al. (2007). Characteristics of hemicellulose, cellulose and lignin pyrolysis. *Fuel* 86 (12-13): 1781–1788.

41 Lu, Y., Lu, Y.-C., Hu, H.-Q. et al. (2017). Structural characterization of lignin and its degradation products with spectroscopic methods. *Journal of Spectroscopy* 2017: 1–15.

42 Methacanon, P., Weerawatsophon, U., Sumransin, N. et al. (2010). Properties and potential application of the selected natural fibers as limited life geotextiles. *Carbohydrate Polymers* 82 (4): 1090–1096.

43 Boopalan, M., Niranjanaa, M., and Umapathy, M.J. (2013). Study on the mechanical properties and thermal properties of jute and banana fiber reinforced epoxy hybrid composites. *Composites Part B: Engineering* 51: 54–57.

44 Chrissafis, K. and Bikiaris, D. (2011). Can nanoparticles really enhance thermal stability of polymers? Part I: an overview on thermal decomposition of addition polymers. *Thermochimica Acta* 523 (1-2): 1–24.

45 Li, X., Tabil, L.G., and Panigrahi, S. (2007). Chemical treatments of natural fiber for use in natural fiber-reinforced composites: a review. *Journal of Polymers and the Environment* 15 (1): 25–33.

46 Sajna, V.P., Mohanty, S., and Nayak, S.K. (2017). A study on thermal degradation kinetics and flammability properties of poly(lactic acid)/banana fiber/nanoclay hybrid bionanocomposites. *Polymer Composites* 38 (10): 2067–2079.

47 Hasan, M.H., Mollik, M.S., and Rashid, M.M. (2017). Effect of nanoclay on thermal behavior of jute reinforced composite. *The International Journal of Advanced Manufacturing Technology* 94 (5-8): 1863–1871.

48 Shahroze, R.M., Ishak, M.R., Salit, M.S. et al. (2019). Sugar palm fiber/polyester nanocomposites: Influence of adding nanoclay fillers on thermal, dynamic mechanical, and physical properties. *Journal of Vinyl and Additive Technology* 26 (3): 236–243.

49 Ramesh, P., Prasad, B.D., and Narayana, K.L. (2019). Effect of MMT clay on mechanical, thermal and barrier properties of treated aloevera fiber/ PLA-hybrid biocomposites. *Silicon* 12 (7): 1751–1760.

50 Mohan, T.P. and Kanny, K. (2015). Nanoclay infused banana fiber and its effects on mechanical and thermal properties of composites. *Journal of Composite Materials* 50 (9): 1261–1276.

51 Hossen, M.F., Hamdan, S., Rahman, M.R. et al. (2016). Effect of clay content on the morphological, thermo-mechanical and chemical resistance properties of propionic anhydride treated jute fiber/polyethylene/nanoclay nanocomposites. *Measurement* 90: 404–411.

52 Eng, C.C., Ibrahim, N.A., Zainuddin, N. et al. (2014). Enhancement of mechanical and dynamic mechanical properties of hydrophilic nanoclay reinforced polylactic acid/polycaprolactone/oil palm mesocarp fiber hybrid composites. *International Journal of Polymer Science* 2014: 1–8.

53 Delhom, C.D., White-Ghoorahoo, L.A., and Pang, S.S. (2010). Development and characterization of cellulose/clay nanocomposites. *Composites Part B: Engineering* 41 (6): 475–481.

54 Kumar, V. and Kumar, R. (2012). Improved mechanical and thermal properties of bamboo-epoxy nanocomposites. *Polymer Composites* 33 (3): 362–370.

55 Mandal, M., Halim, Z., and Maji, T.K. (2020). Mechanical, moisture absorption, biodegradation and physical properties of nanoclay-reinforced wood/plant oil composites. *SN Applied Sciences* 2 (2).

56 Gao, D., Li, R., Lv, B. et al. (2015). Flammability, thermal and physical-mechanical properties of cationic polymer/montmorillonite composite on cotton fabric. *Composites Part B: Engineering* 77: 329–337.

57 Chee, S.S., Jawaid, M., Alothman, O.Y., and Yahaya, R. (2020). Thermo-oxidative stability and flammability properties of bamboo/kenaf/nanoclay/epoxy hybrid nanocomposites. *RSC Advances* 10 (37): 21686–21697.

58 Zanetti, M., Lomakin, S., and Camino, G. (2000). Polymer layered silicate nanocomposites. *Macromolecular Materials and Engineering* 279 (1): 1–9.

59 Hassan, M.A., Kozlowski, R., Obidzinski, B. et al. (2007). The effect of new flame retardant systems containing montmorillonite-butyl acrylate nanoclay on the flammability properties of polyuerethane polymer. *Polymer-Plastics Technology and Engineering* 46 (5): 521–527.

60 Prabhakar, M.N., Shah, A.U.R., and Song, J.-I. (2015). A review on the flammability and flame retardant properties of natural fibers and polymer matrix based composites. *Composites Research* 28 (2): 29–39.

61 Gilman, J. (1999). Flammability and thermal stability studies of polymer layered-silicate (clay) nanocomposites. *Applied Clay Science* 15 (1-2): 31–49.

62 Subasinghe, A., Das, R., and Bhattacharyya, D. (2016). Study of thermal, flammability and mechanical properties of intumescent flame retardant PP/kenaf nanocomposites. *International Journal of Smart and Nano Materials* 7 (3): 202–220.

63 Göcek, İ. (2019). Functionalization of textile materials with nanoclay incorporation for improved characteristics. *Journal of Polytechnic* 22: 509–522.

64 Wang, C. and Liu, C.R. (1999). Transcrystallization of polypropylene composites: nucleating ability of fibres. *Polymer* 40 (2): 289–298.

65 Chen, J. and Yan, N. (2013). Crystallization behavior of organo-nanoclay treated and untreated kraft fiber–HDPE composites. *Composites Part B: Engineering* 54: 180–187.

66 Biswal, M., Mohanty, S., and Nayak, S.K. (2012). Thermal stability and flammability of banana-fiber-reinforced polypropylene nanocomposites. *Journal of Applied Polymer Science* 125 (S2): E432–E443.

67 Sahoo, R.K., Mohanty, S., and Nayak, S.K. (2016). Effect of nanoclay on the nucleation, crystallization and melting behaviour of polypropylene. *Journal of Thermoplastic Composite Materials* 29 (11): 1554–1572.

68 Biswal, M., Mohanty, S., and Nayak, S.K. (2009). Influence of organically modified nanoclay on the performance of pineapple leaf fiber-reinforced polypropylene nanocomposites. *Journal of Applied Polymer Science* 114 (6): 4091–4103.

69 Bumbudsanpharoke, N., Lee, W., Choi, J.C. et al. (2017). Influence of montmorillonite nanoclay content on the optical, thermal, mechanical, and barrier properties of low-density polyethylene. *Clays and Clay Minerals* 65 (6): 387–397.

70 Rahman, M.R., Hamdan, S., Hashim, D.M.A. et al. (2015). Bamboo fiber polypropylene composites: effect of fiber treatment and nano clay on mechanical and thermal properties. *Journal of Vinyl and Additive Technology* 21 (4): 253–258.

71 Bandi, S. and Schiraldi, D.A. (2006). Glass transition behavior of clay aerogel/poly(vinyl alcohol) composites. *Macromolecules* 39 (19): 6537–6545.

72 López-Martínez, E.I., Márquez-Lucero, A., Hernández-Escobar, C.A. et al. (2007). Incorporation of silver/carbon nanoparticles into poly(methyl methacrylate) via in situ miniemulsion polymerization and its influence on the glass-transition temperature. *Journal of Polymer Science Part B: Polymer Physics* 45 (5): 511–518.

73 Arulmurugan, S. and Venkateshwaran, N. (2019). The effect of fiber reinforcement on fracture toughness assessment of nanoclay filled polymer composites. *Surface Review and Letters* 26 (09).

74 Reddy Paluvai, N., Mohanty, S., and Nayak, S.K. (2016). Effect of cloisite 30B clay and sisal fiber on dynamic mechanical and fracture behavior of unsaturated polyester toughened epoxy network. *Polymer Composites* 37 (9): 2832–2846.

75 Lin, Z., Renneckar, S., and Hindman, D.P. (2007). Nanocomposite-based lignocellulosic fibers 1. Thermal stability of modified fibers with clay-polyelectrolyte multilayers. *Cellulose* 15 (2): 333–346.

76 Biswal, M., Mohanty, S., and Nayak, S.K. (2011). Banana fiber-reinforced polypropylene nanocomposites: effect of fiber treatment on mechanical, thermal, and dynamic-mechanical properties. *Journal of Thermoplastic Composite Materials* 25 (6): 765–790.

15

Effect of CNT Fillers on Thermal Properties of the Bamboo Fiber-Based Hybrid Composites

Mohit Hemath[1], *Babu V. Hemath*[2], *Hemath K. Govindarajulu*[2], *Sanjay M. Rangappa*[1], *Suchart Siengchin*[1], *and Suresh N. Sundaram*[3]

[1] *King Mongkut's University of Technology North Bangkok, The Siridhorn International Thai German Graduate School of Engineering, Department of Materials and Production Engineering, Natural Composites Research Group Lab, 1518 Pracharat 1 Road, Wongsawang, Bangsue, Bangkok, 10800, Thailand*
[2] *Composite Research Center, EVR Street, Vinayagapuram, Ambattur, Chennai, 600053, India*
[3] *St. Joseph Engineering College, Department of Electrical and Electronics Engineering, Vamanjoor, Manglore 575028, Karnataka, India*

15.1 Introduction

Polymer-based nanocomposites characterize the most interesting necessary aspect and encouraging methods in the sector of various imminent developed engineering applications [1–3]. They are entailed by adding a nanoform of fibers in the polymer resin to confront the arising orders of the precise characteristics for useful applications [4, 5]. Nanocomposites display enhanced mechanical and physical characteristics than traditional polymer composite-reinforced with microparticles [6]. Indeed, epoxy-based nanocomposites have been broadly investigated in creating material with many functional characteristics that contain remarkable mechanical, thermal, and electrical properties. There are cumulative considerations in using carbon nanotubes as architectural reinforcements in polymer composites due to their remarkable characteristics in growing applications like military hardware, structure of space, and aircraft [7, 8]. Many investigations on interfacial attachment, ruin principle, and mechanical characteristics of carbon nanotubes incorporated polymer-based composites have been studied [8, 9]. The analysis is proceeding to advance polymer composites using absolute environmental friendly agricultural waste fibrils as reinforcement material. Various scientists have investigated plantain-based fibers as a viable source of plant cellulose fiber different utilizations [10–15]. The collection of various kinds of fibers into an exclusive polymer has an outcome in advancing hybrid polymer composites. These hybrid composites are the estimated sum of the different components in which one kind of fiber could control the material's deficiency. In this regard, the similarity in characteristics and achievement could be achieved through an approximate selection of materials [14]. The hybrid composite characteristics depend on the fiber's components,

Natural Fiber-Reinforced Composites: Thermal Properties and Applications, First Edition.
Edited by Senthilkumar Krishnasamy, Senthil Muthu Kumar Thiagamani, Chandrasekar Muthukumar, Rajini Nagarajan and Suchart Siengchin.

the specific length of the fiber, the total number of fibers intermingling, the arrangement of fibers, the failure strain of discrete fibers, and the attachment from fiber to the matrix [15–19]. The study has established that plant cellulose fiber incorporation with other plant fiber or synthetic fibers in polymer resin can confidently affect the polymer composites' thermal, physical, and mechanical characteristics [20–25]. Plant fiber-based polymer composite is presently applied in different automotive components such as boot lining, door panels, headliner panel, internal engine cover, noise insulation panel, engine insulation, bumper, interior insulation roof cover, and wheel box [26]. The results have recommended that the utilization of physical or chemical treatment on plant cellulose fibers is efficient, leading to its supportive impacts on thermal and mechanical characteristics with the influence of time of treatment [27–32]. This investigation targeted on the arrangement of microwave-treated plant cellulose/multiwalled carbon nanotube hybrid nanocomposites possessing 20% of plantain fiber in every condition. The impact of multiwalled carbon nanotubes incorporating thermal conductivity and morphological and mechanical characteristics of the polymer hybrid nanocomposite was studied. Singh and Gaurav [33] determined the dynamic and static mechanical characteristics of woven glass fiber-reinforced polymer composite, where the uniform dispersion of multiwalled carbon nanotubes with lower aspect ratio enhances the fatigue life cycle of the fabricated polymer composites and also impede the generation of fatigue crack in the composite. Qi et al. [34] studied short cellulose fiber's effect with the multiwalled carbon tube chains, and the outcomes displayed excellent and identical sensing capability, which can be applied as smart materials. Tariq et al. [35] established that the utilization of ultrasonication results in the uniform distribution of multiwalled carbon nanotubes in the epoxy resin, specifically under the lower concentration of filler. The investigation observed a significant enhancement in the carbon fiber-based epoxy composite's flexural and tensile strengths incorporated with multiwalled carbon nanotubes. Cividanes et al. [36] studied the impact of varying time and ultrasonication power on the curing characteristics of amino-based carbon nanotube/epoxy hybrid composite. The curing characteristic of carbon nanotube/epoxy composite was examined as the power from ultrasonication. In their consecutive investigation, Cha et al. [37] further observed significant improvement in fracture toughness and tensile characteristics of epoxy composite incorporated with 2 wt% of melamine-functionalized carbon nanotubes. Kharitonov et al. [38] also examined that the addition of fluorinated carbon nanotubes in the epoxy polymer tends to result in a noticeable improvement in flexural and tensile strengths by 58% and 35%, respectively, when compared with neat carbon nanotube epoxy composite. The investigation also signified that the utilization of fluorinated carbon nanotubes does not damage the epoxy composite's thermal characteristics, which leads to little increment in glassy temperature. Salam et al. [39] considered amino and carboxyl-based multiwalled carbon nanotubes to manufacture epoxy composites. The investigation applied the consolidation of three-roll mill and sonication methods to distribute the fillers into the epoxy matrix. The investigation also affirmed that carboxyl-based epoxy composites have better distribution, and a large number of functional regions appeared, which leads to higher thermal, thermomechanical, and flexural characteristics with the effect of increment in cross-linking density. Jiang et al. [40] studied interphase aspects on the glass transition temperature of the epoxy polymer incorporated with hydroxyl, amino, and pure carbon nanotubes. The investigation informs that amino-based carbon nanotubes/epoxy composite has a higher value of glass transition temperature, which is assigned to the production of covalent bonds between the cured polymer and amino-carbon nanotubes, whereas it is not examined in hydroxyl-based carbon nanotubes/epoxy composite.

From the literature review, it is signified that the scientists have created the technologies to improve the thermal and mechanical characteristics of an epoxy polymer by utilizing carbon nanotubes as reinforcement material, which can be appropriately considered in many applications such as automotive, aviation, and aerospace [41–47]. Moreover, in some of the investigations involving pure carbon nanotube-reinforced epoxy composites, lower improvement in characteristics is recorded because of entanglement and accumulation of carbon nanotubes, which leads to weaker interfacial bonding between filler and polymer matrix [48–50]. To avoid these problems, the researchers have introduced different treatments on carbon nanotubes like amino, hydroxyl, fluorination, plasma, carboxyl, hydrogen peroxide, silane, and melamine previously to their reinforcement in the epoxy polymer [51]. The investigation performed by Ma et al. [50] shows that amino-based carbon nanotubes present good wettability and more massive surface energy with the epoxy polymer when compared with pure multiwalled carbon nanotubes. The previous investigations signify an improvement in wear resistance, glass transition temperature, and fracture toughness [52, 53]. It would be impressive to study the influence of the addition of amino-based single-walled carbon nanotubes in epoxy polymer on other essential thermal degradation and mechanical characteristics. The present investigation targets to study the impacts of reinforcing amino-based single-walled carbon nanotubes (SWCNTs) and bamboo fiber (BF) in the epoxy polymer on thermal stability and mechanical characteristics. For the scope of comparison, outcomes are also collected for SWCNT/BF epoxy composites and pure epoxy polymer. The ultrasonication process has been considered to manufacture SWCNT/BF epoxy composite samples to avoid the accumulation of particles within the epoxy polymer. The microstructural characteristics and the fracture structure of different specimens have been observed under the scanning electron microscope (SEM).

15.2 Materials and Methods

15.2.1 Materials Used

Epoxy polymer (LY 556) and hardener (HY 951) were procured from Sakthi Fiber Glass, Inc., Chennai, India. The pure SWCNTs applied in this investigation were purchased from Lab Chemicals, Chennai, India. The chemicals needed for the amino functionalization of SWCNTs such as dimethylformamide, thionyl chloride, acetone, ethanol, and TETA were also supplied by the Lab Chemicals, Chennai, India.

15.2.2 Extraction of Bamboo Microfibers

Bambusa tulda, a type of bamboo, was brought from Madhya Pradesh, India. The BFs were separated into stacks of $2000 \times 10 \times 5\ mm^3$ dimensions. The chemical surface modifications on the BFs were carried out with salt solution (prepared with sodium chloride and fresh tap water in 20 : 1 ratio) by soaking for two days and then treated with 1 N sodium hydroxide (NaOH) solution to remove the residual lignin content by immersing for four hours. After attaining the recommended soaking duration, the BFs were evacuated and soaked with deionized water to decrease the brittleness and dilute the alkaline concentration. Finally, the BFs were manually defibrillated with a pulverizing machine to collect the microparticles of BF. The BF microparticles were cleaned with running deionized water to remove the contaminants and then conditioned in an oven at the temperature of $60 \pm 3\ °C$ for eight hours. All the BFs were then placed in the desiccator with the silica gel to prevent moisture content absorption. The length and diameter of the treated bamboo cellulose fibers were 250–300 and 20 μm, respectively.

15.2.3 Amino Functionalization of SWCNTs

Pure SWCNTs were exposed to amino functionalization, as per the process defined in an earlier investigation by Chandrasekaran et al. [54] and Kashyap et al. [51]. For this scheme, 36 g of SWCNTs was refluxed with 80 ml mixture of nitric and sulfuric acids in 3 : 1 proportion under 42 °C for 420 minutes using a magnetic stirrer in the wooden slab. The balance SWCNTs after the process of refluxing were cleaned with distilled water up to the pH quantity as seven was achieved and dried for eight hours at 100 °C from a vacuum furnace. Two grams of furnace-dried SWCNTs was poured in 40 ml of dimethylformamide and 100 ml of thionyl chloride in inert surroundings under 70 °C for 1440 minutes. The surplus quantity of dimethylformamide and thionyl chloride was eliminated in the fuming glass slab. The obtained SWCNTs were then modified with 200 ml of TETA under 120 °C for four days in the fuming glass slab. The modified SWCNTs were then cleaned with ethanol and distilled water to eliminate the excess quantity of TETA. The SWCNTs hence collected were lastly dried in the vacuum furnace under 110 °C.

15.2.4 Fabrication of SWCNT/Bamboo Fiber (BF)/Epoxy Hybrid Nanocomposites

For the production of SWCNT/BF epoxy hybrid composite samples, consisting of either amino-functionalized SWCNTs (denoted as a-SWCNT/BF) or pure SWCNTs (indicated as p-SWCNT/BF), SWCNTs, and BF, each of 1 wt% was reinforced to the measured quantity of epoxy polymer poured in the cylindrical glass container. The container was located on a magnetic hot plate, and the stirring process was performed for one hour under 2000 rpm. The acetone was occasionally poured to control the mixing process by decreasing the epoxy polymer mixture's viscosity. The epoxy mixture was then sonicated using an ultrasonicator probe (UP400ST,

Hielscher) at 35 °C for two hours, and ultrasonicator was occasionally closed for 10 minutes every 30 minutes of sonication to avoid from overheating. Moreover, the cylindrical container containing the mixture of bamboo microparticles, SWCNTs, and epoxy polymer was also covered with an ice bath to control the mixing temperature below 35 °C during the process of ultrasonication. After the ultrasonication process, the hardener was mixed with the polymer mixture using a mechanical agitator for 10 minutes, and then the mixture was poured into the mild steel molds and permitted to cure under room temperature for six hours. Finally, the fabricated composite samples were dried in a heating furnace at 60 ± 3 °C to remove the internal residual stresses.

15.2.5 Characterization

15.2.5.1 Thermal Properties

To study the impact of reinforcing elements such as p-SWCNT/BF and a-SWCNT/BF in the epoxy polymer on the thermal degradation characteristics, thermogravimetric analysis (TGA) was conducted in TGA 4000 equipment. During the TGA, the fine powdered form of 5–10 mg weight of the specimen was heated from 30 to 600 °C at the heating rate of 10 °C min^{-1}, and weight loss was reported as the function of temperature. The thermal conductivity of epoxy hybrid composites was determined using the transient plane source technique (TPS 2500 S Hot Disk).

15.2.5.2 Fourier Transform Infrared (FTIR) Spectroscopy

Fourier transform infrared (FTIR) examination of p-SWCNT/BF and a-SWCNT/BF was performed on Perkin Elmer Spectrum 2, FTIR spectrometer, by scanning the specimen between 400 and 4000 cm^{-1} with a resolution of 1 cm-1 and a scan rate of 32 scans per minute. The FTIR was performed to confirm the chemical modifications happening on the surface of SWCNTs during the amino functionalization process.

15.2.5.3 Mechanical Properties of SWCNT/BF/Epoxy Composites

Flexural and tensile tests followed the fabricated specimen (pure epoxy, p-SWCNT/BF/epoxy, and a-SWNCT/BF/epoxy) composites. The tensile experiments were conducted at a crosshead speed of 1 mm min^{-1} on Tinius Olsen universal testing machine with 10 kN of a load cell according to ASTM D 3039 standard. The flexural experiments were also performed on the same testing set up according to the ASTM D 790 standard with the crosshead speed of 1 mm min^{-1}, and the span of the sample was kept 60 mm (constant). The flexural and tensile tests were conducted on five similar specimens related to each composition level, and the mean value of these outcomes was applied to present the flexural and tensile strengths of the fabricated sample. The impact test and hardness of the fabricated epoxy hybrid composites were conducted according to the ASTM D 256 and D 785 standard correspondingly with the Izod impact testing machine and Wilson Wolpert Superficial hardness testing machine, respectively. All the mechanical properties were tested under room temperature of 28 °C and relative humidity of 60%.

15.2.5.4 Morphological Characteristics

The dispersion of SWCNTs and BF microparticles in epoxy polymer and morphology of the tensile fractured surfaces were examined under the field emission SEM under 15 kV of accelerating voltage. Before the SEM observations, the sample's tensile fracture surface was coated with gold ion using a plasma sputter coater machine.

15.3 Results and Discussion

15.3.1 Effect on FTIR Spectra

FTIR spectra of neat epoxy, p-SWCNT/BF epoxy, and a-SWCNT/BF epoxy hybrid nanocomposites are presented in Figure 15.1.

The FTIR spectrum of p-SWCNT/BF exhibits vibration bands because of the N—H bonding of amino networks under 3780 cm^{-1}, absorbing stretch band at 1563 cm^{-1} serving N–H scissoring vibrational medium, and C–N stretch bonding related to the range of wavenumber 1230–1300 cm^{-1}, which disappeared in the condition of p-SWCNT/BF epoxy hybrid nanocomposite. Different absorption peaks such as 3721, 2980, 1830, 1498, and 918 cm^{-1} assigned to the O–H stretch, = C—H stretch, C = O asymmetrical vibration, C = C linkage, and C—H bending vibrational medium, respectively, are observed in both a-SWCNT/BF and p-SWNCT/BF epoxy hybrid composites [55]. Hence, it is affirmed from the FTIR spectrum that the structure of a-SWNCT/BF shows perfect architectural changes by producing the amide networks on SWCNT's surface.

15.3.2 Effect on Thermal Degradation Properties

TGA graphs of the pure and amino-functionalized SWCNT particle samples are presented in Figure 15.2a. We conducted a TGA examination to study modifications

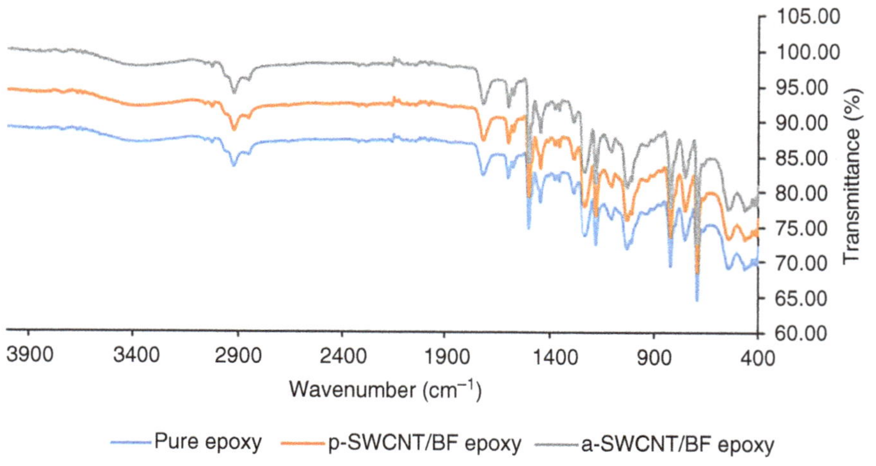

Figure 15.1 FTIR spectrum of SWCNT-based epoxy hybrid nanocomposites.

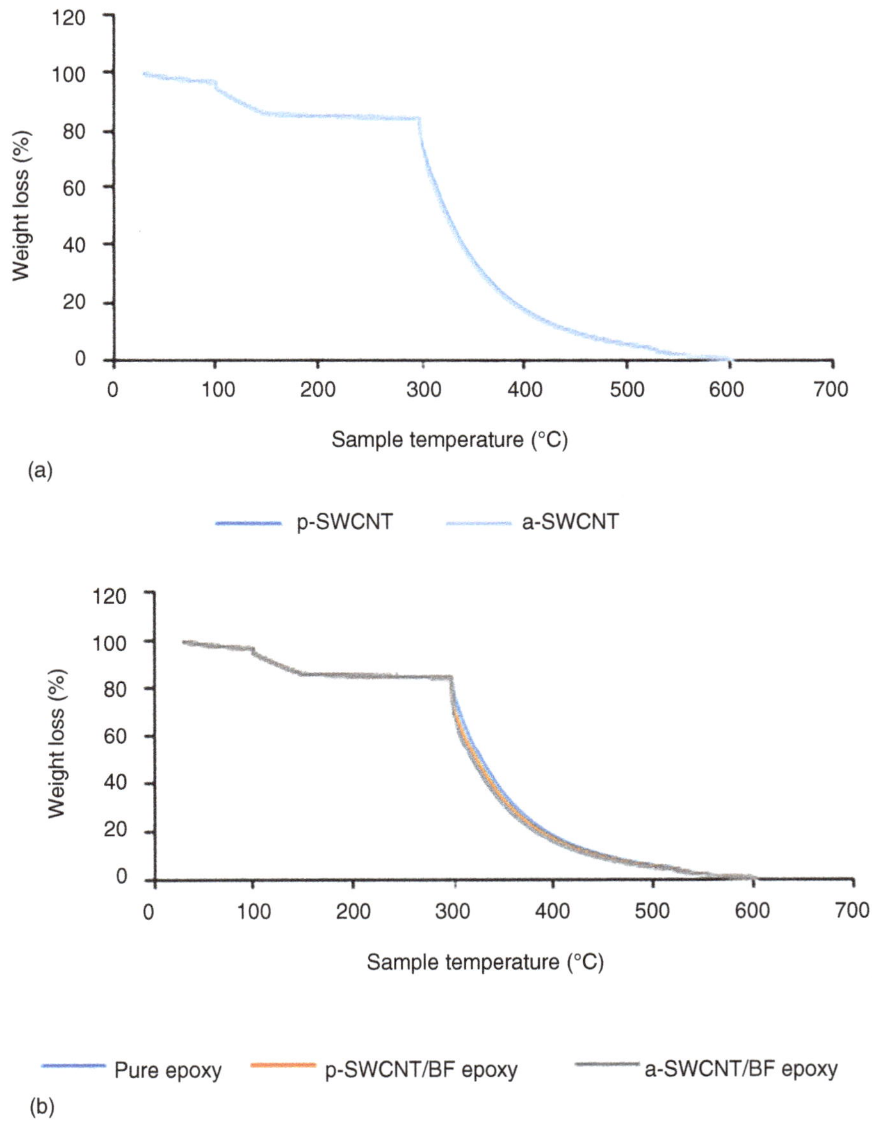

Figure 15.2 TGA curves: (a) modified and unmodified SWNCT and (b) SWCNT-based epoxy nanocomposites.

in the weight of the component vs. temperature under nitrogen gas with a heating rate of 10 °C min^{-1}. There are two thermal deterioration regions per TGA graph for pure and amino-functionalized SWCNT particle samples. In the first thermal decomposition step (between 150 and 300 °C), there is specimen moisture loss, and, in the second part initiated from 300 °C, a layer of char was produced that was degraded under a higher range of temperatures. As observed in TGA graphs, around 320 and 450 °C, the particles possess 78% and 8% of their weight. Furthermore, the numbers collected for amino-functionalized SWCNT under 320 and 450 °C

Table 15.1 Summary of the TGA of epoxy nanocomposites.

Composite type	Onset temperature (°C)	Major degradation temperature (°C)	Residue (%)
Pure epoxy	314.5	335.2	4.32
p-SWCNT/BF epoxy	325.6	341.7	5.12
a-SWCNT/BF epoxy	334.8	360.4	8.36

were 92% and 18%, respectively, which exhibit a relative enhancement in char formation; finally, the particles' higher thermal stability was achieved, owing to their amino-functionalization.

The influence of pure and amino-functionalized SWCNT on the epoxy nanocomposites' thermal stability has been investigated from TGA. The major degradation temperature, onset temperature, and residue amount have been determined to measure nanoparticles' influence. Figure 15.2b presents the TGA weight loss graphs of epoxy nanocomposites filled with p-SWCNT/BF and a-SWCNT/BF. As observed from the graphs, both the major degradation and onset temperature of p-SWCNT/BF and a-SWCNT/BF epoxy were enhanced compared with neat epoxy nanocomposites, as shown in Table 15.1. Hence, the thermal stability of the p-SWCNT/BF and p-SWCNT/BF epoxy nanocomposite was enhanced with the addition of SWCNT nanoparticles. Table 15.1 summarizes the essential TGA outcomes of the epoxy nanocomposites with p-SWCNT/BF and a-SWCNT/BF filler reinforcements. It presents the major degradation, and onset temperature has improved with the reinforcement of fillers. These shows that the epoxy nanocomposites with a-SWCNT/BF are more thermally stable than the pure epoxy polymer. The table also presents that the quantity of residue has improved compared with the pure epoxy polymer sample, which signifies that the fire-retardant capacity of epoxy nanocomposites has improved. The onset temperature has enhanced to up to 20 °C, and the major degradation has increased to up to 25 °C. This is mainly assigned to the higher thermal stability of amino-functionalized SWCNT nanofillers in the epoxy nanocomposites. The a-SWCNT nanofiller, which is not directly connected with the epoxy polymer, protects the heat energy from expeditious transfer to the resin and hence restricts the thermal decomposition [56].

15.3.3 Effect on Thermal Conductivity

From the higher thermal conductivity of SWCNTs, the addition of SWCNTs in the epoxy polymer supports depleting more amount of heat, therefore incrementing the thermal stability of the neat epoxy polymer. Further, uniform dispersion and fine particles of a-SWCNTs/BF in the epoxy polymer, as observed in the preceding section, conclude higher thermal stability than the p-SWCNT/BF epoxy nanocomposites. Among the different specimens, the epoxy nanocomposite

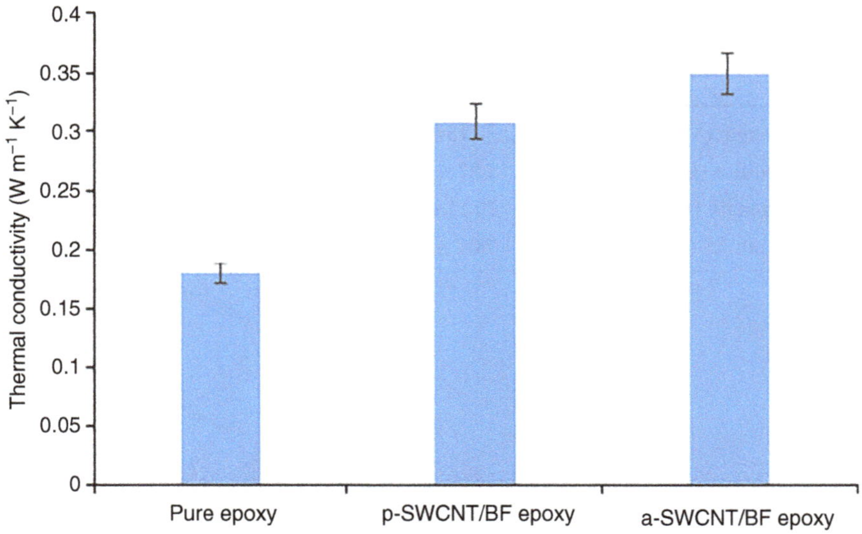

Figure 15.3 Effect of amino functionalization on thermal conductivity of epoxy hybrid nanocomposites.

possessing a-SWCNT/BF exhibits the highest thermal stability characteristics that may be because of stronger interfacial adhesion between the epoxy polymer and a-SWCNT/BF [51]. Depending on the well-established thermal conductivity principles, pure epoxy composite has low thermal conductivity, whereas the thermal conductivity of the composite can be improved by reinforcing the compounds that have a higher range of thermal conductivity value [57]. The fantastic thermal characteristics of SWCNTs are predicted to offer polymers with a higher amount of thermal conductivity. It is also well-known that in polymer-based composites, the transmission of temperature is mainly in vibrational principles. Thus, the complete thermal conductivity of polymer-based composites can be affected by different characteristics like diffusion degree, exterior roughness, crystallinity, inherent fillers, filler-to-matrix interface, aspect ratio, thermal resistance, and segment [58]. Figure 15.3 shows the epoxy hybrid nanocomposites' thermal conductivity reinforced with a-SWCNT/BF and p-SWCNT/BF. From the columns, the addition of SWCNTs probably recovers the thermal conductivity of the composite.

The increasing principles of thermal conductivity incorporating SWCNT in the epoxy composite have been explained by other researchers [58–60]. The addition of SWCNTs in the composite is achieved as a thermal funnel that has effectively strengthen the flow of heat. Improved thermal conductivity of the epoxy hybrid nanocomposites can be assigned to the higher value of SWCNTs' thermal conductivity that can be up to 3000 $W\ m^{-1}\ K^{-1}$ [61]. The thermal conductivity of the epoxy hybrid nanocomposite improved with the aggregate loading of SWCNT. The decrement in the distance among filler could create further thermal networks between them, producing a thermal conducting grid or pathway [62, 63].

Table 15.2 Flexural and tensile characteristics of epoxy hybrid nanocomposites.

Properties	p-SWCNT/BF epoxy	a-SWCNT/BF epoxy
Tensile strength (MPa)	30.15 ± 0.86	37.11 ± 0.87
Tensile modulus (GPa)	2.27 ± 0.9	2.81 ± 0.91
Flexural strength (MPa)	76.11 ± 0.96	93.65 ± 0.95
Flexural modulus (GPa)	7.04 ± 1.01	8.66 ± 0.99

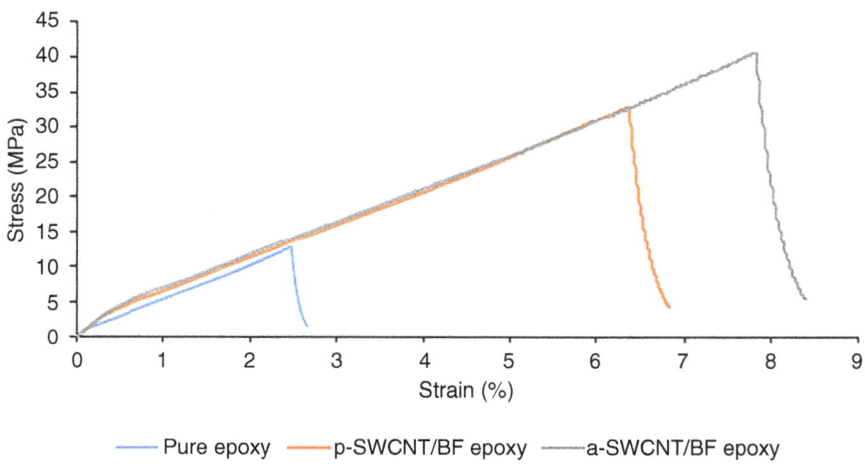

Figure 15.4 Tensile stress–strain graphs of SWCNT/BF epoxy hybrid nanocomposites.

15.3.4 Effect on Mechanical Characteristics

15.3.4.1 Flexural and Tensile Properties

The fluctuation of flexural and tensile characteristics (modulus and strength) of the different specimens, according to the reinforcement, is given in Table 15.2, which is displayed in Figures 15.4 and 15.5, respectively.

The flexural and tensile strength of a-SWCNT/BF epoxy nanocomposites is found to be higher when compared with p-SWCNT/BF and pure epoxy composites (Figures 15.4 and 15.5).

Both strengths (flexural and tensile) are reviewed to increment with the type of reinforcement showing that the addition of SWCNTs in the epoxy polymer improves the proportion of cross-linking and restricts the molecular movement within the polymer. This leads to the improved strengths of the SWCNT epoxy composite when compared with pure epoxy polymer [64]. Similar to the fluctuation of power displayed in Figure 15.4, a-SWCNT/BF epoxy nanocomposite specimen also presents the highest flexural and tensile modulus, whereas the lower quantities of both moduli were examined for pure epoxy composite. The increment discussed in flexural and tensile modulus on incorporating SWCNTs in the epoxy polymer is assigned to higher strength, modulus, and aspect ratio of the SWCNTs [64].

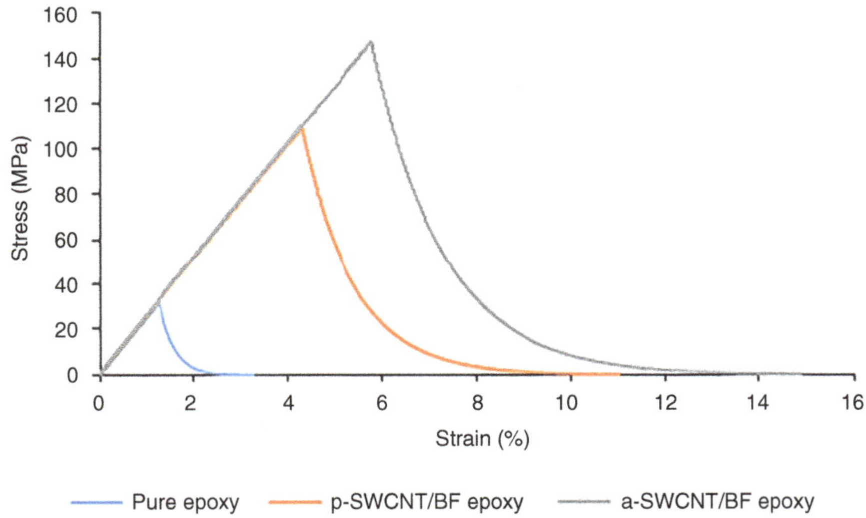

Figure 15.5 Flexural stress–strain graphs of SWCNT/BF epoxy hybrid nanocomposites.

15.3.5 Tensile Fracture

The SEM observation of the tensile fracture surface of different samples was performed to describe the examined fluctuation in other specimens' flexural and tensile characteristics (Figure 15.6a,b). It is affirmed from the SEM micrographs that there is an accumulation of p-SWCNTs and BF in epoxy hybrid composite (Figure 15.6a). Moreover, the dispersion of a-SWCNTs and BF is uniform in epoxy composite (Figure 15.6b). It shows that the amino-functionalized SWCNTs supports decreasing the accumulation and controls the uniform distribution of a-SWCNT in the epoxy polymer. The notified principles are arranged with a similar TEM examination of F-MWCNT and P-MWCNT in the epoxy polymer [51, 64] and

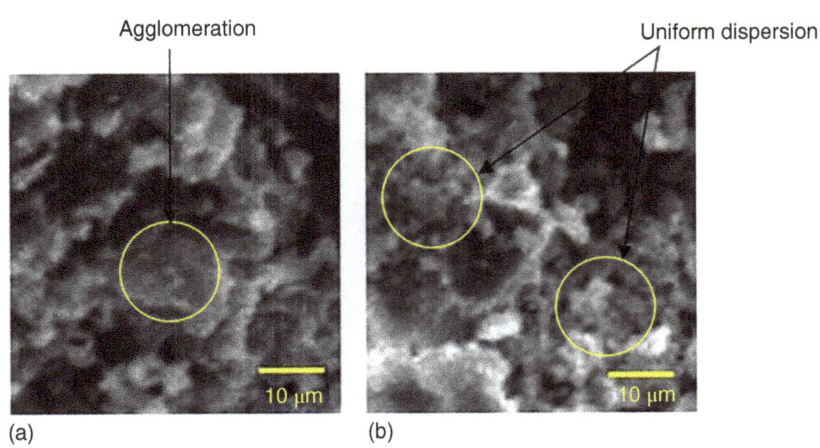

Figure 15.6 Tensile fracture of epoxy hybrid nanocomposites: (a) p-SWCNT/BF epoxy and (b) a-SWCNT/BF epoxy.

in the work described by Ma et al. [50]. The utilization of amino-functionalized SWCNTs also presents to increment the link between SWCNTs and polymer, hence enhancing their interfacial adhesion [50, 64]. During composite production, the interaction of chemical happens between the epoxy molecule and amine functional group composed of the a-SWCNTs, generating ring reactions to cross-chain network reaction. As a result, the production of stronger covalent bond takes place between the epoxy polymer and a-SWCNTs/BF that concludes in the higher transfer of load between fillers and polymer matrix and tends to higher mechanical characteristics of a-SWCNT/BF epoxy hybrid nanocomposites when compared with neat epoxy and p-SWCNT/BF epoxy hybrid nanocomposites.

The lower mechanical characteristics determined from p-SWCNT/BF epoxy may be assigned to the appearance of gaps and accumulation, which is comparably greater than the a-SWCNT/BF epoxy composite. The appearance of gaps examined may be because of the nonuniform dispersion of epoxy polymer, a-SWCNT, and BF, and the accumulation of p-SWCNTs happens owing to their van der Waals attraction forces, π-π bonding, and larger surface area [65].

15.3.5.1 Impact Strength and Hardness

The impact strength for epoxy hybrid nanocomposite is depicted in Figure 15.7, as the resistance showed during high-speed load utilization. As reported in this figure, the increment in impact strength was examined as a-SWCNTs/BF that can be assigned to the reinforcing effect of SWCNT on the polymer region as recorded earlier [8]. Hardness characteristics, as shown in Figure 15.7, is a mean Vickers hardness of 15 observations for every specimen. It can be observed that the reinforcement of a-SWCNT in the BF hybrid composite led to the enhancement of up to 25%. This improvement of the composite hardness is an issue of the stiffing

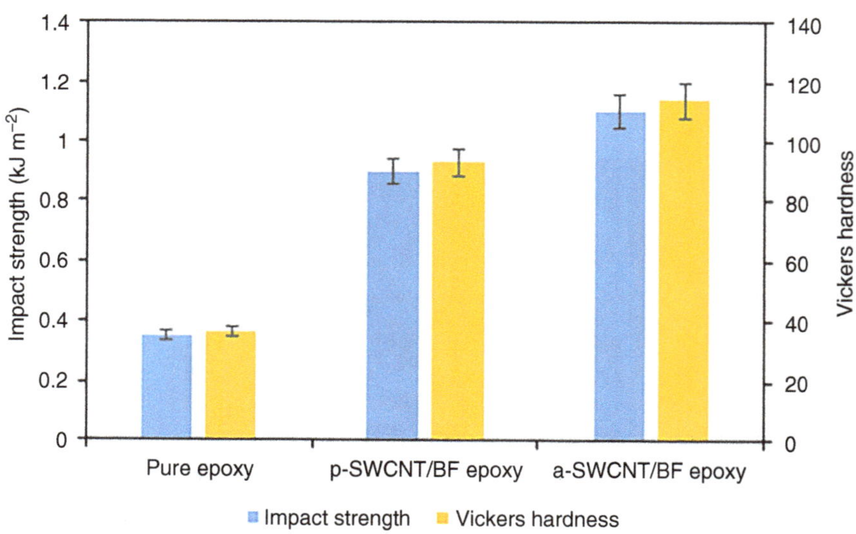

Figure 15.7 Impact strength and Vickers hardness of epoxy hybrid nanocomposite. Source: Based on Gkikas et al.[8].

impact of a-SWCNT on the polymer and the disappearance of gaps observed from the SEM micrographs [62].

15.4 Conclusion

The research studied the thermal and mechanical characteristics of epoxy hybrid composites incorporated with a-SWCNTs (amino functionalized and pure) and BF. The investigation affirms that epoxy hybrid nanocomposite added with a-SWCNTs/BF shows higher flexural, tensile, impact, and hardness characteristics. The flexural and tensile characteristics examined for a-SWCNT/BF epoxy hybrid nanocomposite are found to be more generous, respectively, around 16% and 12.7% when compared with the pure epoxy polymer specimen. Furthermore, flexural and tensile moduli for a-SWCNTs/BF epoxy hybrid nanocomposites are improved by 27% and 14%, respectively, compare with that examined for the pure epoxy polymer. According to mechanical characteristics, the addition of either a-SWCNT or p-SWCNT in the epoxy polymer improves its thermal stability characteristics. The thermal conductivity has displayed significant enhancement up to 41% as amino-functionalized SWCNT, and SEM micrographs recommend that the uniform roughness of surface was attained after the ultrasonication process and better mechanical interlocking of hybrid fillers within the epoxy polymer. The amino-functionalized SWCNTs applied as fillings did not damage the epoxy composites' thermal stability characteristics and tended to the slight increment in glass transition temperature. However, they influence the conductive characteristics of the composites, when compared with the pure SWCNTs. The enhancement in thermal and mechanical characteristics is assigned to the higher interfacial adhesion within the a-SWCNT, BF, and epoxy polymer and uniform distribution of a-SWCNTs/BF compared with those examined for p-SWCNT/BF epoxy composites.

Acknowledgment

This research was supported by the King Mongkut's University of Technology North Bangkok through the Post-Doctoral Program (Grant No. KMUTNB-POST-64-06).

References

1 Rajamohan, A.P.V. and Mathew, A. (2019). Material and mechanical characterization of multi-functional carbon nanotube reinforced hybrid composite materials. *Experimental Techniques* 43 (3): 301–314.

2 Mohit, H., Sanjay, M.R., Vinod, K. et al. (2020). A comprehensive review on mechanical, electromagnetic radiation shielding, and thermal conductivity of fibers/inorganic fillers reinforced hybrid polymer composites. *Polymer Composites* 41 (10): 3940–3965.

3 Kumar, T.S., Senthilkumar, K., Chandrasekar, M. et al. (2020). Influence of fillers on the thermal and mechanical properties of biocomposites: an overview. In: *Biofibers and Biopolymers for Biocomposites* (eds. A. Khan, S.M. Rangappa, S. Siengchin and A.M. Asiri), 111–133. Cham: Springer.

4 Tarfaoui, M., Lafdi, K., and El Moumen, A. (2016). Mechanical properties of carbon nanotubes based polymer composites. *Composites Part B: Engineering* 103: 113–121.

5 Omid, N., Sujan, D., MuhammadEkhlasur, R. et al. (2018). Hybrid composite using natural filler and multiwalled carbon nanotubes (MWCNTs). *Applied Composite Materials* 25 (6): 1323–1337.

6 Ayatollahi, M.R., Shadlou, S., Shokrieh, M.M., and Chitsazzadeh, M. (2011). Effect of multiwalled carbon nanotube aspect ratio on mechanical and electrical properties of epoxy-based nanocomposites. *Polymer Testing* 30: 548–556.

7 Rosca, I.D. and Hoa, S.V. (2011). Method for reducing contact resistivity of carbon nanotube containing epoxy adhesives for aerospace applications. *Composites Science and Technology* 71: 95–100.

8 Gkikas, G., Barkoula, N.-M., and Paipetis, A.S. (2012). Effect of dispersion conditions on the thermos-mechanical and toughness properties of multi walled carbon nanotubes reinforced epoxy. *Composites Part B: Engineering* 43: 2697–2705.

9 Felisberto, M., Aria-Durán, A., Ramos, J.A. et al. (2012). Influence of filler alignment in the mechanical and electrical properties of carbon nanotubes/epoxy nanocomposites. *Physica B: Condensed Matter* 407: 3181–3182.

10 Alvarez, C., Rojas, O.J., Rojano, B., and Ganán, P. (2015). Development of self-bonded fiberboards from a fiber of leaf plantain: effect of water and organic extractives removal. *Bioresources* 10 (1): 672–683.

11 Cadena Ch Edith, M., Manuel Vélez, R.J., Santa, J.F., and Viviana Otálvaro, G. (2017). Natural fibers from plantain pseudostem (*Musa Paradisiaca*) for use in fiber-reinforced composites. *Journal of Natural Fibers* 14 (5): 678–690.

12 Chandrasekar, M., Shahroze, R.M., Ishak, M.R. et al. (2019). Flax and sugar palm reinforced epoxy composites: effect of hybridization on physical, mechanical, morphological and dynamic mechanical properties. *Materials Research Express* 6 (10): 105331.

13 Krishnasamy, S., Thiagamani, S.M., Kumar, C.M. et al. (2019). Recent advances in thermal properties of hybrid cellulosic fiber reinforced polymer composites. *International Journal of Biological Macromolecules* 141: 1–3. (IF: 5.162).

14 Jawaid, M., Khalil, H.P.S.A., and Bakar, A.A. (2010). Mechanical performance of oil palm empty fruit bunches/jute fibres reinforced epoxy hybrid composites. *Materials Science and Engineering* 527: 7944–7949.

15 Dhakal, H.N., Zhang, Z.Y., Guthrie, R. et al. (2013). Development of flax/carbon fibre hybrid composites forenhanced properties. *Carbohydrate Polymers* 96: 1–8.

16 Dinesh, S., Kumaran, P., Mohanamurugan, S. et al. (2020). Influence of wood dust fillers on the mechanical, thermal, water absorption and biodegradation characteristics of jute fiber epoxy composites. *Journal of Polymer Research* 27 (9): 1–13.

17 Jenish, I., Chinnasamy, S.G.V., Basavarajappa, S. et al. (2020). Tribo-mechanical characterization of carbonized coconut shell micro particle reinforced with *Cissus quadrangularis* stem fiber/epoxy novel composite for structural application. *Journal of Natural Fibers*: 1–17.

18 Setty, V.K.S.N., Goud, G., Chikkegowda, S.P. et al. (2020). Characterization of chemically treated *Limonia acidissima* (wood apple) shell powder: physicochemical, thermal, and morphological properties. *Journal of Natural Fibers*: 1–12.

19 Setty, V.K.S.N., Goud, G., Sanjay, M.R., and Siengchin, S. (2020). Raw and chemically treated bio-waste filled (*Limonia acidissima* shell powder) vinyl ester composites: physical, mechanical, moisture absorption properties, and microstructure analysis. *Journal of Vinyl and Additive Technology*: 1–11.

20 Guermazi, N., Haddar, N., Elleuch, K., and Ayedi, H.F. (2014). Investigations on the fabrication and the characterization of glass/epoxy, carbon/epoxy and hybrid composites used in the reinforcement and the repair of aeronautic structures. *Materials & Design* 56: 714–724.

21 Asaithambi, B., Ganesan, G., and Ananda, K.S. (2014). Bio-composites development and mechanical characterization of banana/sisal fibre reinforced poly lactic acid PLA; hybrid composites. *Fibres and Polymers* 15: 847–854.

22 Yashas, G.T.G., Sanjay, M.R., Jyotishkumar, P., and Siengchin, S. (2019). Natural fibers as sustainable and renewable resource for development of eco-friendly composites: a comprehensive review. *Frontiers in Materials* 6: 1–14.

23 Sanjay, M.R. and Siengchin, S. (2021). Editorial corner – a personal view, exploring the applicability of natural fibers for the development of biocomposites. *Express Polymer Letters* 15 (3): 193.

24 Sanjay, M.R., Siengchin, S., and Dhakal, H.N. (2020). Green-composites: eco-friendly and sustainability. *Applied Science and Engineering Progress* 13 (3): 183–184.

25 Sanjay, M.R., Siengchin, S., Jyotishkumar, P. et al. (2019). A comprehensive review of techniques for natural fibers as reinforcement in composites: preparation, processing and characterization. *Carbohydrate Polymers* 207 (1): 108–121.

26 Mohammed, L., Ansari, M.N.M., Pua, G. et al. (2015). A review on natural fiber reinforced polymer composite and its applications. *International Journal of Polymer Science* 2015: 1–15.

27 Imoisili, P.E., Dagogo, I.T., Popoola, A.V., and Okoronkwo, A.F. (2018). Effect of high-frequency microwave radiation on the mechanical properties of plantain (*Musa paradisiaca*) fibre/epoxy biocomposite. *Journal of Physical Science* 29 (3): 23–35.

28 Mohit, H. and Selvan, V.A.M. (2018). A comprehensive review on surface modification, structure interface and bonding mechanism of plant cellulose fiber reinforced polymer based composites. *Composite Interfaces* 25 (5-7): 629–667.

29 Ramesh, M., Deepa, C., Kumar, L.R. et al. (2020). Life-cycle and environmental impact assessments on processing of plant fibers and its bio-composites: a critical review. *Journal of Industrial Textiles*: 1–25.

30 Vinod, A., Sanjay, M.R., Siengchin, S., and Jyotishkumar, P. (2020). Renewable and sustainable biobased materials: an assessment on biofibers, biofilms, biopolymers and biocomposites. *Journal of Cleaner Production* 258: 120978.

31 Yorseng, K., Sanjay, M.R., Tengsuthiwat, J. et al. (2019). Information in United States Patents on works related to 'Natural Fibers': 2000–2018. *Current Materials Science* 12 (1): 4–76.

32 Jagadeesh, P., Yashas, G.T.G., Madhu, P. et al. (2020). Effect of natural filler materials on fiber reinforced hybrid polymer composites: an overview. *Journal of Natural Fibers*: 1–16.

33 Singh, K.K. and Gaurav, A. (2018). Effectiveness of short and straight carbon nanotubes on dispersion state and static/dynamic mechanical properties of woven glass fibre-reinforced polymer laminates. *Proceedings of the Institution ofMechanical Engineers, Part L: Journal of Materials: Design and Applications* 233: 1661–1677.

34 Qi, H., Liu, J., and Mäder, E. (2014). Smart cellulose fibers coated with carbon nanotube networks. *Fibers* 2: 295–307.

35 Tariq, F., Shifa, M., and Baloch, R.A. (2018). Mechanical and thermal properties of multi-scale carbon nanotubes–carbon fiber–epoxy composite. *Arabian Journal for Science and Engineering* 43: 5937–5948.

36 Cividanes, L.D.S., Simonetti, E.A.N., Oliveira, J.I.S.D. et al. (2015). The sonication effect on CNT-epoxy composites finally clarified. *Polymer Composites* 38: 1964–1973.

37 Cha, J., Jun, G.H., Park, J.K. et al. (2017). Improvement of modulus, strength and fracture toughness of CNT/Epoxy nanocomposites through the functionalization of carbon nanotubes. *Composites Part B: Engineering* 129: 169–179.

38 Kharitonov, A.P., Simbirtseva, G.V., Tkachev, A.G. et al. (2015). Reinforcement of epoxy resin composites with fluorinated carbon nanotubes. *Composites Science and Technology* 107: 162–168.

39 Salam, M.B.A., Hosur, M.V., Zainuddin, S., and Jeelani, S. (2013). Improvement in mechanical and thermo-mechanical properties of epoxy composite using two different functionalized multi-walled carbon nanotubes. *Open Journal of Composite Materials* 3: 1–9.

40 Jiang, C., Jiang, D., Zhang, J. et al. (2017). Glass transition temperature of amino groups grafted carbon nanotubes reinforced epoxy resin composites: Role of strong interphase. *Polymer Composites* 39: 1129–1138.

41 Silva, V.A.D. and Rezende, M.C. (2018). Effect of the morphology and structure on the microwave absorbing properties of multiwalled carbon nanotube filled epoxy resin nanocomposites. *Materials Research* 21: 1–9.

42 Arpitha, G.R., Sanjay, M.R., Senthamaraikannan, P. et al. (2017). Hybridization effect of sisal/glass/epoxy/filler based woven fabric reinforced composites. *Experimental Techniques* 41: 577–584.

43 Vinay, S.S., Sanjay, M.R., Siengchin, S., and Venkatesh, C.V. (2020). Effect of Al_2O_3 nanofillers in basalt/epoxy composites: mechanical and tribological properties. *Polymer Composites*: 1–14.

44 Ganapathy, T., Sathiskumar, R., Sanjay, M.R. et al. (2019). Effect of graphene powder on banyan aerial root fibers reinforced epoxy composites. *Journal of Natural Fibers* 18 (7): 1–8.

45 Abhisek, S., Sanjay, M.R., George, R. et al. (2018). Development of new hybrid *Phoenix pusilla*/carbon/fish bone filler reinforced polymer composites. *Journal of the Chinese Advanced Materials Society* 6 (4): 553–560.

46 Ganesan, K., Kailasanathan, C., Sanjay, M.R. et al. (2020). A new assessment on mechanical properties of jute fiber mat with egg shell powder/nanoclay-reinforced polyester matrix composites. *Journal of Natural Fibers* 17 (4): 482–490.

47 Kausar, A., Rafique, I., and Muhammad, B. (2016). Review of applications of polymer/carbon nanotubes and epoxy/CNT composites. *Polymer – Plastics Technology and Engineering* 55: 1167–1191.

48 Ervina, J., Mariatti, M., and Hamdan, S. (2016). Effect of filler loading on the tensile properties of multi-walled carbon nanotube and graphene nanopowder filled epoxy composites. *Procedia Chem* 19: 897–905.

49 Yeh, M.K., Hsieh, T.H., and Tai, N.H. (2008). Fabrication and mechanical properties of multi-walled carbon nanotubes/epoxy nanocomposites. *Materials Science and Engineering A* 483–484: 289–292.

50 Ma, P.C., Mo, S.Y., Tang, B.Z., and Kim, J.K. (2010). Dispersion, interfacial interaction and re-agglomeration of functionalized carbon nanotubes in epoxy composites. *Carbon* 48: 1824–1834.

51 Kashyap, P., Singh, N.P., Arora, S. et al. (2020). Effect of amino-functionalization of MWCNTs on the mechanical and thermal properties of MWCNTs/epoxy composites. *Bulletin of Materials Science* 43: 1–9.

52 Srikanth, I., Kumar, S., Kumar, A. et al. (2012). Effect of amino functionalized MWCNT on the crosslink density, fracture toughness of epoxy and mechanical properties of carbon–epoxy composites. *Composites Part A: Applied Science and Manufacturing* 43: 2083–2086.

53 Cui, L.J., Geng, H.Z., Wang, W.Y. et al. (2013). Functionalization of multi-wall carbon nanotubes to reduce the coefficient of the friction and improve the wear resistance of multi-wall carbon nanotube/epoxy composites. *Carbon* 54: 277–282.

54 Chandrasekaran, V., Santare, M., Krishnan, P., and Advani, S. (2011). Amino functionalization of MWNTs and their effect on ILSS of hybrid nanocomposites. *Composite Interfaces* 18: 339–355.

55 Monthe, G.C. and de Miranda, I.C. (2009). Characterization of sugarcane and coconut fibers by thermal analysis and FTIR. *Journal of Thermal Analysis and Calorimetry* 97: 661–665.

56 Blokhin, A.N., Dyachkova, T.P., Maksimkin, A.V. et al. (2019). Polymer composites based on epoxy resin with added carbon nanotubes. *Materials* 28: 45–49.

57 Zakaria, M.R., Kudus, M.H.A., Akil, H.M., and Thirmizir, M.Z.M. (2017). Comparative study of graphene nanoparticle and multiwallcarbon nanotube filled epoxy nanocomposites based on mechanical, thermal and dielectric properties. *Composites Part B: Engineering* 119: 57–66. http://dx.doi.org/10.1016/j.compositesb.2017.03.023.

58 Wang, F., Drzal, L.T., Qin, Y., and Huang, Z. (2016). Enhancement of fracture toughness, mechanical and thermal properties of rubber/epoxy composites by incorporation of graphene nanoplatelets. *Composites Part A: Applied Science and Manufacturing* 87: 10–22.

59 Yang, S.-Y., Lin, W.-N., Huang, Y.-L. et al. (2011). Synergetic effects of graphene platelets and carbon nanotubes on the mechanical and thermal properties of epoxy composites. *Carbon* 49 (3): 793–803.

60 Martin-Gallego, M., Verdejo, R., Khayet, M. et al. (2011). Thermal conductivity of carbon nanotubes and graphene in epoxy nanofluids and nanocomposites. *Nanoscale Research Letters* 6 (1): 1–7.

61 Vanyorek, L., Kiss, D., Prekob, Á. et al. (2019). Application of nitrogen doped bamboo-like carbon nanotube for development of electrically conductive lubricants. *Journal of Materials Research and Technology* 8 (3): 3244–3250.

62 Rezaur Rahman, M.D., Huque, M., Islam, M.N., and Hasan, M. (2008). Improvement of physic-mechanical properties of jute fiber reinforced polypropylene composite by post treatment. *Composites Part A: Applied Science and Manufacturing* 39: 1739–1747.

63 Imoisili, P.E., Ukoba, K., and Jen, T.-C. (2020). Physical, mechanical and thermal properties of high frequency microwave treated plantain (*Musa paradisiaca*) fibre/MWCNT hybrid epoxy nanocomposites. *Journal of Materials Research and Technology* 9 (3): 4933–4939.

64 Zakaria, M.R., Kudus, M.H.A., Akil, H.M., and Thirmizir, M.Z.M. (2017). Comparative study of graphene nanoparticle and multiwall carbon nanotube filled epoxy nanocomposites based on mechanical, thermal and dielectric properties. *Composites Part B: Engineering* 119: 57–66.

65 Moaseri, E., Hasanabadi, S., Maghrebi, M., and Baniadam, M. (2015). Improvements in fatigue life of amine-functionalized multi-walled carbon nanotube-reinforced epoxy composites: effect of functionalization degree and microwave-assisted precuring. *Journal of Composite Materials* 49: 1961–1969.

16

Effect of Metal Oxide Fillers on Thermal Properties of the Natural Fiber-Based Hybrid Composites

Mohit Hemath[1], Hemath K. Govindarajulu[2], Sanjay M. Rangappa[1], Suchart Siengchin[1], Ruban Ramalingam[3], and Babu V. Hemath[2]

[1] *King Mongkut's University of Technology North Bangkok, The Siridhorn International Thai-German Graduate School of Engineering, Department of Materials and Production Engineering, Natural Composites Research Group Lab, 1518 Pracharat 1 Road, Wongsawang, Bangsue, Bangkok, 10800, Thailand*
[2] *Composite Research Center, EVR Street, Vinayagapuram, Ambattur, Chennai, 600053, India*
[3] *National Institute of Technology, Department of Mechanical Engineering, Tiruchirappalli, Tamil Nadu 620015, India*

16.1 Introduction

In the recent years, polymer-based nanolaminates became an essential area of investigation in different science, engineering, and technology sectors. These types of polymers attracts much attention. The advancement of novel multiple-function polymer-based nanolaminates depends on the metal oxide particles, and the investigation of their characteristics is essential in the physics and chemistry of polymers and the sector of novel radiation protection polymeric materials [1]. Polymer-based nanolaminates are proposed composites for utilization in capacitors, smart microelectronic devices, high-voltage electrical insulation, etc. However, researchers have shown low interest in studies and utilizations of polyethylene types of laminates compared with other polymer resins.

The nano form of the particle is the leading additives to enhance the characteristics of resins. The utilization of the nano form of particles due to their higher surface area is desirable to microparticles. There are higher interfacial interactions within the particles and molecules in nanolaminates because of the larger surface area of nanoparticles than in neat resin. The pure resin cannot obtain such synergies, which develop the complicated interfacial microstructure [2]. Nanofiller-based additives extraordinarily modify the base resin's mechanical, thermal, electrical, and optical characteristics. The inorganic nanoparticles on nanolaminates' attributes are broadly based on the particle's size and shape, features of the fracture surface, particle type, and distribution degree [3, 4]. Composites depend on the metal oxides to have a broad scope of possible utilization in different sectors because of their complicated identical chemical, biological, and physical characteristics. The similar protection of metal oxide materials from radiation comes up with aspects

Natural Fiber-Reinforced Composites: Thermal Properties and Applications, First Edition.
Edited by Senthilkumar Krishnasamy, Senthil Muthu Kumar Thiagamani, Chandrasekar Muthukumar, Rajini Nagarajan and Suchart Siengchin.

for producing their support of absolutely cheap and slightly lower higher strength. These composites can offer efficient radiation shielding to handle spacecraft and specific instruments functioning in a higher range of radiation. Polymer-based materials reinforced with metal oxide nanoparticles are highly consistent and exciting from a budgetary viewpoint because of their functionality and higher strength [5].

Epoxy laminates are classified within engineering polymer composites that are broadly applied in the leading different applications like construction, automobile, aerospace, etc., because of their dimensional stability, resistance from chemical reactions and heat, and better adhesion [6, 7]. Furthermore, the organic behavior of epoxy polymer is constitutionally flammable. The fire catastrophe is generally recorded as obtained from the vapors and detrimental smokes developed by flames. Hence, it is a significant consequence to overcome the fire symptoms of the epoxy polymer. To decrease the flammability properties of epoxy polymers and develop their utilization, extensive impacts have been recommended on advancing new schemes and designs, like chemical budding flame retardants, fillers, and coatings [8, 9]. Nanoscience and technology showed interest from these schemes because of higher flame impedance effectiveness without damaging their mechanical and thermal characteristics [10], which provide a new result for flame-impeding polymeric resins. Presently, different metal or metal oxide nanofillers and their by-product have been accumulated on nanofilms to generate a new type of hybrid composites because of their higher aspect ratio as support components [11]. Such kinds of nanofillers could reduce the accumulation of nanofilms, and their by-products can play as a catalyst to decrease the toxic behavior of gas during the combustion process. In this regard, metal oxide-based hybrid composites are observed for the catalytic reduction of embedded copper, cobalt oxide, iron, and carbon monoxide [12, 13]. Till now, many techniques have been observed to prepare metal renovated nanofiller including sol–gel technique, solution mixing, solvothermal, and hydrothermal technique, etc. [14]. Furthermore, these techniques suggest the utilization of long reaction time and toxic organic solvents. Hence, the advancement of biomethods to collect hybrid composites is highly fascinating.

In the present work, TiO_2 metal oxide nanoparticle-renovated bamboo nanocellulose fiber (BNF) hybrid was prepared through an efficient and biomethodology and consequently reinforced into the epoxy polymer. The fracture surface and structure of these hybrid composites were well represented, and the fire-retardant characteristics and mechanical and thermal stabilities were studied.

16.2 Materials and Methods

16.2.1 Materials

The epoxy polymer (LY 556) and the hardener (HY 951) were applied during the epoxy hybrid nanocomposite fabrication. The pure TiO_2 (metal oxide) nanoparticles were applied with a mean size of particle of 100 nm. Additionally, the Indian bamboo fiber (*Bambusa tulda*), collected from West Bengal, were applied as reinforcement material to modify the polymer resin.

16.2.2 Extraction of Bamboo Nanocellulose Fiber (BNF)

The obtained bamboo fiber was cleaned with a forced fresh tap to remove the dirt and dust particles. The bamboo fiber was then soaked in a salt solution (synthesized from NaCl and water in the ratio of 5 : 1) for two days and then transferred to an industrial grinder to produce fine particles that were segregated from 50 nm nanomesh sieve.

16.2.3 Fabrication of Epoxy Hybrid Nanocomposites

The required quantity of nanofillers (1 wt% of BNF and TiO_2) was distributed in 200 g of epoxy from the ultrasonicator probe for 120 minutes and then composed under higher shear for one hour. The resin mixture was kept up to room temperature before adding 20 g of the hardener and manually containing a mechanical agitator. The beaker was located in the vacuum chamber to evacuate the bubbles and porosities under decreased pressure, and the resin mixture was discharged in the mild steel mold and regasified for 120 minutes. The fabricated specimen was post-cured under 80 °C for six hours in a heating furnace after removed from the mold.

16.2.4 Epoxy Hybrid Nanocomposite Characterization

16.2.4.1 Thermal Properties

The differential scanning calorimetry (DSC) was conducted using DSC 8000 instrument in the heating mode between 20 and 200 °C with a heating rate of 10 °C min^{-1}. The thermal stability of the fabricated epoxy hybrid nanocomposite was investigated under the nitrogen atmosphere on the TG 4000 instrument between 30 and 600 °C with a heating rate of 10 °C utilizing nitrogen gas a purge component.

16.2.4.2 Flame-Retardant Properties

The cone calorimeter experiments were performed on an FTT cone calorimeter equipment according to the ISO 5660-1 standard. The squared epoxy hybrid nanocomposite samples were covered with aluminum foil at the heat flux of 50 kW m^{-2}, and the mean value was measured from an average of three samples.

16.2.4.3 Mechanical Properties

The tensile experiment was applied to study the enhancement in the mechanical characteristics of the epoxy hybrid nanocomposites containing tensile strength and modulus. The tensile test was conducted using the Tinius Olsen universal testing machine, according to the ASTM D 3039 standard. The strain rate was determined with the extensometer and load cell under the 1 mm min^{-1} and 50 kN load cell and crosshead speed.

Uniaxial compression experiments on the epoxy hybrid nanocomposites were evaluated from the Tinius Olsen universal testing machine as per the ASTM D

695 standard. The flexural characteristics are conducted to examine the capacity of epoxy hybrid nanocomposite to withstand the deformation of load, and tests were performed according to the ASTM D 790 standard with a crosshead speed of 6 mm min^{-1}.

The impact strength tests were used to measure the materials' capacity to withstand sudden and shock loading. The impact energy values of the epoxy hybrid nanocomposites were performed as per the ASTM D 256 standard.

16.2.4.4 Morphological Properties

A field emission scanning electron microscope (FE-SEM) was applied to investigate the tensile fracture surface morphology, nanofillers' dispersion, and toughening mechanisms within the laminates. The specimen was sectioned from the broken tensile testing samples and coated with a sputtering machine using plasma ions before the microscopic examination.

16.2.4.5 Fourier Transform Infrared (FTIR) Spectra

Fourier transform infrared (FTIR) was applied to investigate the functional groups and chemical bonding in the fabricated epoxy hybrid nanocomposites and provide their successful functionalization. The spectrum was obtained from adequate particle samples of epoxy hybrid nanocomposites from the potassium bromide (KBr) pellet technique using the Perkin Elmer Spectrum 2, FTIR spectrometer. The spectra were achieved between 4000 and 400 cm^{-1} with a resolution of 1 cm^{-1}.

16.3 Results and Discussion

16.3.1 Effect on FTIR Spectra

The FTIR spectrum of BNF and BNF/TiO_2 epoxy hybrid nanocomposites is presented in Figure 16.1. It is probable to collect the details about the structure of chemicals and provide the appearance of particular chemical bonds and the nanofillers' degree of functionalization.

After the reinforcement of TiO_2 nanofillers in the epoxy polymer, lower vibrations under 1580 cm^{-1} related to aromatic bonds and a strong band under 1611 cm^{-1} were examined due to the stretching vibrations of the COH and aromatic ring functional groups [15]. The peak around 3450 cm^{-1} indicates the O–H stretch bonding to the presence of nanofillers with stronger interfacial adhesion [16]. Furthermore, the peak at 3450 cm^{-1} for BNF/TiO_2 epoxy composites transferred to the higher wavelength as TiO_2 was incorporated that is possibly due to the appearance of NH bonds from the metal oxide. After the reinforcement of BNF and TiO_2 nanoparticles, many new peaks were examined and are general for both nanofillers. Excessive peaks were determined under 1190 and 1004 cm^{-1} because of Si–O stretch bonding. A band at 1556 cm^{-1} can be because of the NH_2 vibrations and NH stretch bond. Furthermore, the peak around 690 cm^{-1} can be assigned to Si–CH_3 vibrations [17]. For BNF/TiO_2, in particular, a peak near 910 cm^{-1} characteristics to epoxy polymer and peaks at 2840 and 2920 cm^{-1} showed the C–H stretching vibration.

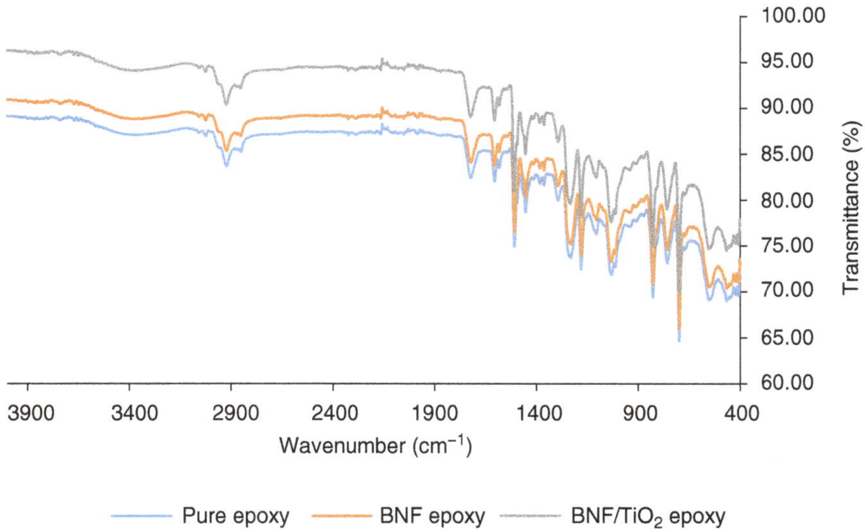

Figure 16.1 FTIR spectra of epoxy hybrid nanocomposites.

16.3.2 Effect on Flammability

The cone calorimeter is a broadly utilized method that could offer a quantitative investigation of the epoxy hybrid nanocomposites' combustion characteristics. The related data indicate the peak of the heat release rate, total heat release, time to peak of heat release rate, total smoke production, fire spread rate index, and ignition time. Figure 16.2 shows the heat release rate as the function of time for the neat epoxy polymer and their hybrid nanocomposites. It can be observed that the peak of heat release rate quantities of the epoxy hybrid nanocomposites consisting

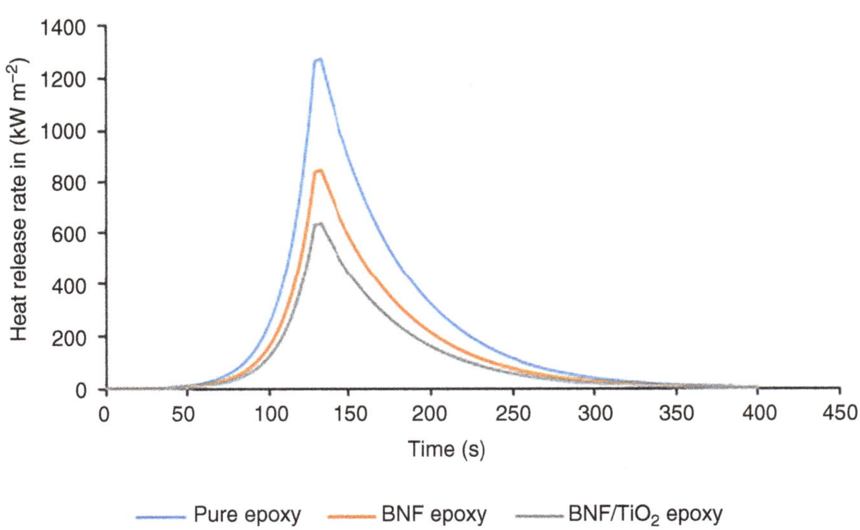

Figure 16.2 Heat release rate as a function of time for the experimented samples.

of BNF and TiO_2 nanoparticles is significantly reduced when compared with the pure epoxy resin. The virgin epoxy resin incinerated severely after the ignition process and attained a peak of heat release rate of 1270 kW m^{-2}. The peak of heat release rate quantity is decreased to 840 kW m^{-2} for the BNF epoxy composite, representing a 33.86% decrement compared with pure polymer. As examined earlier, graphene nanofiller acts as a protective shielding that reduces the evaporation of hazardous products during the heating process [18].

While comparing BNF/TiO_2, epoxy hybrid nanocomposites exhibit a further reduced peak of heat release rate of 635 kW m^{-2} related to the 50% decrement corresponding to the pure polymer. These results signify that BNF/TiO_2 shows higher effectiveness in sustaining the heat release rate assigned to char production's more considerable residue (Table 16.1). However, TiO_2 nanofillers' addition tends to slightly reduce time to ignition but enhanced the heat release rate. The incremented time to peak heat release rate is possibly assigned to char layers' production, which retard the period to the peak of the heat release rate. The fire spread rate index, which is similar to the proportion of peak heat release rate and time to peak of heat release rate, is generally applied to determine the nature of the fire spread of composite materials [19]. As observed in Table 16.1, BNF's reinforcement drastically reduces the fire spread rate index quantities of the fabricated epoxy hybrid nanolaminates. Remarkably, the fire spread index rate quantity of BNF/TiO_2 laminate is decreased to 4.25 kW m^{-2} from the net epoxy polymer (11.78 kW m^{-2} s^{-1}), simplifying the significant abolishment of hazards from the fire. Figure 16.3 provides the total heat release as the function of time for the pure epoxy resin and their hybrid nanolaminates. The pure epoxy polymer has a total heat release quantity of 92.4 MJ m^{-2} at the complete combustion process. The reinforcement of BNF and TiO_2 hybrids initially overcame the heat release rate, but the total heat release rate attains comparable quantities with a pure epoxy polymer.

Table 16.1 Information collected from cone calorimeter experiments for epoxy polymer and their respective composites.

	Time to ignition	Peak of heat release rate	Peak of heat release rate	Time to peak of heat release rate	Total heat release	Fire spread rate index	Total smoke production	Residue
Configuration	**Seconds**	**kW m^{-2}**	**Reduction in %**	**Seconds**	**MJ m^{-2}**	**kW m^{-2} s^{-1}**	**m^2**	**%**
Pure epoxy resin	58	1270	—	114	92.4	11.78	30.57	6.4
BNF epoxy	62	840	—	130	86.8	6.5	29.48	6.6
BNF/TiO_2 epoxy	68	635	33.86–50%	145	81.7	4.25	26.15	6.7

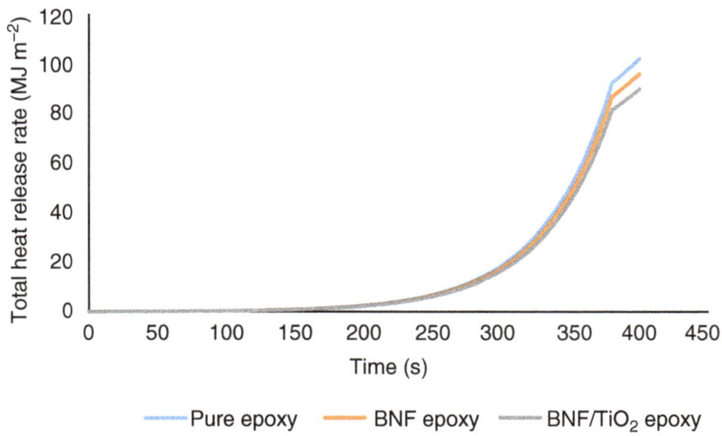

Figure 16.3 Total heat release rate as a function of time for the experimented samples.

This outcome is affected by BNF and TiO_2 nanofillers' addition, which encourages the production of layers of char after the ignition, and the layers of char get collapsed because of its thermal incapability. In this regard, incorporating BNF and TiO_2 nanofillers tends to decrease the total heat release rate to 81.7 MJ m^{-2}. The entire heat release rate scope could be ascribed with a cone calorimeter (Table 16.1). The percentage of residue for the epoxy polymer is 6.4%, and the addition of BNF and TiO_2 hybrids presents a slight modification of the yield of char. Furthermore, the char yield of BNF/TiO_2 increases to 1.7%, recommending the produced chars' enhanced thermal stability. The raised chars state that lower volatile decomposition components is developed to assist the flame and overcome the total heat release.

During accidents, smoke is one of the significant and essential parameters for the catastrophes. It can be observed from Table 16.1 that the adequacy of the total smoke production is similar to the entire heat release. The pure epoxy polymer has a total smoke production of 30.57 m^2 at the final stage of combustion, while the incorporation of TiO_2 reduces the full smoke production quantities. This smoke quench is assigned with the residual weight of the specimen after the combustion. The increment quantity of residual char states that lower decomposition of volatile compounds is produced due to the particles of smoke primarily developed from the volatile components. A similar principle has been examined in the graphene-reinforced poly(methyl methacrylate) laminates [20].

To analyze how char influences the flame retardancy of epoxy laminates and for the neat epoxy polymer, a layer of char with many cracks and holes is examined, and the aluminum foil is present because of the harmful possessive impact of this char content. BNF and BNF/TiO_2 epoxy nanocomposites show a lower residue of char, which has holes and cracks on their surface. In this regard, BNF/TiO_2 shows an intumescent layer of char and can be favorable to the higher quenching effectiveness on the heat release rate. As observed from heat release rate graphs (Figure 16.3), BNF/TiO_2 epoxy hybrid nanocomposites show a broad peak dissimilar to pure epoxy

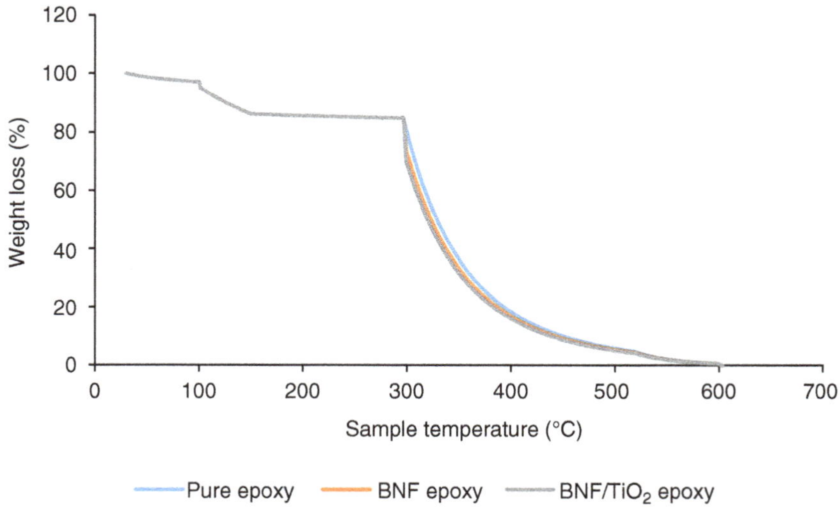

Figure 16.4 TGA curves of epoxy nanocomposites.

polymer, which signifies the development of the production of layers of char and avoids the destruction of layers, which leads to improved flame retardancy [21].

16.3.3 Effect on Thermal Stability

The thermogravimetric curves of pure epoxy, BNF epoxy, and BNF/TiO_2 epoxy hybrid nanocomposites are presented in Figure 16.4. The thermal stability of these epoxy-based nanolaminates is evaluated with the weight loss% as temperature. The examination of these outcomes signifies that incorporating TiO_2 nanofillers in the epoxy polymer resulted in the increment in the thermal stability of the developed hybrid nanolaminate. The onset of deterioration initiates at 370 °C for pure epoxy laminate while it is 375 and 380 °C for BNF and BNF/TiO_2 epoxy nanocomposites, respectively. This shows that the incorporation of TiO_2 nanofillers resulted in enhancing the laminate's thermal stability characteristics. The enhancement in the thermal stability characteristics of epoxy hybrid nanolaminates appears from the physical cooperation of TiO_2 and BNF with the epoxy polymer because of the principle that TiO_2 nanofiller has a greater modulus of elasticity when compared with pure epoxy polymer [22]. This leads to suppressing the deterioration of polymeric networks under a high range of temperatures through the initiation of the second region of TiO_2 nanofillers, which leads to stronger bonding within the polymeric systems.

16.3.4 Effect on Mechanical Properties

16.3.4.1 Tensile Properties

Figure 16.5a shows the tensile stress and strain of the fabricated epoxy nanocomposites. The curve exhibits a linear relation of stress with strain due to the brittle

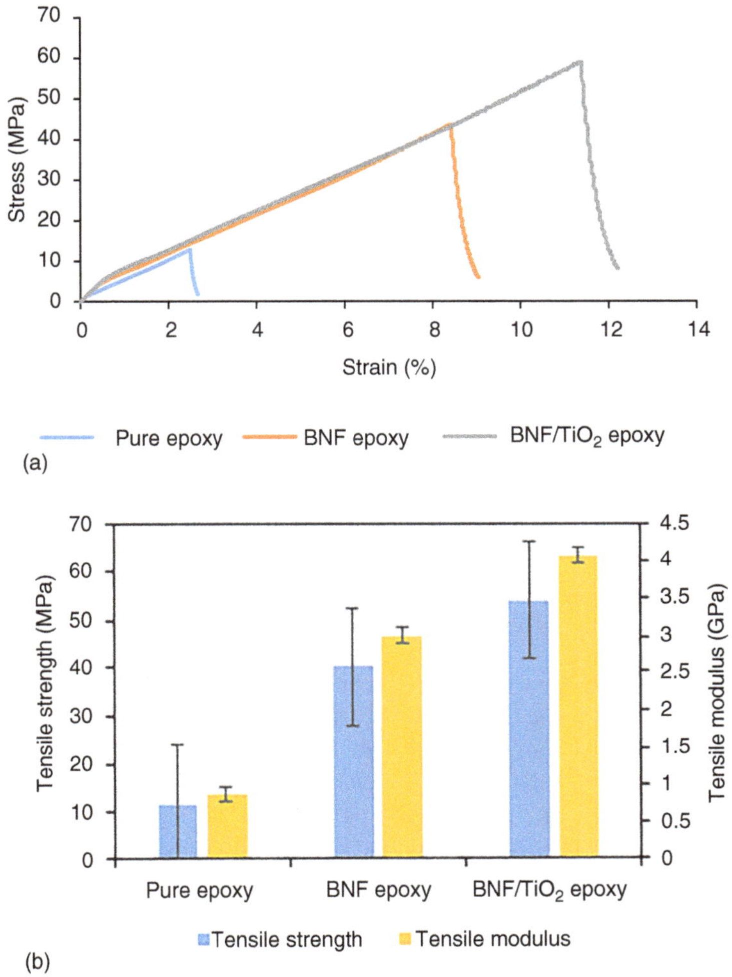

Figure 16.5 Tensile properties of epoxy nanocomposites: (a) stress–strain curves and (b) tensile strength and modulus.

fractural behavior, which is under strain as more than 8% for BNF epoxy laminate and around 12% strain for BNF/TiO_2 epoxy hybrid laminates. This signifies that TiO_2 nanofiller affects the brittleness of the polymer resulting in the increment of tensile characteristics before fracture. The evaluation of the collected outcomes (Figure 16.5b) indicates that through the addition of BNF nanoparticles in the pure epoxy polymer, there is a principle increment in the tensile strength that is around 40 MPa when compared with the tensile strength of pure epoxy of 11.78 MPa.

This outcome incorporates the principle that tensile strength is a filler-dependent characteristic [23]. Hence, there is a better adhesion within the fillers and epoxy matrix, which provides a good transfer of utilized forces from the weaker polymer matrix to the strong fillers, as it is observed from Figure 16.5b. The tensile strength also increases to 54 MPa for BNF/TiO_2 epoxy hybrid nanocomposites due to the

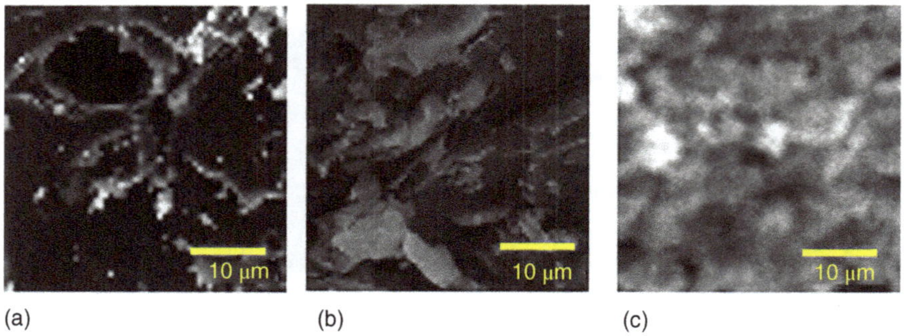

Figure 16.6 SEM micrographs of epoxy nanocomposites: (a) pure epoxy, (b) BNF epoxy, and (c) BNF/TiO_2 epoxy.

reinforcement of TiO_2 nanofillers making the stronger interfacial bonding, which is best among the other two laminates. This increment in tensile properties can be assigned to the principle that there is a uniform distribution of TiO_2 nanofillers in the epoxy matrix that assists in load transfer from fillers to matrix and resists propagating the propagation cracks.

16.3.4.2 Tensile Fracture

The epoxy polymer in the laminate shows better wettability with BNF's dispersion, shielding the damaged fibers from fracture and abrasion. Polymer's wettability with the nanofillers is an essential factor needed to develop the epoxy laminate [24]. This principle can be affirmed from SEM micrographs for pure epoxy laminate with dendrite surface, as depicted in Figure 16.6a.

The shielded fillers hence convey higher tensile characteristics than the neat epoxy polymer. The incorporation of TiO_2 nanofillers has concluded in the increment of tensile characteristics of fiber-reinforced polymer laminate. This increment in the tensile aspects of the epoxy hybrid nanolaminate can be assigned to the principle that the reinforcement of hard TiO_2 nanofillers to the soft polymer will increase stronger bonding within the matrix. This is observed in Figure 16.6b (SEM micrograph for BNF epoxy) exhibiting a pullout of filler with the lower lengths indicating the uniform brittle fracture. Also, the appearance of TiO_2 decreases the active regions for the production of bonds in the interfacial parts within the nanofillers and matrix. This straightly has a manner on the tensile properties that is a filler-dependent characteristic. This can be observed from SEM micrographs of Figure 16.6c, which shows the fiber's clean surface and crucial delamination bare of many polymeric matrix adherence.

16.3.4.3 Flexural Properties

The interpretation of flexural strength and modulus of the developed epoxy hybrid laminate presented in Figure 16.7 demonstrates the principle that the incorporation of TiO_2 nanofillers leads to the improvement in flexural characteristics of the epoxy hybrid nanolaminates because of the homogeneous distribution of TiO_2 nanofillers within the epoxy polymer, providing a more decided impact of hard TiO_2 nanofillers.

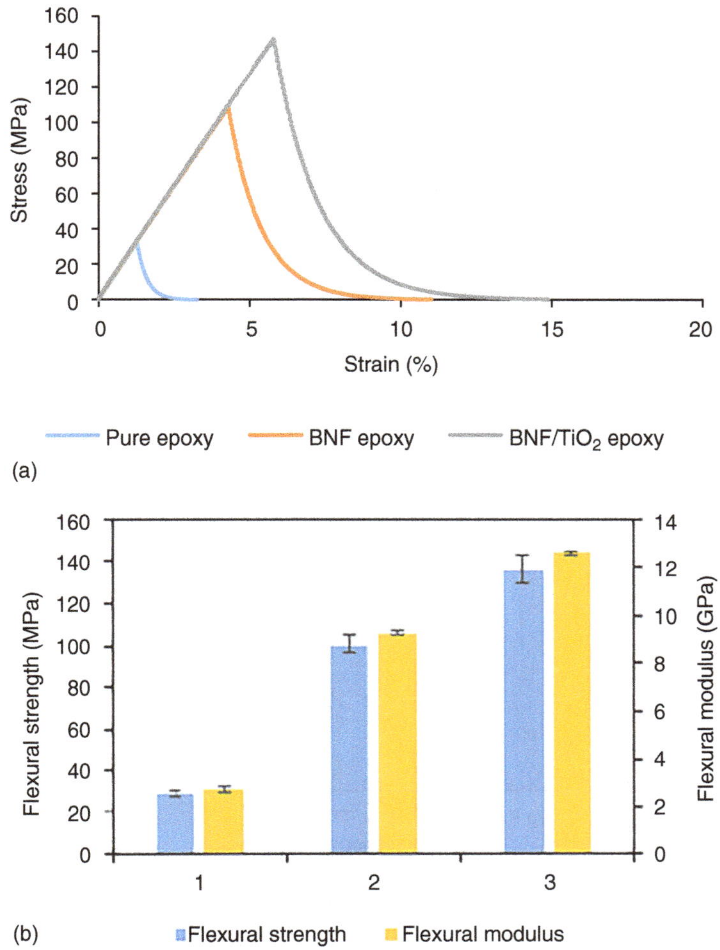

Figure 16.7 Flexural properties of epoxy nanocomposites: (a) stress–strain curves and (b) flexural strength and modulus.

This reason may be assigned in a brittle fracture of the epoxy hybrid laminate without enduring modulus of elasticity under the flexural forces. Examining the results indicates that the flexural properties are larger for BNF/TiO_2 epoxy than the other two types of laminates. This employs the well-known principle that the higher concentration of TiO_2 leads to accumulates, degrading the mechanical characteristics of laminate [25]. Besides, it is reported that the optimum reinforcement of TiO_2 and BNF nanofiller for the homogeneous dispersion within the polymeric resin is nearly lower for similar results [26, 27]. This is proved from the outcomes collected and exhibited here in this chapter.

16.3.4.4 Compression Properties

The compression characteristics of the epoxy hybrid laminates with BNF and BNF/TiO_2 nanofillers were investigated, and the compression stress–strain graphs are shown in Figure 16.8a.

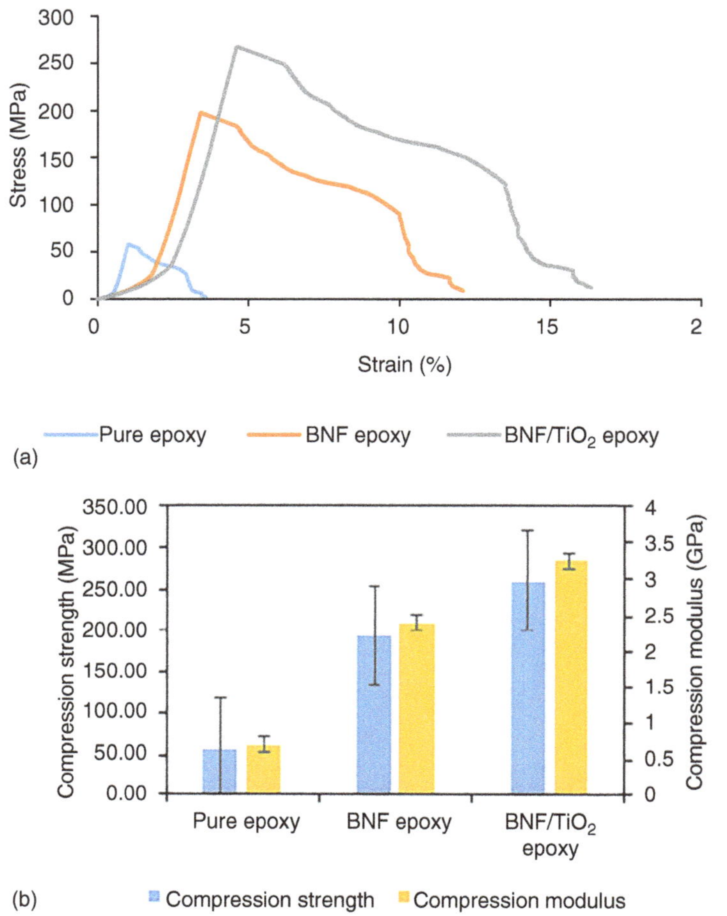

Figure 16.8 Compression properties of epoxy nanocomposites: (a) stress–strain curves and (b) compression strength and modulus.

The specimen's compression strength and modulus were employed as the pressure under failure, whereas the modulus was used identical to the starting slope of the related compression stress–strain graphs crossing through the range. The outcomes are summarized in Figure 16.8b. For neat epoxy polymer, the tensile characteristics had a mean value of 41.15 MPa, and this quantity then incremented to attain a higher value of 51.87 MPa after reinforcing with BNF nanofillers. It was examined that the modification of the TiO_2 nanofillers further enhances the compression characteristics. The utilization of TiO_2 nanofiller increased the tensile properties by 10.1% when compared with BNF epoxy laminate. This state that the addition of nanofillers improves the interfacial adhesion between epoxy polymer and nanofillers. Hence, when the polymer comprises the nanofillers, and the hardener is incorporated, many competing reactions may be considered. The primary reaction is the hardener and epoxy-curing reactivity under a higher range of temperatures to create a large cross-networked polymeric chain. It is sought to combine the epoxy polymeric network with the TiO_2 nanofillers; this base reaction can also be occurring in correspondingly within the reaction of rings of epoxy in the TiO_2 and BNF and the hardener producing a contending reaction with the primary curing method. BNF/TiO_2 nanofillers are also combined with other products in uniform dispersion in different regions of the complete composite. In turn, this uniformity will produce damages in the primary structure and result in the increment of polymer's compression characteristics [28].

16.3.4.5 Impact Strength

The examination of the impact properties (Figure 16.9) signifies that the impact strength of the developed epoxy hybrid laminate improves with TiO_2 and BNF nanofillers' addition. Furthermore, the impact strength of the neat epoxy laminate showed a lower value of impact strength. The impact strength of BNF epoxy nanocomposite was more bass than the laminate with BNF/TiO_2 nanofillers. Impact strength of filler-based polymer nanocomposites is based mainly on the interfacial shear characteristics between the matrix and the filler. A third region's appearance in the form nano-based fillers permits the load transfer from resin to TiO_2 nanoparticles and then onto the BNF [29]. There is an increase in the impact strength of the epoxy hybrid nanocomposite with the BNF and TiO_2 nanofillers when compared with the neat epoxy polymer [30], owing to the uniform dispersion and an enhanced wettability of the TiO_2 nanofillers with the polymer permitting for an efficient transfer of load from the polymer to the nanofillers, leading in the higher energy absorption before the damage. The epoxy polymer is intrinsic brittle and provides a good shielding to crack [31, 32]. The appearance of TiO_2 nanofiller within the frail epoxy polymer concludes in strengthening the polymer [33, 34] that, in turn, enhances the impact characteristics of the epoxy polymer nanolaminates.

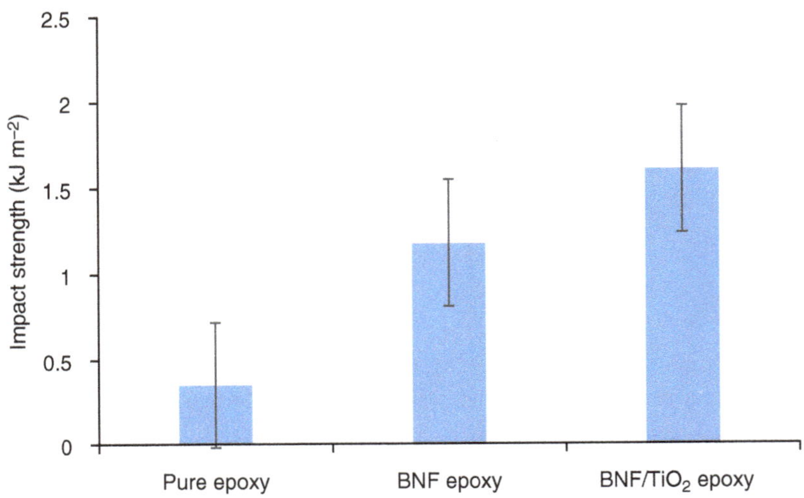

Figure 16.9 Impact strength of epoxy nanocomposites.

16.4 Conclusion

In this chapter, the effect of BNF and TiO_2 nanofiller hybrids on flammability, thermal stability, and mechanical characteristics of epoxy hybrid nanocomposites was studied. Pure epoxy, BNF epoxy, and BNF/TiO_2 epoxy nanocomposites were fabricated from the ultrasonic-assisted wet layup technique. The chemical structure and bonding were determined spectra from FTIR, which affirmed the successful fabrication of epoxy hybrid nanocomposites. The SEM micrographs showed that the TiO_2 and BNF were uniformly reinforced in the epoxy polymer nanocomposites. However, TiO_2 nanofillers' appearance could prevent the accumulation of the BNF while distributing with the epoxy polymer. The incorporation of both the filler's reinforcement slightly influenced the thermal characteristics of the developed epoxy laminates. As forecasted, TiO_2 and BNF nanofillers' addition reduced the flammability of epoxy hybrid nanolaminates examined from cone calorimeter experiments. Among the three types of epoxy composites, the BNF/TiO_2 epoxy hybrid nanocomposite displayed the highest decrement in the peak of heat release rate, total production of smoke, and complete heat release by nearly 50%, 29%, and 14%, respectively, when compared with pure epoxy. The flame-retardant characteristic of the BNF/TiO_2 epoxy hybrid nanocomposite was assigned to the most considerable residual yield value, which was handled as a protection shield to suppress the smoke and heat during the combustion analysis. TiO_2 nanofillers' reinforcement will conclude in the higher thermal stability of the epoxy hybrid nanocomposites, permitting them to be applied in different higher temperature applications. The tensile and flexural characteristics of pure epoxy laminate are lower than the other epoxy nanocomposites, indicating that BNF and TiO_2 nanofillers' reinforcement has a positive impact on these characteristics. The appearance of TiO_2 nanofillers solution in the increment of significant regions for the production of bonds between BNF and epoxy polymer leads to the increased filler-dependent features like flexural and tensile characteristics.

Furthermore, the incorporation of TiO_2 nanofillers in the epoxy laminate transmits good toughness to the polymer by improving the BNF/TiO_2 epoxy hybrid's impact strength nanocomposites. These results indicate that TiO_2 has a higher impact with the epoxy polymer that concludes in an increment in impact strength, which is a polymer-dependent characteristic. Thus, from the overall mechanical and thermal features, it is observed that the addition of BNF and TiO_2 nanofillers leads to higher-performance epoxy hybrid nanocomposites.

References

1 Kumar, S., Purohit, R., and Malik, M.M. (2015). Properties and applications of polymer matrix nano composite materials. *Materials Today: Proceedings* 2: 3704–3711.

2 Rong, M.Z., Zhang, M.Q., and Ruan, W.H. (2006). Surface modification of nanoscale fillers for improving properties of polymer nanocomposites: a review. *Materials Science and Technology* 22: 787–796.

3 Ghafoor, B., Mehmood, M.S., Shahid, U. et al. (2016). Influence of γ-ray modified MWCNTs on the structural and thermal properties of high-density polyethylene. *Radiation Physics and Chemistry* 125: 145–150.
4 Amin, M. and Ali, M. (2015). Polymer nanocomposites for high voltage outdoor insulation applications. *Reviews on Advanced Materials Science* 40: 276–294.
5 Riggs, B.C., Adireddy, S., Rehm, C.H. et al. (2015). Polymer nanocomposites for energy storage applications. *Materials Today: Proceedings* 2: 3853–3863.
6 Kong, Q.H., Wu, T., Zhang, J.H., and Wang, D.Y. (2018). Simultaneously improving flame retardancy and dynamic mechanical properties of epoxy resin nanocomposites through layered copper phenylphosphate. *Composites Science and Technology* 154: 136–144.
7 Zhang, J.H., Kong, Q.H., and Wang, D.Y. (2018). Simultaneously improving the fire safety and mechanical properties of epoxy resin with Fe-CNTs via large-scale preparation. *Journal of Materials Chemistry A* 6: 6376–6386.
8 Shabestari, M.E., Kalali, E.N., Gonzalez, V.J. et al. (2017). Effect of nitrogen and oxygen doped carbon nanotubes on flammability of epoxy nanocomposites. *Carbon* 121: 193–200.
9 Hsiue, G.H., Liu, Y.L., and Liao, H.H. (2001). Flame-retardant epoxy resins: an approach from organic–inorganic hybrid nanocomposites. *Journal of Polymer Science, Part A: Polymer Chemistry* 39: 986–996.
10 Wang, X., Kalali, E.N., Wan, J.T., and Wang, D.Y. (2017). Carbon-family materials for flame retardant polymeric materials. *Progress in Polymer Science* 69: 22–46.
11 Perivoliotis, D.K. and Tagmatarchis, N. (2017). Recent advancements in metal-based hybrid electrocatalysts supported on graphene and related 2D materials for the oxygen reduction reaction. *Carbon* 118: 493–510.
12 Song, E.H., Wen, Z., and Jiang, Q. (2011). CO catalytic oxidation on copper-embedded graphene. *Journal of Physical Chemistry C* 115: 3678–3683.
13 Li, Y., Zhou, Z., Yu, G. et al. (2010). CO catalytic oxidation on iron-embedded graphene: computational quest for low-cost nanocatalysts. *Journal of Physical Chemistry C* 114: 6250–6254.
14 Hu, C., Lu, T., Chen, F., and Zhang, R. (2013). A brief review of graphene–metal oxide composites synthesis and applications in photocatalysis. *Journal of the Chinese Advanced Materials Society* 1: 21–39.
15 Mrówczynski, R., Jurga-Stopa, J., Markiewicz, R. et al. (2016). Assessment of polydopamine coated magnetic nanoparticles in doxorubicin delivery. *RSC Advances* 6: 5936–5943.
16 Ma, Y.R., Zhang, X.L., Zeng, T. et al. (2013). Polydopamine-coated magnetic nanoparticles for enrichment and direct detection of small molecule pollutants coupled with MALDI-TOF-MS. *ACS Applied Materials & Interfaces* 5: 1024–1030.
17 Chang, B.P., Akil, H.M., Nasir, R.M., and Nurdijati, S. (2011). Mechanical and antibacterial properties of treated and untreated zinc oxide filled UHMWPE composites. *Journal of Thermoplastic Composite Materials* 24: 653–667.
18 Pour, R.H., Soheilmoghaddam, M., Hassan, A., and Bourbigot, S. (2015). Flammability and thermal properties of polycarbonate/acrylonitrile–butadiene–styrene nanocomposites reinforced with multilayer graphene. *Polymer Degradation and Stability* 120: 88–97.

19 Wang, X., Zhou, S., Guo, W.W. et al. (2017). Renewable cardanol-based phosphate as a flame retardant toughening agent for epoxy resins. *ACS Sustainable Chemistry & Engineering* 5: 3409–3416.

20 Huang, G.B., Chen, S.Q., Song, P.G. et al. (2014). Combination effects of graphene and layered double hydroxides on intumescent flame-retardant poly(methyl methacrylate) nanocomposites. *Applied Clay Science* 88-89: 78–85.

21 Kalali, E.N., Guo, W., Wang, X. et al. (2020). Effect of metal-based nanoparticles decorated graphene hybrids on flammability of epoxy nanocomposites. *Composites Part A: Applied Science and Manufacturing* 129: 105694.

22 Ramezanzadeh, B., Attar, M.M., and Farzam, M. (2011). A study on the anticorrosion performance of the epoxy–polyamide nanocomposites containing ZnO nanoparticles. *Progress in Organic Coating* 72: 410–422.

23 Gojny, F.H., Wichmann, M.H., Fiedler, B. et al. (2004). Carbon nanotube-reinforced epoxy-composites: enhanced stiffness and fracture toughness at low nanotube content. *Composites Part A: Applied Science and Manufacturing* 36: 2363–2371.

24 Chang, C.J., Tsai, M.H., Hsu, Y.H., and Tuan, C.S. (2008). Morphology and optoelectronic properties of ZnO rod array/conjugated polymer hybrid films. *Thin Solid Films* 516: 5523–5526.

25 Peng, G., Li, Q., Yang, Y. et al. (2008). Degradation of nano ZnO-glass fiber-unsaturated polyester composites. *Polymers for Advanced Technologies* 19: 2128–2133.

26 El Saeed, A.M., El-Fattah, M.A., and Azzam, A.M. (2015). Synthesis of ZnO nanoparticles and studying its influence on the antimicrobial, anticorrosion and mechanical behavior of polyurethane composite for surface coating. *Dyes and Pigments* 121: 282–289.

27 Halder, S., Prasad, T., Khan, N.I. et al. Superior mechanical properties of poly vinyl alcohol-assisted ZnO nanoparticle reinforced epoxy composites. *Materials Chemistry and Physics* 192: 198–209.

28 Mohit, H. and Selvan, V.A.M. (2020). Effect of Al-SiC nanoparticles and cellulose fiber dispersion on the thermomechanical and corrosion characteristics of polymer nanocomposites. *Polymer Composites* 41 (5): 1878–1899.

29 Jancar, J., Douglas, J.F., Starr, F.W. et al. (2010). Current issues in research on structure–property relationships in polymer nanocomposites. *Polymer Journal* 51: 3321–3343.

30 Sanes, J., Carrión, F.J., and Bermúdez, M.D. (2010). Effect of the addition of room temperature ionic liquid and ZnO nanoparticles on the wear and scratch resistance of epoxy resin. *Wear* 268: 1295–1302.

31 Turk, M., Hamerton, I., and Ivanov, D.S. (2017). Ductility potential of brittle epoxies: thermomechanical behaviour of plastically-deformed fully-cured composite resins. *Polymer Journal* 120: 43–51.

32 Barroso, A., Correa, E., Freire, J., and París, F. (2018). A device for biaxial testing in uniaxial machines. Design, manufacturing and experimental results using cruciform specimens of composite materials. *Experimental Mechanics* 58: 49–53.

33 Gu, X., Zhang, Z., Yuan, L. et al. (2016). Developing high performance cyanate ester resin with significantly reduced postcuring temperature while improved toughness, rigidity, thermal and dielectric properties based on manganese-Schiff base hybridized graphene oxide. *Chemical Engineering Journal* 298: 214–224.

34 Hulugappa, B., Achutha, M.V., and Suresha, B. (2016). Effect of fillers on mechanical properties and fracture toughness of glass fabric reinforced epoxy composites. *Journal of Minerals and Materials Characterization and Engineering* 4: 1–14.

17

Influence of Chemical Treatments on the Thermal Properties of Natural Fiber-Reinforced Hybrid Composites (NFRHC)

Rafael de Avila Delucis[1] and José Humberto S. Almeida Jr[2,3]

[1] *Federal University of Pelotas, Engineering Centre, Rua Benjamin Constant 989, Pelotas 96010-020, Brazil*
[2] *Aalto University, Department of Mechanical Engineering, Otakaari 4, Espoo, 02150, Finland*
[3] *Advanced Composites Research Group, School of Mechanical and Aerospace Engineering, Queen's University Belfast, Belfast, UK*

17.1 Introduction

Natural fibers have been widely used for composite materials since ancient times when straws and other vegetable fibers were used allied to bricks and other construction and building materials [1]. There is a growing awareness regarding the need to restrict the use of non-renewable resources, which implies several concerns, like the feasibility of certain environmentally friendly composite materials [2, 3]. Current environmental laws of several countries have tighten restrictions of several industrial activities, and thus researchers and producers have sought to optimize the exploitation of plant matrices to extract natural fibers capable of replacing traditional synthetic fibers in order to reduce the consumption of oil derivatives [4–6]. Therefore, sustainable solutions can reduce negative impacts caused by the overuse of oil products, such as the depletion of natural resources and landfill abilities.

Composites reinforced with natural fibers are progressively used in engineering applications, but mainly in automotive, civil fields, furniture, sporting goods, but in non-primary structural components [7, 8]. Regarding automotive industries, recent projects have been developed on natural fiber-reinforced polymer composites. Depending on the type of applications, proper selections of fibers, matrix, additives, and manufacturing processes can result in tailored properties for these composites [9]. In addition, the pretreatment process of fiber plays a key role as it controls the interfacial bonding properties and it can positively affect both mechanical and thermal responses of the composite system [5]. Natural fibers can be derived from mineral, vegetal, or animal resources. Among the plant-based ones, there are fibers extracted from leaves (curaua, pineapple, sisal, and abaca), cores (hemp, jute, and kenaf), grasses and reeds (wheat, corn, and rice), seeds (cotton, kapok, and coir), and basts (flax, jute, hemp, ramie, and kenaf) [4, 6].

Natural Fiber-Reinforced Composites: Thermal Properties and Applications, First Edition.
Edited by Senthilkumar Krishnasamy, Senthil Muthu Kumar Thiagamani,
Chandrasekar Muthukumar, Rajini Nagarajan and Suchart Siengchin.

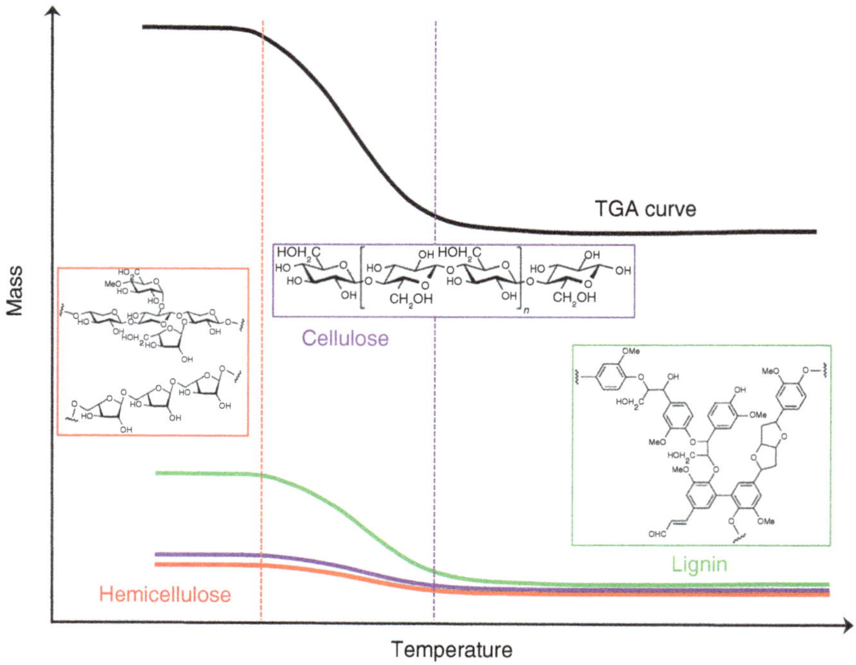

Figure 17.1 Typical thermogravimetric curves showing the main components of lignocellulosic fibers and their identification.

These fibers are also called as lignocellulosic fibers since they are majorly composed of cellulose, hemicellulose, and lignin, although there are some minor components, including waxes, pectins, and extractives [10]. Cellulose is a linear homopolysaccharide composed of glucose units connected to each other by β-1,4-glycosidic bonds, producing large chains that tightly pack in crystalline microfibrils [11]. Hemicelluloses are composed of different monosaccharides, including anhydro-β-(1-4)-D-xylopyranose, glucopyranose, mannopyranose, and galactopyranose units [12]. Lignin is an amorphous polymer formed by aromatic and aliphatic components derived from an enzyme-initiated dehydrogenative polymerization of three primary precursors: trans-coniferyl, *trans-p*-coumaryl, and *trans*-sinapyl. Lignin is found in the internal spaces of the plant cell wall, such as coupled to pectin, hemicellulose, and cellulose [13]. The pectin is an important component for its key role in binding the fibers to form fiber bundles. The amount of plant tissues represents the variable pectin contents in their cell wall, especially fruit peels and gums [14]. Waxes are rich in water-soluble alcohols, which are also soluble in some acids, for instance, phenolic, oleaginous, and stearic acid [13]. Nevertheless, more than 95% of the composition of the majority of natural fibers is hemicellulose, cellulose, and lignin, as shown in Figure 17.1.

Independently on the fiber, it can be generalized that the three major components of the natural fibers drive the behavior of the entire system. For instance, the mechanical performance of vegetable fibers is mainly dependent on cellulose content. The hemicelluloses are the least crosslinked macromolecule and, because of that, has the lowest molecular weight, the highest density of polar functional groups, and the smaller thermal stability. This means that, in order to modify the thermal behavior and stability of natural fibers, the hemicellulose should be the targeted component. Hemicelluloses are also hydrophilic compounds, that is, the higher the hemicellulose content in the natural fiber, the earlier the decomposition temperatures and the higher the moisture uptake [11, 15]. On the other hand, the lignin is mainly responsible to improve biological degradation and thermal stability.

Comparing to synthetic fiber composites (mostly glass-fiber-reinforced polymers, GFRP), composites incorporated with natural fibers have some benefits, such as biodegradability, low density, low cost, and both low thermal and acoustic conductivities [16]. On the other hand, they have lower thermal and mechanical properties. Therefore, a natural synergetic comes up when replacing, at least partially, glass fibers with natural-based ones in order to find a balance looking for exploring the potentialities of each fiber toward more environmentally friendly materials yet functional in terms of thermal and mechanical aspects.

In terms of material resources, thermoplastic composites reinforced with natural fibers can be classified into two major groups: the non-degradable but recyclable synthetic thermoplastics and the biodegradable thermoplastics, or the "green" composites [11]. The most common synthetic commodity thermoplastic materials are acrylonitrile butadiene styrene (ABS), polyethylene (PE), polycarbonate (PC), polyamide (nylon), high impact polystyrene (HIPS), and polypropylene (PP). Among them, PP has the highest consumption since it is inexpensive, light, highly processable, resistant to high temperature, and dimensionally stable and can have high stiffness, strength, impact, and electrical properties [17]. Commonly used biodegradable thermoplastics include polylactide (PLA), polyglycolide or poly(glycolic acid) (PGA), poly-β-hydroxyalkanoates, and poly(ε-caprolactone) (PCL). Glass transition temperature (T_g) and melting temperature (T_m) of some polymers are shown in Table 17.1. All the mentioned polymeric matrices can be processed with natural fibers without degrading the fibers, which may ensure suitable fibers/matrix interfacial adhesions since they have melting temperatures below 200 °C.

Table 17.1 Thermal and physical properties of some thermoset and thermoplastic polymers.

Polymer	T_g (°C)	T_m (°C)	Density (kg m^{-3})
PP	−10 to −23	160–175	0.90–0.92
HDPE	−110	126–135	0.95
PS	100	110–135	1.04–1.09
PLA	55-65	120–175	1.21–1.25
PCL	−60 to −65	58-65	1.11–1.15
PHB	5–15	168–182	1.18–1.26
Epoxy	170–300	—	1.20–1.30
Polyester	130–160	—	1.11–1.40
Phenolic	175	—	1.20–1.30

Sources: Based on Crosky et al. [18]; Gurunathan et al. [13]; Panthapulakkal et al. [11].

Thermoset polymers are resins that generate crosslinks during curing. Crosslinking is induced either by the addition of curing agents or by heating. Consequently, thermoset polymers usually have higher strength and stiffness than thermoplastics, which make them attractive even though their crosslinking chains cannot be broken. The most common thermosets are polyester, epoxy, vinyl ester, and phenolic. Similar to thermoplastic polymers, some biodegradable resins have been recently developed, which are biobased polyurethane (PU), biobased phenolic, and lactic acid (LA) [18]. Epoxy resins are the most employed as a matrix for hybrid composites with natural fibers, but their being widespread in commercial products is unlikely due to their high cost. Thereby, less costly resins, such as polyester and vinyl ester, are more normally employed in natural fiber hybrid composites. Phenolic resins have excellent fire resistance, and their cost is in between epoxies and polyesters, but their processability is very difficult.

Several studies have evaluated both thermoset and thermoplastic polymer matrices with natural and synthetic fibers. In those that specially focused on those hybrid composites with natural fibers, it is common to apply chemical treatments onto natural fibers to enhance their compatibility with both the glass fiber and mainly the polymeric matrix. This can be achieved through chemical treatments or coupling agents onto the fiber [19, 20]. For example, when hybridized with glass fibers, their composites with the following fibers present higher thermal stability: bamboo fiber [21], hemp fiber [17], kenaf [22], curaua [4], and sisal [23]. In general, the thermal performance of hybrid composites is driven by fiber content, fiber length, fiber orientation, fiber/matrix interface, stacking sequence, and the failure strain of each fiber [24].

The major drawbacks related to a more widespread dissemination of natural fibers in commercial products are mainly related to their high moisture absorption and heterogeneity in their microstructure, which generates a quite high scattering in their properties. Moreover, the hydrophilic fibers and the hydrophobic matrix may have an inherent incompatibility, which is a key limitation and provides an impoverished matrix/fiber stress transfer stress [1].

Furthermore, pectin and waxes may cover the fiber cell wall and some of its reactive functional groups, which hinders its wetting with the matrix when liquid composite molding is processed. Once some important chemical groups from the fiber cell wall are inaccessible, a poor adhesion across the phase boundary may be developed and the matrix-reinforced coupling becomes injured.

Fiber treatment is, indeed, the preferred route looking for overcoming these issues. Chemical, physical, and biological approaches are utilized to improve the fiber/matrix interface. For instance, there are biological methods that use bacteria that can induce enzymatic mechanisms onto the fiber cell wall, leading to increases in the fiber/matrix interfacial adhesion. There are also physical methods that can modify the surface properties of the fibers. Although these treatments do not modify structural components of the fibers, enhancements in the fiber/matrix adhesion may be achieved. Stretching, calendering, cold plasma, electric discharge, and thermos-treatments are the most known physical procedures. However, physical treatments are far the most spread and efficient method to treat natural fibers.

Chemical treatments change the structure of the fibers, improving fiber/matrix interfacial adhesion. Several chemicals have been used, for instance, alkali, silane, acetic anhydride, and peroxides. Changes in crystalline structure and removal of amorphous and low-molecular-weight compounds (such as hemicelluloses) from the fiber surface are intensively attributed to these treatments in the literature. The called alkalinization is the most common treatment. Several alkali solutions have been investigated, but sodium hydroxide (NaOH) is clearly the main treatment solution since it is a cheap, simple-to-prepare, and an effective pathway to induce those known effects ascribed to chemical treatments. According to Da Silva et al. [5], the NaOH can also be combined with sodium borohydride ($NaBH_4$), which may act as a reducing agent for the end-group (i.e. aldehyde) present in the C-1 free polysaccharides chain at the high pHs reached during this treatment. This modified alkaline treatment yielded increases in the interfacial strength of the natural fiber hybrid composite.

17.2 Chemical Modifications for Natural Fibers Applied in Hybrid Composites

17.2.1 Chemical Treatments

The main goal of a chemical modification onto natural fibers is at improving the interfacial strength between the fiber and the matrix. The different chemical agents for the treatments include alkali, acetylation, acrylation, benzoylation, isocyanate, sodium chlorite, permanganate, peroxide, and stearic acid [25]. The influence of surface treatments onto natural fibers in hybrid composites is presented next.

17.2.1.1 Chemically Treated Natural Fibers in Hybrid Thermoset Composites

Chemical modifications mostly change the microfibrillar structure of natural fibers, removing some amorphous compounds from the fiber surface, which changes the entire bulk behavior. Brittleness of the treated fibers, for instance, can be significantly reduced if compared to pristine natural fibers [26]. Grafted oil palm fibers (OPF) showed such thermal behavior that indicated that grafting leads to increased thermal stability.

Mishra et al. [27] subjected sisal and pineapple leaf fiber (PALF) to 5% of alkali treatment for one hour and then manufactured hybrid composites with sisal/pineapple leaf/glass fiber-based polyester and found that water uptake was considerably lower than that of the individual fiber-based composite. Venkata Reddy et al. [28] analyzed the thermal behavior of alkali-treated kapok/sisal hybrid composites using differential scanning calorimetry (DSC). Their results indicated that, compared to untreated ones, the hybrid composites reinforced with treated fibers showed a slight increase in melting temperature.

Dhakal et al. [29] evaluated the thermal stability of hemp fiber-reinforced composites using NaOH-treated fibers and hybrid composites incorporated with glass fibers. This research also dealt with surface energy features of polyester-based composites molded by hot compression, as well as the effects of fiber treatment and glass fiber hybridization on the wetting behavior. Compared to the untreated hemp/polyester composites, that hybrid composite reinforced with NaOH-treated hemp fibers and glass fibers presented both increased thermal stability and increased char yield, as well as a lower thermodecomposition rate. Moreover, the hybrid composited presented only one step in its thermal degradation profile, and meanwhile, that single hemp fiber-reinforced composite showed a three-step decomposition process.

Bavan and Kumar [30] investigated maize stalk fibers (pulverized form) and jute fibers (mat form) with a biopolymer of epoxy soya bean oil as the used matrix to produce a biocomposite material. The composite was processed through hot compression molding. Thermal degradation analysis was studied by thermogravimetric analysis (TGA). An alkali chemical treatment with 5% of NaOH wt was applied onto the fibers. They separated the decomposition process into two stages: the initial stage comprises the removal of moisture and few cellulose elements and the second stage consists of the removal of cellulose, hemicelluloses, lignin, pectin, waxes, and alkali segments on the fiber surface. The initial (stage I) and final (stage II) degradation temperature for raw fibers was 190–310 and 310–410 °C, respectively. For the hybrid biocomposites, the initial and final degradation starts from 180–280 and 280 to 400 °C, respectively. In summary, 5% of NaOH w.t. improved the thermal properties of both fiber and composite.

Shanmugam and Thiruchitrambalam [31] investigated thermomechanical features of a hybrid composite of Palmyra palm leaf stalk (PPLSF) and jute fibers. The PPLSF fibers were alkali-treated and arranged as two continuous unidirectional layers. Moreover, an unsaturated polyester resin was used as a matrix. Compared to the untreated fibers, the treated ones yielded an improved interfacial adhesion, which led to a lower tan delta peak in a dynamic mechanical profile.

Shanmugam et al. [32] further investigated PPLSF fibers treated using different treatment solutions, including alkali, benzoyl chloride, and potassium permanganate. These fibers were incorporated into two kinds of polyester-based composites. In the first, continuous PPLSF were unidirectionally aligned. The other one was manufactured by hybridizing non-woven glass fiber mats and bilayered continuous unidirectional alkalinized PPLSF. Compared to that composite incorporated with untreated fibers, both the storage and loss moduli were higher for those hybrid composites with chemically treated fibers.

Jayamani et al. [33] applied an unsaturated polyester resin as a matrix for the manufacturing of jute/bamboo hybrid composites. Moreover, a sodium hydroxide (NaOH) solution was used as a treatment solution for both these fibers, which promoted a surface modification. These authors reported both increase in interfacial adhesion and decreases in void content, dielectric constant, and water uptake, which were ascribed to the chemical reactions that took place onto the fiber surface. As expected, these results indicate that the alkalinization was able to confer an increased hydrophobicity to both the studied lignocellulosic fibers.

Jayamani et al. [34] investigated betelnut fiber-reinforced polyester composites. Before the processing of the composites, these fibers were alkali-treated. As a result, the treated fibers had higher thermal stability than the untreated ones. Jawaid et al. [35] studied some thermal features of oil palm empty fruit bunch (EFB) and jute fibers, which were pretreated and then incorporated as reinforcements in epoxy-based hybrid composites. The TGA revealed that the thermal stability of the hybrid composite produced with treated fibers overcame that same property obtained for composites incorporated with pristine fibers. This treatment was carried out using a simple immersion in 2-hydroxy ethyl acrylate (2-HEA).

Mohan and Kanny [36] studied nanoclays into banana fibers treated by alkalinization (NaOH). These authors addressed some effects of the incorporated of these reinforcements on thermomechanical features of epoxy hybrid composites. That composite with pristine banana fibers presented a weight loss of about 12% when the temperature reached 100 °C. In the same condition, the other composites did not overcome the weight loss of 8%, which included those composites incorporated with NaOH-treated banana fibers and both alkali-treated banana fibers and nanoclays. This probably occurred due to the reduction in moisture content and removal of low-molecular-weight compounds since both these mechanisms were ascribed to the studied chemical treatment. In a comparison between the hybrid nanoclay-reinforced composite and that traditional composite reinforced with alkali-treated banana fiber, the nanocomposite showed a smaller mass loss rate. This may be ascribed to the clay nanolayers, which probably acted as a barrier medium.

Anand and Anbumalar, [37] investigated thermal characteristics of epoxy composites reinforced with both cellulose and hemp fibers. Fiber volume fraction and fiber surface modification were taken into account as the main factors in this study. An aqueous sodium hydroxide (NaOH) treatment was applied onto the hemp fiber, which induced a certain mechanism, in which hydroxyl groups were ionized to the alkoxide. A water–ethanol solution (4 : 1) was prepared (6 wt% of NaOH) and used to perform the treatments on fiber mats, which were successively soaked one by one. After being immersed, the fiber mats were then left undisturbed for 180 minutes. Then, the mats were washed and oven-dried. The final composites with treated showed higher thermal stability than the non-untreated ones.

Anand and Anbumalar [37] also employed a benzoylation treatment onto the fibers. Benzoyl chloride can be considered one of the most important treatment solutions since it promotes remarkable changes in organic synthesis. Benzoyl chloride is comprised of benzoyl ($C_6H_5C = O$), which contributes to the increased hydrophobic nature when a natural fiber is chemically modified. The alkalinization was performed to induce activation of hydroxyl groups from some macromolecules (including cellulose and lignin) and then this fiber was immersed in a benzoyl chloride solution with 10% NaOH. The isolated fibers were then soaked in ethanol for one hour to remove the benzoyl chloride, which was then washed and oven-dried. Compared to untreated natural fibers, this treatment may promote better compatibility of treated reinforcements and hydrophobic polymeric matrices. The benzoylation is capable of improving some adhesion features, as well as strength, water absorption, and thermal stability of different kinds of polymer composites. According to these authors, this treatment yielded more thermally stable composites than those composites manufactured with pristine fibers.

Neto et al. [38] investigated the thermal performance of three hybrid composites produced with different combinations of jute, sisal, ramie, and curauá fibers. These fibers were alkalized and also separately alkalized prior to being silanized. In all, more thermally stable composites were produced with the aid of the chemically treated natural fibers. Arulmurugan et al. [39] developed hybrid composites using aloe vera, flax, within epoxy resin treating the fibers with barium sulfate ($BaSO_4$). The influences of $BaSO_4$ on the thermal properties were assessed. The chemical treatment with $BaSO_4$ led to more thermally stable hybrid composites than that composited made with untreated fibers.

17.2.1.2 Chemically Treated Fibers in Hybrid Thermoplastic Composites

Gyoung et al. [40] evaluated some changes in the thermal properties of PP-based composites after alkalization treatment on wood fibers. Also, hybrid composites were obtained by incorporating variable weight fractions (from 0.5 to 3 wt%) of a talc mineral power. This treatment with alkali washed off some natural impurities from the fiber surface, such as segments of hemicelluloses, lignin, and extractives. The hybrid composites presented higher crystallization temperatures, which was ascribed to the presence of the talc particles, acting as nucleating agents. Regarding the chemical modification, there was an accelerated crystallization mechanism of the entire composite in relation to the neat PP. Both crystallinity and melting enthalpy decreased due to the incorporation of the studied fillers, which was attributed to the reduced amount of resin used in these cases and the aforementioned interferences of fillers on the crystallization process.

Asaithambi et al. [41] treated both banana and sisal fibers with alkalization after benzoylation (reaction with benzoyl peroxide), both using simple soaking procedures, and produced PLA-based composites. DSC studies indicated that the hybrid fiber-reinforced composites (HNFRC) made with untreated fibers showed glass transition temperature (T_g), melting temperature (T_m), and degradation temperature (T_d) of 61, 151, 217 °C, respectively. These temperature peaks were followed by an oxidation peak (T_0) at 304 °C. These values shifted to slightly higher values for those HNFRC with treated natural fibers, namely 63, 153, 301 °C for T_g, T_m, and T_d, respectively. These thermal events were accompanied by a crystallization peak (T_c) at 109 °C. The increase in T_g indicates a change in the degree of plasticization, which is explained by the smaller mobility of the overall PLA chains caused by better compatibility and stronger anchoring of the treated reinforcement on the polymer. The increase in T_m indicates that the treatment brought a more effective heat transfer in this case. The crystallization peak (T_c) may be due to the reorganization of amorphous domains into crystalline regions of the decreased macromolecular flexibility and mobility. These findings revealed that the benzoylated natural fibers acted as nucleating agents and led to a higher crystallization mechanism since the PLA chains were able to wet, interdiffuse, and covalently bond to the surface of the treated fibers. These chemical interactions yielded less mobile polymer chains.

There were recent reports addressing the effects of benzoylation in the surface of natural fibers, which indicated that remarkable changes in surface roughness.

Regarding the processing extrusion followed by the injection molding process is commonly mentioned [41]. The removal of amorphous segments from hemicellulose and cellulose yielded composites with treated fibers often presented a crystallinity index of about 40%, while the composite with untreated fibers presented 26% for this property. TGA demonstrated that the composites with untreated fibers showed a final weight loss of 95%, which corresponded to an onset peak ($T_{d\,min}$) and a $T_{d\,max}$ from 342 and at 373 °C, respectively. This $T_{d\,min}$ began at 349 °C and finished at a $T_{d\,max}$ of 378 °C, which corresponds to a 98% mass loss. Benzoylation after to alkalization normally dissolves amorphous segments from hemicelluloses and lignin through mercerization followed by peroxidation, which significantly affects the whole cellulose macromolecule and delays its onset temperature of thermal degradation. Hydrogen groups from the matrix may further chemically bond themselves to the surface of the treated natural fiber.

17.2.2 Chemical Coupling Agents

Chemical coupling agents are widely used to improve the reinforcement–matrix interaction of both thermosetting and thermoplastic composites. They are also called as compatibilizers. The most common mechanisms reported in the literature can be briefly described as the elimination of low-molecular-weight compounds from the fiber surface, creation of a thin and formable outer layer, the formation of crosslinked interphase between fiber and matrix, decreasing in hydrophobicity of the fiber surface, increasing in wettability of the matrix on the fiber surface, and development of covalent bonds with both fiber and matrix [25].

17.2.2.1 Maleic Anhydride

Maleated compounds are the main coupling agent and when used to produce PP-based composites, in which a grafted version of the maleic anhydride grafted polypropylene (MAPP) is often used. The graft copolymerization of maleated agents normally results in the formation of a coating on the fiber surface, in which chemical linkages are prone to occur. This may lead to increases in surface energy and decrease in both surface polarity and hydrophilicity, which may positively affect the interfacial bonding with polymeric resins or matrices [42]. The bridging between the maleated coupling agent and the polymeric matrix may be formed either by van der Waals interactions or covalent bonds [43–45]. Covalent bonding between maleic anhydride attached to natural fibers and PP matrix may occur by the opening of C = C groups from maleic anhydride due to a subsequent reaction with dicumyl peroxide when the liquid polymer is mixed. Because of that, a copolymer formation may take place since unsaturated linkages (C = C) are opened and hydrogen groups are removed from PP due reaction of free radicals from the dicumyl peroxide [44].

Samal et al. [45] studied thermal characteristics of hybrid banana/glass composites and used MAPP as a chemical coupling agent. The composites incorporated with MAPP were more thermally stables than single banana fiber-reinforced polymer composite. These results were ascribed to SiO groups from the glass fibers, which were able to interlink themselves to anhydride group from the MAPP, causing a synergic action of both banana and glass fibers.

Nayak et al. [46] carried out a thermal analysis on hybrid banana/glass-reinforced PP composites and reported that compared to the traditional natural fiber-reinforced PP, hybrid composites showed improvements in melting point, crystallization temperature, and onset thermal degradation temperature. A further investigation was performed using DSC and thermogravimetric TG analyzes, which revealed that MAPP treatment on banana and glass fibers increased both the crystallization temperature and the thermal stability of the neat PP. In another study, [47]; Samal et al. [45, 48] produced PP-based composites reinforced with sisal, banana, and bamboo, all of them processed using MAPP as a chemical coupling agent and hybridized with glass fibers. Detailed results on each studied fiber are drawn below:

- Sisal fiber: DSC and TGA results indicated increases in both thermal stability and crystallization temperature, ascribed to the presence of MAPP in the hybrid composites. Thermomechanical profiles showed that both the use of MAPP and hybridization with glass fibers yielded increases in storage modulus. This indicates that the studied hybrid composite had a higher capacity of store mechanical energy when it was plastically deformed. However, the incorporation of MAPP led to a decrease in the damping properties of the hybrid composites. The obtained β and γ transition peaks indicated a significant influence of the fibers/PP interface, which was improved by adding MAPP during the processing of the composites, causing a clear segmental immobilization of the matrix.
- Banana fiber: As for the other natural fibers, the insertion of MAPP conferred increases in melting point, crystallization temperature, and onset thermal degradation temperature.
- Bamboo fiber: The MAPP's presence induced an increase in storage modulus, which indicates a more efficient transformation of mechanical energy into internal energy. Nevertheless, as for the sisal fibers, there was a decrease in damping properties due to the addition of fibers and MAPP in this case. Both DSC and TGA thermograms allowed us to observe that the composites became more thermally stables and the crystallization temperature of PP matrix increased due to both the addition of the mineral fibers and the MAPP. Regarding the melting temperature, similar levels were found in a comparison between the hybrid composite and neat PP. Moreover, the introduction of both fibers and MAPP broke the linear crystallizable sequence of the neat thermoplastic polymer and, consequently, reduced its degree of crystallization.

17.2.2.2 Silanization

According to Siakeng et al. [49], HNFRC made of silanized PALF and PLA matrix showed increased thermal stability when these natural fibers were silanized (reacted with an organosilane). Asim et al. [22] produced hybrid composites reinforced with silane-treated and -untreated PALF/kenaf. As a matrix, phenolic resin was used. The presence of hemicelluloses, lignin, pectin, and glycosidic linkages from cellulose led to a broad mass loss event ranging from 34% to 45% at a 278–306 °C temperature range. This also induced increased thermal stability to the treated hybrid composites in comparison to those composites made of pristine fibers. The hybrid composites displayed two major peaks in the derivative thermos-gravimetric analysis (DTG) thermograms, in a way that the first one was below 100 °C ascribed to hydroxyl groups from the fibers. The subsequent peak was attributed to the degradation of cellulose and hemicellulose, as well as voids and disrupted fibers, which were found throughout the composites. From the thermomechanical results, those hybrid composites made of untreated and treated (50% PALF + 50% kenaf) showed a steep down curve at the rubbery stage, which was attributed to the aforementioned low crosslinked segments from the natural fibers. Both untreated and treated (70% PALF + 40% kenaf) hybrid composites presented T_g signs at about 60 °C.

Regarding the flammability behavior, the pristine hybrid PALF/kenaf composites presented a total combustion time of less than 30 seconds. Regarding the hybrid composites made of treated fibers, smaller flammabilities were reached and combustion times less than 10 seconds were reported in most cases. The reached findings on fire retardancy were ascribed to good compatibility between the treated fibers and the matrix since, because of that, the thermosetting matrix involved the entire fiber surface and thus protected it.

Sajna et al. [50] produced hybrid bionanocomposites out of PLA utilizing silane-treated banana fibers and Closite 30B nanoclays as reinforcing fillers. The TGA thermograms of biocomposites made of banana fibers exhibited both slight high initial (T_i) and final (T_f) of 286 and 373 °C, respectively. Moreover, a percentage of char residue of 6.93% was found. Both these results were considered as effects of the prior silanization performed on the fibers immersed in a water–ethanol medium, which leads to hydrolysis in epoxy groups from the aminopropyltriethoxysilane (APS), forming reactive silanol. One end of the silanol formed a hydrogen bond with a free hydroxyl group of fiber and the other end formed a covalent bond with the hydroxyl group from the PLA, contributing to an improved fiber/matrix interfacial adhesion and resulting in increased thermal stability. Regarding the flammability behavior, the insertion of silanized banana fibers promoted reduction in both the burning rate and dripping of the neat matrix. These results were associated with the above described bonds developed due to the silane treatment.

Eng et al. [51] studied oil palm mesocarp fibers treated with methacrylate silane on PLA/PCL/clay/natural fibers hybrid composites. The thermomechanical analysis showed that the insertion of silane-treated fibers promoted an increase in the storage modulus. Also because of that, two T_g values shifted and became closer to each other, and tan δ values became smaller, which was associated with the well-known increase in the reinforcing/matrix adhesion.

17.2.3 Two-Step Treatments

Neto et al. [38] treated natural fiber fabrics with alkalization and then silanization. The alkalization was responsible of change the surface roughness of the fiber and for removing waxes and other minor impurities. The silane groups were added to the fiber surface to act as chemical coupling agents. Both single- and two-step treatments provided higher thermal stability in the whole 20–400 °C temperature range. Nevertheless, the single-step treatment was more effective until 300 °C, on the removal of certain compounds with low thermal stability via mercerization. The two-step treatment performed by alkalinization prior to silanization became more effective after this level of temperature due to the presence of silane groups, which created a thermally stable barrier around the natural fibers. Compared to the simple silanization, the two-step treatment yielded both higher storage modulus and loss modulus and high $\tan\delta$ in all cases. However, these authors did not address the effect of a single treatment with silane.

Shih et al. [52] produced thermoplastic composites based on PLA and poly (butylene succinate) (PBS) and reinforced these composites with PALF and bamboo fibers. These fibers were cleaned using a detergent solution and treated by alkalinization and then treated these fibers by silanization. They also proposed different PALF/bamboo weight proportions and found that the treated fibers yielded higher thermal stability and heat deflection temperatures in all cases.

Arifuzzaman Khan et al. [53] hybridized banana stem with coir fibers and produced PP-based composites using MAPP as a coupling agent. Prior to the molding, they treated the banana fibers by acetylation, alkalinization, and bleaching. The MAPP/low-density polyethylene (LDPE) composites and the acetylated and alkalized banana stem fiber (BSF)/MAPP/LDPE composites presented similar thermal degradation profiles. This indicates that bleached fibers were successfully bonded to the matrix in the acetylated and alkalized banana fiber/MAPP/LDPE cases. Regarding the thermograms, drying mechanisms on the fibers caused an endothermic peak at the 114–120 °C and then another endothermic peak ranging from 390 to 420 °C was associated with cleavage in chemical groups from lignin and cellulose.

Chalapathi et al. [54] studied bamboo nonwoven fabric-based composites, in which a vinyl ester resin was used as a matrix. Both NaOH and silane treatments were applied to bamboo fiber in order to modify the surface chemistry. In addition, three flame retardants (namely, chitosan, ammonium polyphosphate, and zinc borate) were incorporated at variable percentages (from 3 to 6 wt%). The composites with treated fiber presented significantly improved thermal stability, burning rate, and heat release peak rate compared to pure bamboo fiber composites.

17.3 Concluding Remarks

Hybrid polymeric composites reinforced with natural fibers are promising alternatives to GFRP synthetic composites. Chemical treatments, chemical coupling agents, or both have been applied to obtain natural fibers with better compatibility with conventional polymeric matrices used in polymer composites. The majority of the studies reported that these treatments mainly work on the fiber/matrix interfacial strength, achieved via a better fiber/matrix compatibility. Hence, these chemical modifications and additives play an important role in the thermal features of the composites. Improved thermal stability was reported in many studies cited in this chapter. Of the chemical modifications, alkalinization stands out as the most applied treatment; maleic anhydride and silanization are the most used coupling agents. There is a gap related to the use of other chemical treatments since some of them showed good potential to improve the thermal stability of fiber-reinforced hybrid composites. The biobased coupling agents should be developed in order to replace maleic anhydrides and silanes in order to retain the ecological appeal.

Many chemical treatments were reported in the literature for hybrid composites with fibers either treated by alkalinization or benzoylation chemicals. As future perspectives, novel chemical treatments might appear for improving the compatibility of natural fibers with resins, in which promising candidates include $NaClO_2$, H_2O_2, or acetic anhydride (mercerization). Maleic anhydride should be replaced with alternative coupling agents with faster biodegradation and obtained from vegetable sources. Some carboxylic acids and tannins should be taken into account since they are renewable alternatives. The combination of chemical treatments and coupling agents for the two-step treatments seems to be an effective way to produce hybrid composites with suitable fiber/matrix interface, although it needs to be further investigated. Graft copolymerization of plant fibers with vinyl monomers (such as styrene, methacrylates, and acrylates) is another promising route to manufacture hybrid polymer composites with improved thermal performances.

References

1. Amiandamhen, S.O., Meincken, M., and Tyhoda, L. (2020). Natural fibre modification and its influence on fibre-matrix interfacial properties in biocomposite materials. *Fibers and Polymers* 21 (4): 677–689, 23 abr.
2. Almeida, J.H.S. Jr., Amico, S.C., Botelho, E.C. et al. (2013). Hybridization effect on the mechanical properties of curaua/glass fiber composites. *Composites Part B: Engineering* 55: 492–497.
3. Lotfi, A. et al. (2019). Natural fiber-reinforced composites: A review on material, manufacturing, and machinability. *Journal of Thermoplastic Composite Materials* 34: 1–47.
4. Almeida, J.H.S. Jr., et al. (2012). Study of hybrid intralaminate curaua/glass composites. *Materials & Design* 42: 111–117.
5. da Silva, L.V., José Humberto, S.A.J., Angrizani, C.C. et al. (2013). Short beam strength of curaua, sisal, glass and hybrid composites. *Journal of Reinforced Plastics and Composites* 32 (3): 197–206.
6. Faruk, O. et al. (2012). Biocomposites reinforced with natural fibers: 2000–2010. *Progress in Polymer Science* 37 (11): 1552–1596.
7. Ornaghi, J.R., L, H., Zattera, A.J., and Amico, S.C. (2014). Thermal behavior and the compensation effect of vegetal fibers. *Cellulose* 21 (1): 189–201, fev.
8. Silva, H.S.P. et al. (2014). Mechanical behavior and correlation between dynamic fragility and dynamic mechanical properties of curaua fiber composites. *Polymer Composites* 35: 1078–1086.
9. Ho, M. et al. (2012). Critical factors on manufacturing processes of natural fibre composites. *Composites Part B: Engineering* 43 (8): 3549–3562.
10. Ornaghi, H.L.J. et al. (2012). Dynamic mechanical properties of curaua composites. *Journal of Applied Polymer Science* 116 (5): 2658–2667.

11 Panthapulakkal, S. et al. (2017). Natural fiber and hybrid fiber thermoplastic composites: Advancements in lightweighting applications. In: *Green Composites: Waste and Nature-based Materials for a Sustainable Future*, 2e. [s.l.] (eds. C. Baillie and R. Jayasinghe), 39–72. Elsevier Inc.

12 Ornaghi, H.L. et al. (2020). Mechanisms involved in thermal degradation of lignocellulosic fibers: a survey based on chemical composition. *Cellulose* 27 (9): 4949–4961.

13 Gurunathan, T., Mohanty, S., and Nayak, S.K. (2015). A review of the recent developments in biocomposites based on natural fibres and their application perspectives. *Composites Part A: Applied Science and Manufacturing* 77: 1–25.

14 John, M.J. and Thomas, S. (2008). Biofibres and biocomposites. *Carbohydrate Polymers* 71 (3): 343–364.

15 Sathishkumar, T.P., Naveen, J., and Satheeshkumar, S. (2014). Hybrid fiber reinforced polymer composites – a review. *Journal of Reinforced Plastics and Composites* 33 (5): 454–471.

16 Mochane, M.J. et al. (2019). Recent progress on natural fiber hybrid composites for advanced applications: a review. *Express Polymer Letters* 13 (2): 159–198.

17 Panthapulakkal, S. and Sain, M. (2007). Injection-molded short hemp fiber/glass fiber-reinforced polypropylene hybrid composites–mechanical, water absorption and thermal properties. *Journal of Applied Polymer Science* 103 (4): 2432–2441, 15 fev.

18 Crosky, A. et al. (2013). Thermoset matrix natural fibre-reinforced composites. In: *Natural Fibre Composites: Materials, Processes and Applications*. [s.l.] (eds. A. Hodzic and R. Shanks), 233–270. Elsevier Inc.

19 Ramesh, P., Prasad, B.D., and Narayana, K.L. (2020). Effect of MMT clay on mechanical, thermal and barrier properties of treated aloevera fiber/PLA-hybrid biocomposites. *Silicon* 12 (7): 1751–1760.

20 Reale Batista, M.D. et al. (2020). Hybrid cellulose-inorganic reinforcement polypropylene composites: Lightweight materials for automotive applications. *Polymer Composites* 41 (3): 1074–1089.

21 Lee, S.-H. and Wang, S. (2006). Biodegradable polymers/bamboo fiber biocomposite with bio-based coupling agent. *Composites Part A: Applied Science and Manufacturing* 37 (1): 80–91.

22 Asim, M. et al. (2018). Thermal, physical properties and flammability of silane treated kenaf/pineapple leaf fibres phenolic hybrid composites. *Composite Structures* 202: 1330–1338.

23 Ornaghi, H.L. et al. (2010). Mechanical and dynamic mechanical analysis of hybrid composites molded by resin transfer molding. *Journal of Applied Polymer Science* 118 (2), out.: 887–896.

24 Jawaid, M. and Abdul Khalil, H.P.S. (2011). Cellulosic/synthetic fibre reinforced polymer hybrid composites: a review. *Carbohydrate Polymers* 86 (1): 1–18.

25 Karthi, N. et al. (2020). An overview: natural fiber reinforced hybrid composites, chemical treatments and application areas. *Materials Today: Proceedings* 27: 1–7.

26 Raju, G. et al. (2007). Graft copolymerization of methyl acrylate onto oil palm empty fruit bunch (OPEFB) fiber. *Polymer-Plastics Technology and Engineering* 46 (10): 949–955, out.

27 Mishra, S. et al. (2003). Studies on mechanical performance of biofibre/glass reinforced polyester hybrid composites. *Composites Science and Technology* 63 (10): 1377–1385.

28 Venkata Reddy, G. et al. (2009). Flexural, compressive, and interlaminar shear strength properties of kapok/glass composites. *Journal of Reinforced Plastics and Composites* 28 (14): 1665–1677.

29 Dhakal, H.N., Zhang, Z.Y., and Bennett, N. (2012). Influence of fibre treatment and glass fibre hybridisation on thermal degradation and surface energy characteristics of hemp/unsaturated polyester composites. *Composites Part B: Engineering* 43 (7): 2757–2761.

30 Bavan, D.S. and Kumar, G.C.M. (2013). Significant aspects on thermal degradation of hybrid biocomposite material. *AIP Conference Proceedings* 1536: 1318–1319.

31 Shanmugam, D. and Thiruchitrambalam, M. (2013). Static and dynamic mechanical properties of alkali treated unidirectional continuous Palmyra palm leaf stalk fiber/jute fiber reinforced hybrid polyester composites. *Materials & Design* 50: 533–542, set.

32 Shanmugam, D., Thiruchitrambalam, M., and Thirumurugan, R. (2014). Continuous unidirectional palmyra palm leaf stalk fiber/glass–polyester composites: static and dynamic mechanical properties. *Journal of Reinforced Plastics and Composites* 33 (9): 836–850, maio.

33 Jayamani, E. et al. (2014). Comparative study of dielectric properties of hybrid natural fiber composites. *Procedia Engineering* 97: 536–544.

34 Jayamani, E. et al. (2014). Investigation of fiber surface treatment on mechanical, acoustical and thermal properties of betelnut fiber polyester composites. *Procedia Engineering* 97: 545–554.

35 Jawaid, M. et al. (2015). Effect of fibers treatment on dynamic mechanical and thermal properties of epoxy hybrid composites. *Polymer Composites* 36 (9): 1669–1674, set.

36 Mohan, T. and Kanny, K. (2016). Nanoclay infused banana fiber and its effects on mechanical and thermal properties of composites. *Journal of Composite Materials* 50 (9): 1261–1276, abr.

37 Anand, P. and Anbumalar, V. (2017). Investigation on thermal behavior of alkali and benzoyl treated hemp fiber reinforced cellulose filled epoxy hybrid green composites. *Cellulose Chemistry and Technology* 51 (1–2): 91–101.

38 Neto, J.S.S. et al. (2019). Effect of chemical treatment on the thermal properties of hybrid natural fiber-reinforced composites. *Journal of Applied Polymer Science* 136 (10): 1–13.

39 Arulmurugan, M. et al. (2020). Effect of barium sulphate on mechanical, DMA and thermal behaviour of woven aloevera/flax hybrid composites. *Bulletin of Materials Science* 43 (1): 58, dez.

40 Gyoung, J. et al. (2010). Effects of chemical treatments of hybrid fillers on the physical and thermal properties of wood plastic composites. *Composites Part A Applied Science and Manufacturing* 41 (10): 1491–1497.

41 Asaithambi, B., Ganesan, G.S., and Ananda Kumar, S. (2017). Banana/sisal fibers reinforced poly(lactic acid) hybrid biocomposites; influence of chemical modification of BSF towards thermal properties. *Polymer Composites* 38 (6): 1053–1062.

42 Khan, G.M.A., Yilmaz, N.D., and Yilmaz, K. (2019). Effects of chemical treatments and degumming methods on physical and mechanical properties of okra bast and corn husk fibers. *Journal of the Textile Institute* 111: 1–18.

43 Hong, C.K. et al. (2008). Mechanical properties of maleic anhydride treated jute fibre/polypropylene composites. *Plastics Rubber and Composites* 37 (7): 325–331.

44 Rozman, H.D., Saad, M.J., and Ishak, Z.A.M. (2003). Flexural and impact properties of oil palm empty fruit bunch (EFB) – polypropylene composites – the effect of maleic anhydride chemical modification of EFB. *Polymer Testing* 22: 335–341.

45 Samal, S.K., Mohanty, S., and Nayak, S.K. (2009). Banana/glass fiber-reinforced polypropylene hybrid composites: fabrication and performance evaluation. *Polymer-Plastics Technology and Engineering* 48 (4): 397–414.

46 Nayak, S.K., Mohanty, S., and Samal, S.K. (2009). Influence of interfacial adhesion on the structural and mechanical behavior of PP-banana/glass hybrid composites. *Polymer Composites* 16 (2): 1247–1257.

47 Nayak, S.K. and Mohanty, S. (2010). Sisal glass fiber reinforced PP hybrid composites: effect of MAPP on the dynamic mechanical and thermal properties. *Journal of Reinforced Plastics and Composites* 29 (10): 1551–1568.

48 Samal, S.K., Mohanty, S., and Nayak, S.K. (2009). Polypropylene–bamboo/glass fiber hybrid composites: fabrication and analysis of mechanical, morphological, thermal, and dynamic mechanical behavior. *Journal of Reinforced Plastics and Composites* 28 (22): 2729–2747.

49 Siakeng, R. et al. (2018). Effects of surface treatments on tensile, thermal and fibre-matrix bond strength of coir and pineapple leaf fibres with poly lactic acid. *Journal of Bionic Engineering* 15 (6): 1035–1046.

50 Sajna, V.P., Mohanty, S., and Nayak, S.K. (2017). A study on thermal degradation kinetics and flammability properties of poly(lactic acid)/banana fiber/nanoclay hybrid bionanocomposites. *Polymer Composites* 38 (10): 2067–2079, out.

51 Eng, C.C. et al. (2015). Chemical modification of oil palm mesocarp fiber by methacrylate silane: effects on morphology, mechanical, and dynamic mechanical properties of biodegradable hybrid composites. *Bioresources* 11 (1): 861–872.

52 Shih, Y.F. et al. (2014). Pineapple leaf/recycled disposable chopstick hybrid fiber-reinforced biodegradable composites. *Journal of the Taiwan Institute of Chemical Engineers* 45 (4): 2039–2046.

53 Arifuzzaman Khan, G.M. et al. (2013). Influence of chemical treatment on the properties of banana stem fiber and banana stem fiber/coir hybrid fiber reinforced maleic anhydride grafted polypropylene/low-density polyethylene composites. *Journal of Applied Polymer Science* 128 (2): 1020–1029.

54 Chalapathi, K.V., Song, J.I.L., and Prabhakar, M.N. (2020). Impact of surface treatments and hybrid flame retardants on flammability, and thermal performance of bamboo fabric composites. *Journal of Natural Fibers* 00: 1–11.

18

Physical, Mechanical, and Thermal Properties of Fiber-Reinforced Hybrid Polymer Composites

Subramanian Ravichandran[1], *Suresh Sagadevan*[2], *and Md Enamul Hoque*[3]

[1] *Sathyabama Institute of Science and Technology, Department of Physics, Chennai, 600119, India*
[2] *University of Malaya, Nanotechnology& Catalysis Research Centre, Kuala Lumpur, 50603, Malaysia*
[3] *Military Institute of Science and Technology (MIST), Department of Biomedical Engineering, Mirpur Cantonment, Dhaka, 1216, Bangladesh*

18.1 Introduction

The recent increase in the usage of composite technology in the materials science and engineering sector is due to the possibility of creating novel architectures with hybrid properties compared with that of the single components. For the successful use of this technology principles, the role and selection of reinforcing phase are critical as it provides the strength and stiffness to the composite, and so the reinforcement material forms help to generate stronger, harder, and stiffer composites. The commonly used reinforcing materials are applied in the form of fibers/particles and are generally originated from polymers, glass, plastic, carbon, ceramics, Kevlar, and natural plant-based ones like kenaf, jute, silk, cotton, etc. [1–7]. For example, the fiber-reinforced polymer (FRP) composites are the structures that can be applied as the major load-bearing substituents and also face many challenges to explore toward its applications in many different sectors [1]. Similarly, the glass-reinforced polymers (GRPs) are the strongest and the most resistive ones to the deforming forces, and the extent of its restiveness is decided by the orientation of the fibers. When the force is exerted parallel to the orientation of the fiber, the composites serve as the strongest while the weakest in the case of the force applied perpendicular to the fiber [8]. Taking into consideration the fiber orientation-based resistive properties offered by the GRP-mediated nanocomposites, they are used mostly for the manufacturing of artificial human parts, prosthesis, and dental implants [9]. Similarly, the GRP-based nanocomposites when used in the manufacturing of machine parts for the automobile and other engineering sectors, although their mechanical properties are indicated to be resistant against fractures, often involve in wear and tear and are due to the failures of generating the homogeneous phase composites [10]. Hence, it is a difficult task to fully hybridize the reinforcement matrix's unique features with that of the recycling and natural

Natural Fiber-Reinforced Composites: Thermal Properties and Applications, First Edition.
Edited by Senthilkumar Krishnasamy, Senthil Muthu Kumar Thiagamani,
Chandrasekar Muthukumar, Rajini Nagarajan and Suchart Siengchin.

degradable properties of polymers to form the nanocomposites that are widely involved in the production of automobile body parts with much higher efficiency at all conditions of temperature, pressure, and humidity.

In the recent years, much of the research related to polymer nanocomposites is being focused on the development of mechanical, thermal, and biodegradation properties where the commonly involved sectors for these composites include the construction, engineering, automotive, aerospace, food, and packaging industries. To test the suitability of formed composites in those sectors, the commonly used techniques involve the testing of mechanical properties including interlaminar shear strength (ILSS), dynamic mechanical analysis (DMA), etc. For example, Ravichandran et al. [11] developed hybrid polymer composites based on low-density polyethylene (LDPE), and the composites include glass, carbon, and Kevlar fibers as the reinforcements. The tests of DMA indicated that the LDPE composites reinforced with Kevlar fiber have the lowest impact energy against the carbon fiber and have lower energy than the glass fiber ones and the observation of this behavior can be linked to the occurrence of morphological changes to the final composite due to the fiber type. Also, the storage modulus of the composite followed the order of LDPE–carbon > LDPE–glass > LDPE–Kevlar and a minimum value of $\tan\delta$ (damping factor) were observed for the LDPE–glass composite against the other two [11]. In a different study, Zhang et al. studied the DMA of high-density polyethylene (HDPE)-reinforced rice husk biochar composites, where the morphological analysis supported for the successful embedding of rice husk biochar within the HDPE and the same is responsible for an enhancement in the storage modulus and decrease of loss modulus for the final composite when studied by the DMA [12]. Similarly, Jesuarockiam et al. studied the mechanical and thermodynamic properties of synthetic/natural hybrid composites containing the graphene nanoplatelets, where the analysis showed that the replacement of Kevlar fiber with that of *Cocos nucifera* resulted in an effective cross-linking that can affect the structural applications [13]. The results of the study concluded that the thermogravimetric analysis (TGA) and DMA of hybrid composites S3G3 and S4G3 can effectively replace Kevlar/epoxy (S1G0) polymeric structure at conditions of elevated temperature [13]. Also, the DMA of delaminated flax fiber-reinforced composites were studied by Shen et al. and the analysis of results indicated that the delamination area of the composite increases in both 12.8% laminate and 12% loss factor. Based on the overall analysis, they concluded that the effect of delamination influences the transverse position, and also the detectability of delamination was more notable than the delamination size [14]. Karvanis et al. reported the DMA of glass or carbon fiber/polymer composite by measuring the storage modulus, loss modulus, $\tan\delta$, and glass transition temperature. The analysis of results has provided very useful information for the effective utilization based on the orientation of various fibers with different architectures like the placing of unidirectional fibers in the longitudinal or transverse direction, while the orientation of the woven fabric is at 0° (fibers 0°/90°) or 45° (fibers 45°/+45°) [15]. Fan et al conducted the ILSS of stationary glass fiber mats reinforced with that of multiwalled carbon nanotube (MWCNT)–epoxy suspensions. The results indicated that the shearing strength of the composite is getting increased by 33%, following the MWCNT–epoxy loading and is due to an increase in the multiwall nanotubes (MWNT) orientational thickness [16]. The ILSS is the stress that gets acted on the interface of two adjacent laminates in layered composites, and these stresses cause relative deformation between the consecutive laminate and in some cases may cause total failure along the midplane if sufficiently high stress gets between the two adjacent laminates. Therefore, it is highly essential to investigate the ILSS nature of any newly formed composites as the test investigates the failures in composite laminates that get initiated in a shear stress mode (delamination mode) [17, 18].

In a general tradition, the materials used for the aerospace, automobile, and marine industries possess many advanced characteristics and in that view, the LDPE is a well-known commercial thermoplastic polymer that can offer many industrial applications like food and juice containers, waterproof surfaces, computer components, etc. The industrial applications of this LDPE material are due to its unique combination of properties like insulating behavior for the electrical devices, mechanically strong nature, and at the same time the capacity of thermal resistances [19]. When using the LDPE material for the composite formation with other reinforcements and/or fillers, the generated hybrid composites possess the additional characteristics of high specific strength and modulus, excellent toughness against the fractures, fatigue properties, good resistance to corrosion, etc. [19–21]. This unique combination of properties, in particular, the high strength to weight ratio, makes the LDPE polymers and their composites very potential candidates for transport applications. Therefore, to take advantage of the inbuilt properties of LDPE material, the present study deals with the LDPE composites containing epoxy resin and various reinforcements such as carbon, glass, and Kevlar fibers. For the formation of the composites, we followed the compression molding method where the formed hybrid composites were characterized thoroughly and further studied toward the change in the mechanical, thermal, and water absorption properties.

18.2 Preparation of Composite Material

To fabricate the FRP composites, we used the compression molding method and for that, a mixture of LDPE and toluene solvent was taken in the ratio of 1 : 3 and heated to a temperature of 190 °C for 45 minutes. After cooling to room temperature, the reaction mixture was washed with degassed ethanol, centrifuged, separated the precipitate, and dried the product to see the formation of solidified smooth LDPE in its powdered form. The obtained powdered was transferred to a round bottom flask and subjected to mixing on a magnetic stirrer with that of epoxy resin and fiber by taking in a ratio of 1 : 2 : 3 for LDPE powder, epoxy resin, and fiber, respectively. The mixture was allowed to stir for about 30 minutes or more until all the contents get blended well to form a homogenous mixture. After that, 10% of hardener was added to this mixture. The spacer of 5 mm thickness and $300 \times 300\ mm^2$ (inner dimension) was mounted on the compression molding machine. Then layers of glass fibers were placed one after another by applying the LDPE and resin mixture.

Both sides of the spacer were covered by a polythene sheet coated with wax. The fabricated composites were finally cured at a constant pressure of 0.2 (3 psi) bar and a temperature of 80 °C for two hours.

18.3 Characterization of Composite Material

18.3.1 Tensile Test

The tensile test samples were prepared (as per the ASTM D638 standard) by cutting the fabricated bulk composite into the required dimension by a suitable cutter and finishing the edges using an emery paper. The test involved mounting the sample on a universal testing machine and then subjected to the various tensions until it fractured. The tensile force was recorded as a function of an increase in gauge length. During the application of different tensions, the elongation of the gauge section was also recorded against the applied force.

18.3.2 Flexural Test

The three-point flexural test was performed using a universal testing machine to measure the flexural strength that is reasonably significant with a tensile test. For the flexural test, the specimen was rested onto two knife supports and the opposite probe produced a flexural effect repeatedly until it was broken. The flexural specimens were prepared as per the ASTM D790 standard where the three-point flexure test is commonly used to for testing the flexural properties of the composites. The force was applied to the test sample until it was fractured. The tests are carried out at 50% average relative humidity conditions.

18.3.3 Test of Water Absorption

Good matrix wetting of the fiber and adequate bonding of the fiber matrix can reduce the rate and quantity of water absorbed in the composite interphone region [22]. To determine the amount of water absorbed under the testing conditions, the water absorption tests of fiber-reinforced composites were done as per ASTM 570 by immersing the samples in distilled water at room temperature. The dried specimen's weight (w_1) and after submerging the same material in distilled water for a 24 hours period (w_2) were recorded. The factors that affect water absorption characteristics like plastic type, additives used, exposure temperature, and duration of soaking were also investigated. Due to the nature of fibers, the water absorption of composites containing various reinforcement fibers (glass, carbon, and Kelvar) may be a limiting parameter for several composite applications. All the samples are dried completely before being immersed in water until we confirm the constant weight on a four-digit balance. The percentage of water consumption is calculated by Eq. (18.1):

$$\text{WA}\,(\%) = [(w_2 - w_1)/w_1] \times 100 \tag{18.1}$$

where is w_1 is the initial weight of dried specimen (gr) and w_2 is the specimen weight after N hours of water soaking (gr).

18.4 Results and Discussion

18.4.1 Mechanical Properties

To investigate the effects of type of reinforcement material on the LDPE's internal characteristics, the synthesized composites were first subjected to a wide range of mechanical tests. In general, the tensile properties of the composites depend on the type of materials, preparation methods, specimen preparation condition, and percentage of reinforcement [23]. In Table 18.1, it can be observed that the carbon fiber reinforcement of LDPE led to a substantial improvement of the stiffness to 448 MPa, followed by the glass fiber of 298 MPa and Kevlar fiber of 207 MPa. Table 18.1 compares the tensile strength, breaking load (or breaking strength and the tensile strength at its break), energy absorption (maximum force applied), and impact energy of LDPE–epoxy resin with that of other fiber reinforcements like glass, carbon, and Kelvar fiber.

Figure 18.1a,band the respective data for both are shown in Table 18.1. From the analysis of flexural strength data shown in Figure 1a, all the composites are exhibiting values greater than the pure LDPE–epoxy-laminated composite. As similar to the earlier tests of tensile and impact studies, the carbon fiber-reinforced composites have a high flexural strength of 37 N M^{-1}, followed by the glass fiber of 22 N M^{-1} and Kevlar fiber of 16 N M^{-1}. The observation of such a higher flexural value for the carbon fiber composite supports the high brittle nature as compared to the other

Table 18.1 Ultimate tensile strength, maximum force, breaking load, and energy absorption of fiber-reinforced LDPE composites.

LDPE composites	Ultimate tensile strength (Mpa)	F_{max} (KN)	Breaking load (KN)	Energy absorbed (J)
LDPE–epoxy resin laminate	120	7.2	0.05	0.01
LDPE–glass fiber	298	20.6	1.3	8.6
LDPE–carbon fiber	448	35.9	1.5	9.7
LDPE–Kevlar fiber	207	15.5	1.2	3.5

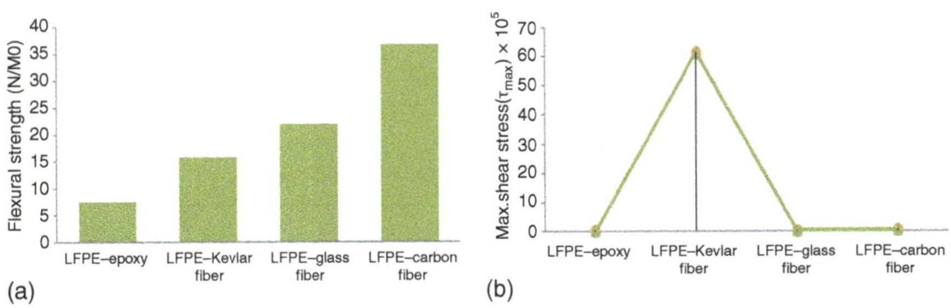

Figure 18.1 Comparison of variations in the flexural strengths (a) and shear stress (b) for the LDPE fiber-reinforced composites.

fibers and pure LDPE. This may be related to the higher stiffness of carbon fibers and the ability of components within the LDPE–carbon fiber to redistribute across the fiber so as to withstand the load. As a result, the flexural strength of carbon fiber reinforced composite is 2.3 times greater than the Kevlar and 1.7 times greater than the glass fiber reinforcement. This remarkable improvement in the flexural strength of LDPE–carbon fiber composite against the other fibers may be attributed to the effect C–C coupling effect of fiber with that of the LDPE polymer, thereby making the composite to be more homogenous. Turla et al. [24] investigated the flexural strength of carbon and glass fiber-reinforced epoxy based hybrid composites where the results indicated higher values for the carbon composite compared with the glass ones. Based on the literature studies and from our data, it can be concluded that the flexural strength is dependent on the fiber material type and a significant enhancement in those properties can be achieved by reinforcing the polymers with that of fibers [25, 26]. Similarly, the comparison of shear stress values for the LDPE composites shown in Figure 18.1b and Table 18.2, it was observed that the LDPE–Kevlar fiber composite has the highest value of 62, while all other composites exhibited the negligible values of <0.5. The higher value of shear stress (tangential stress) indicates that the LDPE–Kevlar composite maintains greater slippage capacity all along the planes in parallel to the applied force while no such capacity is seen for the other fiber-reinforced composites.

Hardness is also one of the main characteristics of a polymer material where it determines the degree of material deformation and is generally accepted as a significant property. This is defined as the indentation resistance and determined by measuring the indentation's permanent depth. In the Wilson hardness testing of LDPE–fiber composites, a load is applied from 10 to 60 kg to the specimen for a dwell time (time spent in the same position) of five seconds and the hardness values are measured. The results obtained from the Wilson hardness test with our LDPE–resin polymer composites of different fiber reinforcements are also summarized in Table 18.2. From the comparison of values, it indicates that the hardness of the composites is getting varied with respect to the nature of fiber reinforcement. The composites with glass and Kevlar fiber have high hardness values because of the alignment and the length of the fibers, leading to the contribution of the loads along the fibers. However, the LDPE–carbon fiber composite is maintaining the

Table 18.2 Comparison of flexural strength, maximum shear stress, and hardness for the LDPE–fiber composites.

Different LDPE composites	Flexural strength (N/M)	Max. shear stress (τ_{max}) × 10^5 (N/M)	Hardness values
LDPE–epoxy resin	7.5	0.1	80
LDPE–glass fiber	22	0.3	107
LDPE–carbon fiber	37	0.4	67
LDPE–Kevlar fiber	16	62	111

lower values of hardness and is because of a homogeneous mixture that allowed for the improved interaction among the groups of the LDPE matrix and the carbon fiber [27, 28].

18.4.2 Water Absorption Studies

The water absorption may adversely affect several mechanical properties and also can increase moisture content in the fiber walls and interphase regions of the fiber matrices and so the test of water absorption is a very important one for determining the outdoor work potential of fiber-reinforced composites. These performances of the composites may suffer if they are exposed to various environmental conditions and physical properties of the fiber material like the changes in surrounding temperature, fiber size and volume fraction, fiber type and orientation, total exposed area, diffusivity, interface bonding, void space, viscosity and surface protection, water-matrix interaction, etc. In general, good wetting of the fiber with the matrix and adequate interfacial bonding can reduce the water absorption amount and rate in the interphone region of the composite [29, 30]. Therefore, to investigate the extent of interaction between the epoxy resin, fiber, and LDPE groups, the LDPE–epoxy resin and LDPE–fiber reinforced composites were tested for the water absorption and the results are provided in Figure 18.2a. From the data shown in Figure 18.2a, it is indicated that the composite with Kevlar fiber has high water absorption, while the composite with carbon fiber has a less water absorption coefficient. Furthermore, this value is even less than the LDPE–epoxy resin. The observation of very less water absorption for the LDPE–carbon fiber-reinforced composite may be attributed to the absence of voids and cavities in the composite. However, for the Kevlar fiber composite, there may be much space at the interface region between the fiber and matrix which supported more water intake.

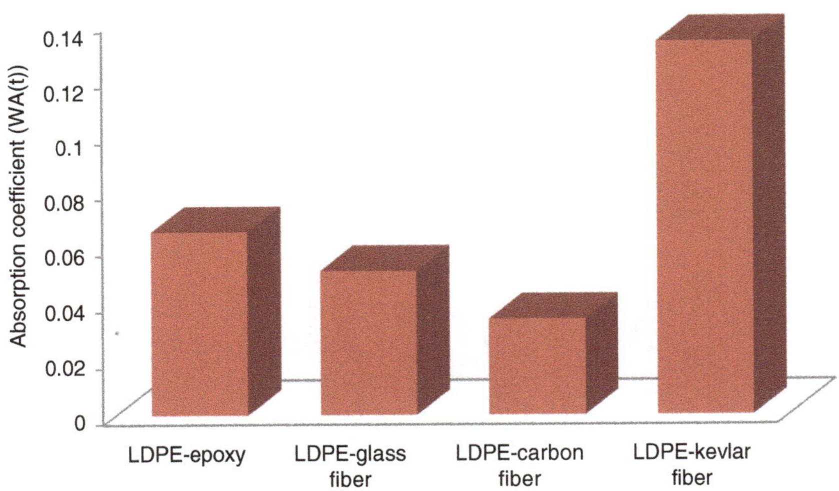

Figure 18.2 Variation of the water absorption coefficient of LDPE-fiber composites.

18.4.3 Thermal Properties

TGA and differential thermal gravimetric (DTG) are used to characterize the decomposition and thermal stability under various conditions and to analyze the degradation kinetics of the physicochemical processes of the samples. Figure 18.3a–d shows the TGA and DTG for the LDPE–epoxy and LDPE fiber-reinforced composites where, for the testing, the initial sample weights were taken in the range of 2–5 mg and were heated at a heating rate of 20 °C min^{-1} from room temperature to higher levels. The weight loss usually rate indicates the of decomposition of polymers and often provides mechanisms for degradation and other important information related to physicochemical changes in different environments [31]. From Figure 18.3a, the TGA of LDPE–epoxy resin represents a dynamic weight loss profile, and for the LDPE–fiber composites, the major degradation is occurring in two steps, one due to the fiber degradation while the other from the polymer degradation. We observed the initial decomposition of material in the temperature range of 200–300 °C where the weight losses appeared to be very less and may be due to the loss of adsorbed water/moisture. The second weight loss stage occurred in the range of 400–750 °C and can be linked to the degradation of the polymer present in the composites. From the TGA curves, we observed the significant weight losses in the range of 170–800 °C in the glass-reinforced fiber (Figure 18.3b), 195–978 °C for the carbon fiber (Figure 18.3c), and 156–470 °C for the Kevlar fiber (Figure 18.3d). The complete decomposition of the composite material was observed around 1000 °C in LDPE with carbon fiber, at 800 °C for the glass fiber, and at 470 °C for the Kevlar fiber-reinforced composites. Overall, we

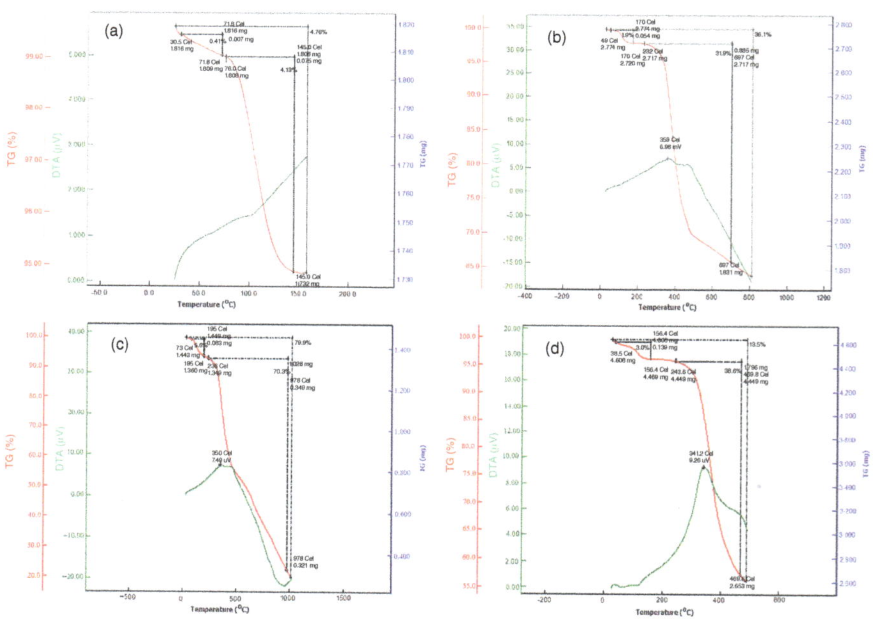

Figure 18.3 TGA and DTA of LDPE with resin composite (a), LDPE–glass fiber (b), LDPE–carbon fiber (c), and LDPE–Kevlar (d) fiber-reinforced composites.

observed a 65.8% residue at 697 °C with the glass fiber, 22% at 978 °C for the carbon fiber, and 57% residue at 469 °C for the Kevlar fiber-reinforced composites. This indicates that the LDPE–carbon fiber composite has significant thermal stability as it is able to withstand the heating temperature of 978 °C with 22% residual weight.

Similarly, in the differential thermal analysis (DTA) of LDPE fiber-reinforced composites, we observed the polymer degradations around 359 °C for the glass fiber (Figure 18.2b), 350 °C for the carbon fiber (Figure 18.2c), and 341 °C for the Kevlar fiber (Figure 18.2d) reinforced composites. Since the LDPE is a crystalline polymer and it is obvious to observe such degradation peaks is obvious and can be linked to the phase changes, where the obtained results are consistent with the literature studies [32, 33]. The curve also shows that thermal degradation only started to occur after absorbing certain amounts of heat energy by the composite materials. Under the heat absorbed conditions, the degradation processes get initiated and help for the breakdown of the fiber structure and polymer matrix by causing ruptures in the molecular chain. Rezaei et al. [34] and Zabihzadeh et al. [35] reported similar results of thermal degradation of fibers and polymer matrices occurring due to the molecular chain ruptures. Thus, from the comparison of peaks found in the TGA and DTG curves, the physical structure of the LDPE polymer matrix seems to be changed after the fiber reinforcement and which significantly affected the thermal degradation process of composites.

18.5 Conclusions

In conclusion, we studied the thermal, mechanical, and water absorption characteristics of LDPE matrices with epoxy resin and other composites reinforced with glass, carbon, and Kevlar fibers. From the studies of mechanical properties, we observed a significant improvement in the fiber-reinforced composites as compared to the LDPE with epoxy resin. Within the fiber-reinforced composites, the LDPE–carbon fiber has high tensile strength and impact energy, while the LDPE–Kevlar composite has the highest hardness only. On testing the behavior of composites toward the water absorption, we found that the LDPE–Kevlar fiber composite is very sensitive to swelling as it showed the maximum absorption coefficient and the least for the LDPE–carbon fiber composite. This can be attributed to the voids in the polymer matrix and the nature of reinforcing fibers. Further, the thermal studies confirmed that with the reinforcement of carbon fiber, the thermal stability of LDPE composite increased significantly. Overall, as compared to the non-reinforced polymer matrices, the reinforcement of the LDPE matrix with the fiber has the significant properties of mechanical, thermal, and water absorption, and within the different fibers, the carbon fiber reinforcement is the most influential one due to the modifications occurring in its morphology. Further, these composites can be incorporated directly for various industrial applications by taking advantage of the limited/no water absorption property of LDPE–carbon fiber composite and high water absorption of LDPE–Kevlar fiber, in addition to the thermal and mechanical stabilities of LDPE–carbon fiber. Finally, the enhancement in the mechanical and other physical properties due to the fiber reinforcement and morphological changes can be explored to form composites with other polymer matrices that have a significant effect on industrial product development.

Conflicts of Interest

The authors declare no conflict of interest.

References

1 Rajak, K., Pagar, D.D., Menezes, P.L., and Linu, E. (2019). Fiber-reinforced polymer composites: manufacturing, properties, and applications dipen. *Polymers* 11: 1667.

2 Golieskardi, M., Satgunam, M., Ragurajan, D. et al. (2020). Microstructural, tribological, and degradation properties of Al_2O_3- and CeO_2-Doped 3 mol.% yttria-stabilized zirconia bioceramic for biomedical applications. *Journal of Materials Engineering and Performance* 29: 2890–2897.

3 Ragurajan, D., Golieskardi, M., Satgunam, M. et al. (2018). Advanced 3Y-TZP bioceramic doped with Al_2O_3 and MnO_2 particles potentially for biomedical applications: study on mechanical and degradation properties. *Journal of Materials Research and Technology* 7 (4): 432–442.

4 Golieskardi, M., Satgunam, M., Ragurajan, D. et al. (2019). Advanced 3Y-TZP bioceramic doped with Al_2O_3 and CeO_2 potentially for biomedical implant applications. *Materials Technology* 34 (8): 480–489.

5 Hoque, M.E., Sakinah, N., Chuan, Y.L., and Ansari, M.N.M. (2014). Synthesis and characterization of hydroxyapatite bioceramic. *International Journal of Scientific Engineering and Technology* 3 (5): 458.

6 Haniffah, W.H., Sapuan, S.M., Abdan, K. et al. (2015). Kenaf fibre reinforced polypropylene composites: effect of cyclic immersion on tensile properties. *International Journal of Polymer Science* 2015, Article ID: 872387: 1–6.

7 Fong, T.C., Saba, N., Liew, C.K. et al. (2015). Yarn flax fibres for polymer-coated sutures and hand lay-up polymer composite laminates. In: *Manufacturing of Natural Fibre Reinforced Polymer Composites* (eds. M. Jawaid, M. Sapuan Salit, N.B. Yusoff and M.E. Hoque), 155–175. Switzerland: Springer-Verlag.

8 Etcheverry, M. and Barbosa, S.E. (2012). Glass fiber reinforced polypropylene mechanical properties enhancement by adhesion improvement. *Materials (Basel)* 5 (6): 1084–1113.

9 Wang, X. (2019). Bioartificial organ manufacturing technologies. *Cell Transplantation* 28 (1): 5–17.

10 Veličković, S., Stojanović, B., Ivanović, L. et al. (2019). Application of nanocomposites in the automotive industry. *Mobility & Vehicle Mechanics* 45 (3): 51–64.

11 Ravichandran, S., Vengatesan, E., and Ramakrishnan, A. (2019). Synthesis and dynamic mechanical analysis of fiber-reinforced low-density polyethylene hybrid

polymer composites. *Materials Today: Proceedings* https://doi.org/10.1016/j.matpr.2019.09.216.

12 Zhang, Q., Cai, H., Ren, X. et al. (2017). The dynamic mechanical analysis of highly filled rice husk biochar/high-density polyethylene composites. *Polymers* https://doi.org/10.3390/polym9110628.

13 Jesuarockiam, N., Jawaid, M., Zainudin, E.S. et al. (2019). Enhanced thermal and dynamic mechanical properties of synthetic/natural hybrid composites with graphene nanoplateletes. *Polymers* https://doi.org/10.3390/polym11071085.

14 Shen, Y., Tan, J., Fernandes, L. et al. (2019). Dynamic mechanical analysis on delaminated flax fiber reinforced composites. *Materials* https://doi.org/10.3390/ma12162559.

15 Karvanis, K., Rusnakova, S., Zaludek, M., and Alexandercapka (2018). Preparation and dynamic mechanical analysis of glass or carbon fiber/polymer composites. *IOP Conference Series: Materials Science and Engineering* https://doi.org/10.1088/1757-899X/362/1/012005.

16 Fan, Z., Santare, M.H., and Advani, S.G. (2008). Interlaminar shear strength of glass fiber reinforced epoxy composites enhanced with multi-walled carbon nanotubes. *Composites Part A: Applied Science and Manufacturing* 39 (3): 540–554.

17 Davis, D.C. and Whelan, B.D. (2011). An experimental study of interlaminar shear fracture toughness of a nanotube-reinforced composite. *Composites Part B: Engineering* 42 (1): 105–116.

18 Bowles, K.J. and Frimpong, S. (1992). Void effects on the interlaminar shear strength of unidirectional graphite-fiber-reinforced composites. *Journal of Composite Materials* 26 (10): 1487–1509.

19 Jordan, J.L., Casem, D.T., Bradley, J.M. et al. (2016). Mechanical properties of low-density polyethylene. *Journal of Dynamic Behavior of Materials* 2: 411–420.

20 Khanam, P.N. and AlMaadeed, M.A.A. (2015). Processing and characterization of polyethylene-based composites. *Advanced Manufacturing: Polymer & Composites Science* 1: 63–79.

21 Dintcheva NTz, Filippone, G., Arrigo, R., and La Mantia, F.P. (2017). Low-density polyethylene/polyamide/clay blend nanocomposites: effect of morphology of clay on their photooxidation resistance. *Journal of Nanomaterials* 2017: 3549475, 9 pp.

22 Wang, Q., Liu, C., and Chen, Y. (2001). Studies on PA6-PP-wollastonite composite compatibilized by PP-*graft*-malefic anhydride prepared via pan milling. *Plastics, Rubber and Composites* 30: 363–369.

23 Nassar, A. and Nassa, E. (2013). Study on mechanical properties of epoxy polymer reinforced with nano SiC particles. *Nanoscience and Nanoengineering* (2): 89–93.

24 Turla, P., Kumar, S.S., Reddy, H.P. et al. (2014). Processing and flexural strength of carbon fiber and glass fiber reinforced epoxy-matrix hybrid composite. *International Journal of Engineering Research and Technology* 3: 394–398.

25 Amuthakkannan, P., Manikandan, V., Winowlin Jappes, J.T. et al. (2013). Effect of fiber length and fiber content on mechanical properties of short basalt fiber

reinforced polymer matrix composites. *Materials Physics and Mechanics* 16: 107–117.

26 Jang, J. and Han, S. (1999). Mechanical properties of glass-fiber mat/PMMA functionally gradient composite. *Composites Part A: Applied Science and Manufacturing* 30: 1045–1053.

27 Joseph, P.V., Kuruvilla, J., and Thomas, S. (1999). Effect of processing variables on the mechanical properties of sisal-fiber-reinforced polypropylene composite. *Composites Science and Technology* 59: 625–1640.

28 Emad, A.L. and Hasani, S. (2007). Study of tensile and hardness property for epoxy reinforced with glass fiber layers. *Engineering and Technology* 25: 988–997.

29 Nitz, H., Reichert, P., Ro Mling, H. et al. (2000). Influence of compatibilizers on the surface hardness, water uptake and the mechanical properties of poly(propylene) wood flour composites prepared by reactive extrusion. *Macromolecular Materials and Engineering* 276/277: 51–58.

30 Huner, U. (2015). Effect of water absorption on the mechanical properties of flax fiber reinforced epoxy composites. *Advanced Science and Technology Letters* 9: 1–6.

31 Niranjana Prabhu, T., Demappa, T., and Harish, V. (2012). Thermal degradation of hdpe short fibers reinforced epoxy composites. *Journal of Applied Chemistry* 2: 39–44.

32 Chung, D.D.L. (2000). Thermal analysis of carbon fiber polymer-matrix composites by electrical resistance measurement. *Thermochimica Acta* 364: 121–132.

33 Liufu, S.-C. and Han-NingXiao, Y.-P.L. (2005). Thermal analysis and degradation mechanism of polyacrylate/ZnO nanocomposites. *Polymer Degradation and Stability* 87: 103–110.

34 Rezaei, F., Unus, R., and Ibrahim, R.A. (2009). Effect of fiber length on thermomechanical properties of short carbon fiber reinforced polypropylene composites. *Materials and Design* 30: 260–263.

35 Zabihzadeh, S.M., Dastoorian, F., and Ebrahimi, G. (2010). Tensile properties and thermal analysis of chemimechanical pulp-reinforced HDPE composites. *Journal of Reinforced Plastics and Composites* (12): 1814–1819.

Index

Natural Fiber-Reinforced Composites: Thermal Properties and Applications, First Edition.
Edited by Senthilkumar Krishnasamy, Senthil Muthu Kumar Thiagamani,
Chandrasekar Muthukumar, Rajini Nagarajan and Suchart Siengchin.

www.ingramcontent.com/pod-product-compliance
Lightning Source LLC
Chambersburg PA
CBHW080932240926
51849CB00003BA/437
* 9 7 8 3 5 2 7 3 4 8 8 3 1 *